Three

D0119543

Modern Cell Biology

MODERN CELL BIOLOGY

SERIES EDITOR

Birgit H. Satir
Department of Anatomy and Structural Biology
Albert Einstein College of Medicine
1300 Morris Park Avenue
Bronx, New York 10461

ADVISORY BOARD

NONINVASIVE TECHNIQUES IN CELL BIOLOGY

NONINVASIVE TECHNIQUES IN CELL BIOLOGY

Editors

J. Kevin Foskett
Sergio Grinstein

Division of Cell Biology
The Hospital for Sick Children
Toronto, Canada

WILEY-LISS

A JOHN WILEY & SONS, INC., PUBLICATION
New York • Chichester • Brisbane • Toronto • Singapore

Address all Inquiries to the Publisher
Wiley-Liss, Inc., 41 East 11th Street, New York, NY 10003

While the authors, editors, and publisher believe that drug selection and dosage and the specifications and usage of equipment and devices, as set forth in this book, are in accord with current recommendations and practice at the time of publication, they accept no legal responsibility for any errors or omissions, and make no warranty, express or implied, with respect to material contained herein. In view of ongoing research, equipment modifications, changes in governmental regulations and the constant flow of information relating to drug therapy, drug reactions and the use of equipment and devices, the reader is urged to review and evaluate the information provided in the package insert or instructions for each drug, piece of equipment or device for, among other things, any changes in the instructions or indications of dosage or usage and for added warnings and precautions.

Library of Congress Cataloging-in-Publication Data

Noninvasive techniques in cell biology / edited by J. Kevin Foskett
and Sergio Grinstein.
    p.   cm. — (Modern cell biology series : 9)
    Includes bibliographical references.
    Includes index.
    ISBN 0-471-56809-0
    1. Cytology—Methodology. 2. Fluorescence microscopy.
3. Confocal microscopy. 4. Flow cytometry. I. Foskett, J. Kevin.
II. Grinstein, Sergio, 1950- . III. Series: Modern cell biology :
v. 9.
    [DNLM: 1. Cytological Technics. 2. Fluorescent Dyes.
3. Microscopy. Fluorescence. 4. Molecular Biology. W1 MO124T v. 9
/ QH 585 N813]
QH573.M63 vol. 9
[QH585]
574.87 s—dc20
[574.87′028]
DNLM/DLC
for Library of Congress

·12 MAR 1991

90-12359
CIP

Contents

Contributors

Daniel Axelrod, Biophysics Research Division and Department of Physics, University of Michigan, Ann Arbor, MI 48109 [93]

Robert S. Balaban, Laboratory of Cardiac Energetics, National Heart, Lung, and Blood Institute, National Institutes of Health, Bethesda, MD 20892 [213]

Steven M. Block, Rowland Institute for Science, Cambridge, MA 02142, and Department of Cellular and Developmental Biology, Harvard University, Cambridge, MA 02138 [375]

John C. Cambier, Division of Basic Sciences, Department of Pediatrics, National Jewish Center for Immunology and Respiratory Medicine, Denver, CO 80206 [353]

W.A. Carrington, Program in Molecular Medicine, Physiology Department, University of Massachusetts Medical School, Worcester, MA 01605 [53]

Kenneth W. Dunn, Department of Pathology, Columbia University College of Physicians and Surgeons, New York, NY 10032 [153]

F.S. Fay, Program in Molecular Medicine, Physiology Department, University of Massachusetts Medical School, Worcester, MA 01605 [53]

Claus Fittschen, Division of Basic Sciences, Department of Pediatrics, National Jewish Center for Immunology and Respiratory Medicine, Denver, CO 80206 [353]

K.E. Fogarty, Program in Molecular Medicine, Physiology Department, University of Massachusetts Medical School, Worcester, MA 01605 [53]

J. Kevin Foskett, Division of Cell Biology, The Hospital for Sick Children, Toronto, Ontario M5G 1X8 Canada [237]

David J. Gross, Department of Bio-chemistry and Program in Molecular and Cellular Biology, University of Massachusetts, Amherst, MA 01003 [21]

Richard P. Haugland, Molecular Probes, Inc., Eugene, OR 97402 [1]

Louis B. Justement, Division of Basic Sciences, Department of Pediatrics, National Jewish Center for Immunology and Respiratory Medicine, Denver, CO 80206 [353]

Lazaro J. Mandel, Division of Physiology, Department of Cell Biology, Duke University Medical Center, Durham, NC 27710 [213]

Frederick R. Maxfield, Department of Pathology, Columbia University College of Physicians and Surgeons, New York, NY 10032 [153]

Akwasi Minta, Molecular Probes, Inc., Eugene, Oregon 97402 [1]

The numbers in brackets are the opening page numbers of the contributors' articles.

x   Contributors

Richard Nuccitelli, Zoology
Department, University of California,
Davis, CA 95616 **[273]**

Gerald H. Pollack, Division of
Bioengineering, University of Washington, Seattle, WA 98195 **[311]**

Ger T. Rijkers, Department
of Immunology, University Hospital
for Children and Youth, "Het
Wilhelmina Kinderziekenhuis," 3501
Utrecht, The Netherlands **[353]**

Mitchell C. Sanders, Cell Biology
Group, Worcester Foundation for
Experimental Biology, Shrewsbury,
MA 01545 **[177]**

Bruce Seligmann, Department of
Inflammation and Osteoarthritis
Research, Ciba-Geigy Pharmaceuticals
Division, Summit, NJ 07901 **[327]**

Ernst H.K. Stelzer, Confocal Light
Microscopy Group, Physical
Instrumentation Programme, European
Molecular Biology Laboratory (EMBL),
D-6900 Heidelberg, Federal Republic
of Germany **[73]**

James Thomas, Department of Physics,
Cornell University, Ithaca, NY
4853 **[129]**

Yu-Li Wang, Cell Biology Group,
Worcester Foundation for Experimental
Biology, Shrewsbury, MA 01545 **[177]**

Watt W. Webb, Department of Applied
Physics, School of Applied and
Engineering Physics, Cornell University,
Ithaca, NY 14853 **[129]**

Preface

It is the ultimate goal of cell biology to understand and integrate the physiology, biochemistry, and biophysics of unperturbed, living cells. The attainment of this objective requires the development and application of technical approaches that can noninvasively probe a variety of dynamic cellular processes. Whereas no technique is truly noninvasive, several recently developed, innovative methods approach this objective. The purpose of the present volume is to review a number of these technologies and their applications to cell biological problems. The reader will notice that the emphasis is on optical techniques, reflecting our view that such methods are among the most versatile, informative, and least invasive. Due to space limitations, other, nonoptical spectroscopic techniques (e.g., NMR) are not included.

The design of novel optical probes, which enable specific biochemical and structural components to be observed in living cells, is discussed in Chapter 1. Other chapters describe methodologies developed for spectroscopic measurements in populations of cells (Chapters 13 and 14) and at the single cell level (Chapter 2). Related chapters discuss applications of these methods to biological problems, such as cell metabolism (Chapter 9), endocytosis (Chapter 7), cell surface and cytoskeletal dynamics (Chapter 6 and 8), contractile processes (Chapter 12), and ion and water transport (Chapter 10 and 11), demonstrating the power and promise of noninvasive optical and electrophysiological approaches. Chapters 3, 4, and 5 describe emerging technologies that allow subcellular processes to be imaged with high temporal and spatial resolution. Finally, an exciting new area of development, the application of noninvasive optical techniques to manipulate cells and subcellular organelles within intact cells, is introduced in Chapter 15.

Although such a volume cannot be comprehensive, we hope that the assembled chapters, contributed by a selection of leading experts, will provide an updated, representative sample of recent and exciting developments in the area of noninvasive measurements of biological processes at the cellular level. We anticipate that the future will bring refinement of many of the techniques discussed in this volume as well as the development of new technologies and applications that will enable biology to be understood within the context of the intact, living cell. Perhaps the availability of the present volume will contribute to this goal.

Finally, we would like to thank Dr. R.S. Balaban for his helpful suggestions of possible contributors to this volume and Ms. T. Fisher for her diligent secretarial help.

J. Kevin Foskett
Sergio Grinstein

Noninvasive Techniques in Cell Biology: 1–20
© 1990 Wiley-Liss, Inc.

1. Design and Application of Indicator Dyes

Richard P. Haugland and Akwasi Minta

Molecular Probes, Inc., Eugene, Oregon 97402

I. INTRODUCTION

Fluorescence is the most popular optical tool for measuring ion properties inside living cells. Bright and Taylor [1986] have listed at least five reasons for this. The first and foremost is *sensitivity.* It is possible to detect a very small number of fluorescent molecules and this number should be improved with advances in probe chemistry and instrumentation. The second reason is *specificity.* Single classes of cells can be studied in a mixture of many other molecules by selecting proper probes and filter combinations. The third is *spectroscopy*. Information about molecules and their environment can be obtained from spectroscopic measurements. The fourth is *spatial resolution*. Fluorescence permits the analysis of fluorescent parameters at the limit of resolution of light microscopy. Two- and three-dimensional maps of molecular distribution [Agard, 1984] and molecular activity are thus possible in cells. The fifth is *temporal resolution*. The time dependence of processes such as membrane potential and ligand binding can be measured by fluorescence.

Fluorescent indicator dyes are generally used to sense concentrations of small molecules or ions in organelles, cells or tissues. For these dyes to be effective they must meet the following criteria [Tsien, 1989]: i) They must operate at suitable wavelengths. Visible wavelengths are generally preferred but some near UV wavelengths are usable. ii) They must have good quantum efficiencies. The greater the quantum efficiency the less the amount needed to load into cells to give a useful signal. iii) They must resist bleaching. iv) They must localize in the appropriate cellular or tissue compartments. v) There must be minimum disturbance of cell function during use. vi) They must respond to their intended stimulus with minimum interfering influences. This usually means high selectivity for the specific ion in the presence of competing ions.

Presently there are fluorescent indicators for almost all the important ions in cell biology. However, none of these meet all of the stringent criteria described above. A summary of available fluorescent indicators for the various ions can be found in Table I, and their design and applications are described under individual ions in the text. Several of these indicators such as Fluo-3 and Rhod-2 for Ca^{2+}, the SNARF and SNAFL pH indicators, SBFI for Na^+, PBFI for K^+, and all of the Mg^{2+} indicators have been described so recently that almost no references are published on their applications.

Most of these dyes are introduced into the cell noninvasively as a lipophilic acetoxymethyl (AM) ester [Tsien, 1981]. Cells loaded with this ester release the dye inside the cell through an enzymatic hydrolysis by nonspecific esterases. The spectral response properties of the dye after hydrolysis tend to be similar to its in vitro properties, but may be somewhat influenced by cytoplasmic viscosity, protein binding or other cell-specific properties.

II. DESIGN AND APPLICATION OF INDICATOR DYES
A. Introduction

The design of the appropriate fluorescent dyes varies from ion to ion. Some are based on known models, or created models, while others are arrived at fortuitously. In all cases the invention of the first dye marks only the beginning of the difficult task of structural modifications to make it usable or to meet most of the criteria mentioned earlier. This has accounted for the exis-

TABLE I. Fluorescent Indicator Dyes for Various Ions*

Ion	Dye	Low ion concentration			High ion concentration			
		EX	$(\epsilon \times 10^{-3})$	EM	EX	$(\epsilon \times 10^{-3})$	EM	K_d
Ca^{2+}	Quin-2	352	(5)	492		——		126 nM
	Fura-2	362	(30)	512	335	(35)	505	224 nM
	Indo-1	349	(34)	485	331	(34)	410	250 nM
	Fluo-3	506	(78)	526		——		450 nM
	Rhod-2	556	(80)	576		——		1,000 nM
	Rhod-1	556	(80)	578		——		2,300 nM
Na^+	SBFI	346	(42)	551	334	(47)	525	18 mM
K^+	PBFI	345	(42)	545	336	(47)	515	107 mM
Mg^{2+}	Mag-Fura-2	376	(30)	506	344	(27)	492	1.5 mM
	Mag-Indo-1	354	(30)	475	349	(30)	419	2.7 mM
	Mag-Quin-1	349	(5.0)	499		——		6.7 mM
	Mag-Quin-2	353	(4.2)	487		——		0.8 mM

		Acid solution			Base solution			
		EX	$(\epsilon \times 10^{-3})$	EM	EX	$(\epsilon \times 10^{-3})$	EM	pK_a
H^+	Carboxyfluorescein		——		489	(74)	518	6.4
	BCECF		——		508	(77)	531	6.9
	SNAFL-2	485,514	(25)	546	547	(48)	630	7.65
	SNARF-1	518,548	(26)	587	579	(53)	640	7.50
	DHPN	349	(6.3)	400	409	(7.2)	484	8.0
Cl^-	SPQ	344	(4.2)	450		——		——

*Compiled from Molecular Probes' Handbook of Fluorescent Probes and Research Chemicals [Haugland, 1989].

tence of more than one dye for various ions. The search for the ideal indicators continues in a few laboratories. The applications of these dyes will be presented in terms of their general use with the appropriate references to help first-time users.

B. Calcium

The study of the function of Ca^{2+} ions inside cells is one of the most dynamic areas of modern cell biology. Ca^{2+} is known to control neurosynaptic transmission, secretion of hormones, muscle contraction, and a myriad of physiological functions. It is suspected to be involved in cell division, movements of non-muscle cells and memory and learned patterns of nervous system. In all cases, localized fluctuations in cytosolic free Ca^{2+} levels inside cells are believed to control these functions. Techniques for the measurement and manipulation of Ca^{2+} are therefore crucial and have been advancing rapidly largely as a result of inventions of fluorescent Ca^{2+} indicators in Professor Roger Tsien's laboratories.

1. Design. Free Ca^{2+} levels in resting cells are extremely low (about 10^{-7} M), necessitating an indicator with high sensitivity. Furthermore the Ca^{2+} transients to the range of 1 μM must be measurable in the presence of relatively high concentration (about 1 mM) of the similar divalent cation Mg^{2+}. Although other indicators for calcium such as arsenazo III, chlortetracycline, and aequorin have been described, the most successful of the current fluorescent Ca^{2+} indicators (Table I) are derived from ethylene glycol bis (2-aminoethyl ether) -N,N,N',N'-tetraacetic acid (EGTA) [Ammann et al., 1975]. The organic synthesis of the analogous BAPTA series of Ca^{2+}-selective indicators [Tsien, 1980] takes advantage of the high selectivity of EGTA for Ca^{2+} ions over Mg^{2+} ions while eliminating its major disadvantages such as a) high pKa which results in Ca^{2+} binding by EGTA that is pH dependent in the physiological range, b) slow uptake and release of Ca^{2+} ions which limits the time resolution of the response that can be measured, and c) lack of suitable optical properties of absorption and fluorescence. All of the BAPTA derivatives have the tetracarboxylate structure and the nitrogen-oxygen chelating-structure of EGTA. Unlike EGTA, carbon atoms between the nitrogen and oxygen in BAPTA are hybridized into an aromatic system. Hybridization lowers the basicity of the nitrogen from a pKa between 8 to 9 down to an acceptable 5 to 6. BAPTA has an affinity for Ca^{2+} in the range required for measurements in the cytosol, tenfold better Ca^{2+}-Mg^{2+} selectivity than EGTA and negligible pH sensitivity of Ca^{2+} binding above pH 7. Its absorbance, however, is in the ultraviolet, overlapping protein absorption. BAPTA and its derivatives are useful for buffering Ca^{2+} both intra- and extracellularly. Although BAPTA derivatives undergo Ca^{2+}-dependent absorption shifts, they have not found any application as ion indicators in cellular studies because of the possibility of a UV-induced cell damage and possible interference from cell protein and nucleotide absorbance. Fluorinated versions of BAPTA are used in fluorine-nuclear magnetic resonance (^{19}F-NMR) studies of Ca^{2+} flux inside whole cells [Marban et al., 1987].

The prototypical fluorescent calcium indicator is called Quin-2 (3) (Fig. 1). It has three nitrogens and one oxygen hybridized into a quinoline nucleus and a benzenoid aromatic nucleus. The other major Ca^{2+} indicators are Fura-2 (4) and Indo-1 (5) [Grynkiewicz et al., 1985], Fluo-3 (6), and Rhod-2 (7) [Minta et al., 1989]. These are all obtained by structural modification of BAPTA. They all preserve the useful properties of BAPTA while eliminating some of the negative properties of Quin-2 that result from its relatively weak UV absorption. It has taken 8 years since the invention of Quin-2 and a multitude of synthetic compounds [Tsien, 1983] of varying usefulness to obtain these dyes. The key BAPTA-derived synthetic precursors for the useful dyes are (8), (9), and (10). Reaction of precursor (8) with compounds of the type (11) leads to a benzofuran fluorophore. Similarly (9) and (12) give an indole fluorophore. Treatment of (10) with t-butyl lithium gives an intermediate which reacts with

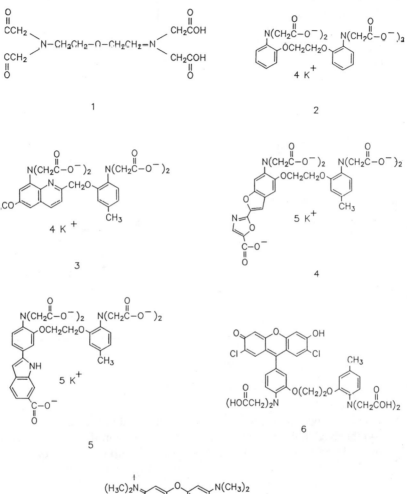

Fig. 1.   Structures of fluorescent indicator dyes.

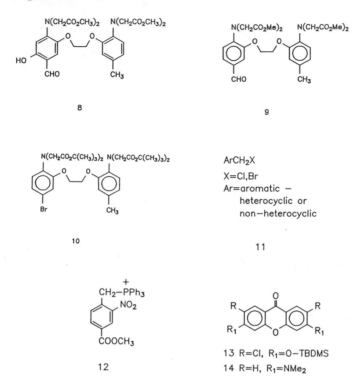

Fig. 1.   (Continued)

(13) to give a fluorescein-like fluorophore, and with (14) to give a rhodamine-like fluorophore.

Two types of spectral response have been observed in the many Ca^{2+} indicators synthesized in the Tsien lab. The first kind, exemplified by Fura-2 (Fig. 2), shows a large shift in the wavelength of maximum excitation on binding Ca^{2+}. Indo-1 belongs to the same family of compounds but shows a shift in emission wavelength rather than excitation wavelength. The spectral shift permits the use of excitation or emission intensity ratio measurements. In fluorescence ratio measurements, the determination of ion concentration is *independent* of the indicator concentration as long as the signal is sufficiently above background. Variable effects such as degree of cell loading, photobleaching, detector sensitivity, cell thickness, and optical path are eliminated. So far the available ratioable dyes have excitation maxima in the near UV and have a K_d between 100 and 250 nM. These high affinities mean they become saturated and cannot distinguish Ca^{2+} levels above a few micromolar. The second type of spectral response to Ca^{2+}, as exemplified by Fluo-3 (Fig. 3), Quin-2, and Rhod-2, is a remarkable enhancement of fluorescence yield on Ca^{2+} binding

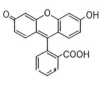

15

16

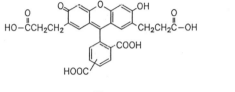

17

18

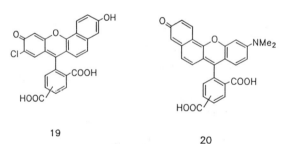

19

20

Fig. 1.   (Continued)

without an appreciable shift in either excitation or emission wavelength. Ratio measurements are not possible with this type of indicator. So far it has not been possible to predict whether a dye is going to exhibit shifting in its spectrum on complexation but there has been an attempt to explain why some of the dyes do not exhibit this useful phenomenon [Minta et al., 1989]. The non-ratioable dyes show excitation and emission maxima from the ultraviolet for Quin-2 to the visible for Fluo-3 and Rhod-2 and have dissociation constants in the range of 400 nM to 2,300 nM. The ideal indicator will therefore be ratioable, have excitation and emission maxima in the visible range to reduce artifacts of

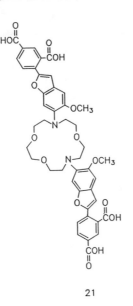

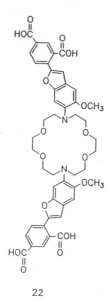

21   22

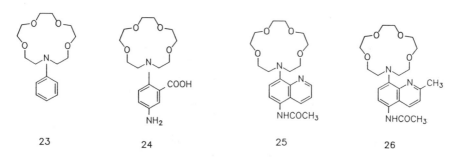

23   24   25   26

Fig. 1.   (Continued)

autofluorescence and have affinities in the 1 μM range. Until such a dye is found, one may have to select dyes to fit particular needs.

   2. Application.   The key to the breakthrough of ion measurement techniques in cell biology is not just the invention of these dyes but finding a means to introduce noninvasively the dyes into cells. The most common method of introducing the indicators is through conversion of the polar carboxylates to lipophilic membrane-permeable acetoxymethyl (AM) esters which can be hydrolyzed by intracellular esterases to regenerate the free dye in the cytosol [Tsien, 1981]. The cells are incubated with the AM form of the dye for 30 min to 1 h or longer and washed to eliminate the unloaded dye. Since the AM form of the dye is usually not water soluble, it is commonly dissolved in anhydrous dimeth-

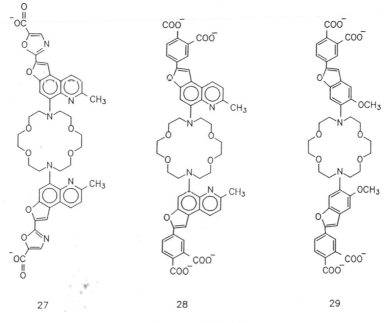

27   28   29

Fig. 1.   (Continued)

ylsulfoxide as a stock solution for the incubation. Adding "Pluronic F-127" and/or fetal calf serum [Poenie et al., 1986] may aid in the loading process. Problems associated with loading cells have been discussed expertly in Cobbold and Rink [1987]. The dyes may be used in cell suspensions but it is essential to have the solution in the cuvette stirring. This prevents localized illumination of cells in the beam and reduces photobleaching. Leakage of the indicator dye from the cell into a high Ca^{2+} medium is a potential problem. Problems associated with measurements in cell suspension have been discussed by Tsien [Tsien et al., 1982]. The signal obtained from measurements in cell suspensions represents the *average* Ca^{2+} level inside the cell. The low response of unhealthy cells to a stimulus can grossly perturb estimation of the *peak* Ca^{2+} level in suspensions. This is less of a problem in single cell measurements by microscopy or flow cytometry where non-responding cells can be detected.

The most beneficial use of the ratioable dyes has been in single cells. This research has been greatly facilitated by using ratioable dyes. These dyes permit i) dual-wavelength measurement of single cells or clusters of cells [Poenie et al., 1985] and ii) digital imaging fluorescence microscopy to study Ca^{2+} in whole fields of cells and different locations within single cells [Tsien and Poenie, 1986]. Digital imaging methods are particularly suitable for studying what goes on in a single cell, however, the cost of setting up and the training required may restrict its usage to relatively few laboratories. These dyes may

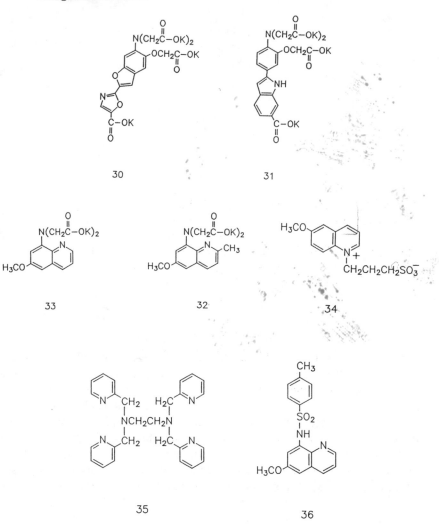

Fig. 1. (Continued)

also be employed in studying Ca^{2+} in monolayers, tissues and organs, and in stopped-flow fluorescence [Cobbold and Rink, 1987].

So far Indo-1 has been the most popular indicator for measurement of Ca^{2+} by flow cytometry, where it is more feasible to do laser excitation with a single wavelength while measuring Ca^{2+}-dependent emission at two wavelengths. Because Indo-1 absorbs only at short wavelengths, excitation requires use of the ultraviolet output of the multi-line argon laser. The data obtained by irradiating with the laser in a fraction of a second can be calculated on a single

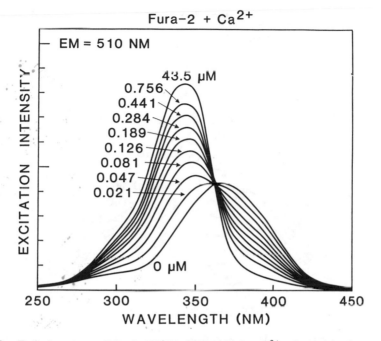

Fig. 2. Excitation spectra of Fura-2 at 22°C in buffers with free Ca^{2+} values ranging from <1 nM to 5 μM. The titration was done in a buffer containing 100 mM KCl, 10 mM K-MOPS, 10 mM K_2H_2 EGTA, and 1 μM Fura-2 at pH 7.0. The Ca^{2+} values were obtained by replacement of 1 ml of the above buffer with 1 ml of a buffer containing 100 mM KCl, 10 mM K-MOPS, 10 mM K_2Ca EGTA, and 1 μM Fura-2 at pH 7.0 The emission was set at 510 nm for the excitation measurement.

cell basis using ratios in intensities near the emission maxima (490 nm without Ca^{2+}, 405 nm with Ca^{2+}).

In order to obtain $[Ca^{2+}]$ using these dyes in cells it is necessary to know the theory and practice of calibration of these fluorescent indicators. For non-ratioable dyes, this is provided by Tsien for Quin-2 [Tsien et al., 1982] and Kao for Fluo-3 [Kao et al., 1989], and for ratioable dyes by Poenie et al. [1985]. The difference in the practice of calibration is that with non-ratioable dyes it is always necessary to perform a post-experiment calibration with a heavy metal, usually Mn^{2+}, to obtain the value for the autofluorescence. The $[Ca^{2+}]$ is then obtained from Equation 1:

$$[Ca^{2+}] = K_d (F - F_{min}) / (F_{max} - F) \qquad (1)$$

where F is any given fluorescence recorded, F_{max} the fluorescence of the Ca^{2+} saturated dye, and F_{min} the fluorescence of the free dye. K_d is the effective dissociation constant.

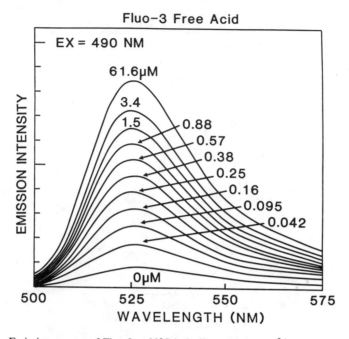

Fig. 3. Emission spectra of Fluo-3 at 22°C in buffers with free Ca^{2+} values ranging from 4 nm to 62 μM. The titration was done in a buffer containing 100 mM KCl, 10 mM K-MOPS, 10 mM K_2H_2 EGTA, and 10 μM Fluo-3 at pH 7.03. The Ca^{2+} values were obtained by replacement of 1 ml of the above buffer with 1 ml of a buffer containing 100 mM KCl, 10 mM K-MOPS, 10 mM K_2Ca EGTA, and 10 μM Fluo-3 at pH 7.03. The excitation was set at 490 nm for emission measurement.

The value of F_{min} must be adjusted for known influence by other ions or experimental controls for each individual dye. In the case of Quin-2 it is corrected for its significant Mg^{2+} binding using Equation 2:

$$F_{min} = F_{Mn2+} + 1/6 (F_{max} - F_{min}) \qquad (2)$$

where F_{Mn2+} is the fluorescence signal after the indicator fluorescence is mostly quenched by Mn^{2+} and 1/6 is an empirical factor correcting for Mg^{2+} binding to Quin-2 [Hesketh et al., 1983].

Fluo-3 does not bind to Mg^{2+} significantly and operates at longer wavelength so the correction is essentially for the background as shown in Equation 3:

$$F_{min} = F_{max} - F_{bkg}/40 + F_{bkg} \qquad (3)$$

where 1/40 is the ratio of the fluorescence of metal-free Fluo-3 compared to the Ca^{2+} complex and F_{bkg} is the signal obtained after lysis in the presence of Mn^{2+}.

Ratioable dyes on the other hand give $[Ca^{2+}]$ values according to Equation 4:

$$[Ca^{2+}] = K(R - R_{min}) / (R_{max} - R)$$
$$K = K_d (F_o/F_s) \tag{4}$$

R_{min} and R_{max} are the fluorescence intensity ratio (e.g., at 340 and 380 nm for Fura-2) obtained in zero and saturating Ca^{2+} respectively. K_d is the effective dissociation constant. F_o is the excitation signal in the absence of Ca^{2+} and F_s is the excitation signal at saturating Ca^{2+}. The ratio of the signal is independent of the dye concentration. Therefore, if one is certain about the autofluorescence background, one can measure $[Ca^{2+}]$ directly.

C. pH

A knowledge of cytoplasmic pH is necessary to study intracellular mechanisms, from organelle function to enzyme kinetics. There are now quite a few indicators for studying pH (Table I), the most popular of which is BCECF [Rink et al., 1982]. It has a pKa of 6.97–6.99, which is almost ideal because nearly all cells have a cytosolic pH which is tightly regulated to be near 7.0, generally varying in the range of a few tenths of a unit.

1. Design. 6-Carboxyfluorescein (15) can be incorporated into cells by uptake of its diacetate with a subsequent hydrolysis to generate the original dye [Thomas et al., 1979]. Attempts at using it in lymphocytes [Rink et al., 1982] resulted in an undesirable leakage. The pKa of carboxyfluorescein at 6.3 is also somewhat low for studying pH ranges around 7. 5,6-Dicarboxyfluorescein diacetate (16) [Rink et al., 1982] was synthesized in the hope that the extra carboxy group would prevent leakage. The leakage was slightly suppressed but the pKa remained the same. BCECF, bis(carboxyethyl)carboxyfluorescein (17), which has two extra carboxylate functions via short alkyl chains, was synthesized with the hope of preventing leakage. It not only slowed considerably the leakage but also increased the pKa to a more desirable value of 6.99. Although the fluorescence emission spectra of BCECF show only quenching without a spectral shift, its fluorescence excitation spectra have an isosbestic point near 438 nm (Fig. 4). Estimation of the intracellular pH using BCECF has usually been made by ratio measurements of the excitation spectra at 450 and 500 nm. Microscopic imaging using BCECF fluorescence has been done in single cells [Paradiso et al., 1984]. BCECF has been introduced into cells as the AM esters rather than acetates.

Molecular Probes has recently introduced new dual emission or dual exci-

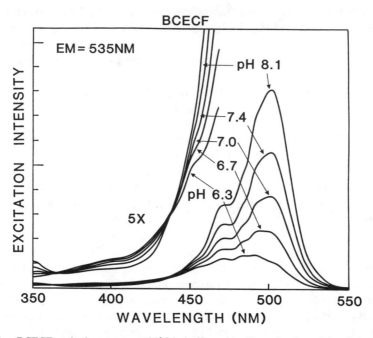

Fig. 4.   BCECF excitation spectra at 22°C in buffers with pH ranging from 6.3 to 8.1. The pH titration was done in 50 mM K_2HPO_4 at pH indicated, containing 1 μM BCECF. The emission was set at 535 nm for the excitation measurement.

tation intracellular pH indicators called SNARF (19) and SNAFL (20) [Whitaker et al., 1988]. These indicators were found while searching for longer wavelength dyes and resulted from incorporation of a naphthalene moiety into fluorescein and rhodamine systems. They have pKa values (Table I) ranging from 7.50 to 7.90. Unlike the fluorescein indicators, SNARF and SNAFL pH indicators undergo *both* pH dependent emission and excitation shifts with clear, isosbestic points in both the excitation and emission spectra. The long and dual wavelengths of these dyes allow a facile excitation by argon laser lines used in flow cytometry. The emission and excitation shifts of wavelength with pH also make them ideal for digital ratio imaging in single cells. Like BCECF, these indicators have AM forms for non-invasive introduction into cells.

There is yet another pH indicator, 1,4-dihydroxyphthallonitrile (DHPN) [Kurtz and Balaban, 1985], which is an emission shifting indicator. This is usually loaded as the acetate and has a pKa of 8.0 [Brown and Porter, 1977]. It tends to leak out of cells quite readily, presumably because it has only one or two negative charges which are partly localized.

2. Application. All pH indicators are loaded noninvasively through the permeant esters (AM or acetates) just as described for Ca^{2+} indicators. The big difference is the method for calibration. Calibration curves for pH indicators are constructed using a variety of permeabilizing agents e.g. nigericin, K^+/H^+ ionophore [Thomas et al., 1979] or monensin for ouabain-treated cells in Na^+-containing buffers [Paradiso et al., 1984].

D. Sodium and Potassium

There is a large gradient between intracellular Na^+ concentrations (typically 10–40 mM) and extracellular Na^+ concentrations (typically 120–450 mM) in animals. This gradient is important for biological processes such as action potentials, synaptic depolarization, active uptake of nutrients, neurotransmission, epithelial transport, and regulation of cell volume. Indeed, cells devote a major fraction of their metabolic energy to maintaining intracellular Na^+ at levels much lower than extracellular. Fluorescent Na^+ and K^+ indicators are a very recent development. These have come from two laboratories. Smith [1988] and his group have reported a cryptand-based sodium indicator FCryp-2 with excellent Na^+ affinity (K_d 6 mM). The results inside live cells would be of great interest but have not been reported. Tsien and collaborators [Minta et al., 1987] have also recently reported a crown-ether-based Na^+ indicator, SBFI (21). Applications have been reported in fibroblasts, hepatocytes and lymphocytes [Harootunian et al., 1988], smooth muscle cells [Moore et al., 1988] and gastric glands [Negulescu et al., 1988]. The K^+ equivalent of SBFI is called PBFI (22) [Minta and Tsien, 1989] and differs from SBFI only by the size of the crown ether.

1. Design. Unlike Ca^{2+} indicators, there was no suitable model molecule with reasonable Na^+ to K^+ selectivity at pH 7 in water. The search therefore started with a basic structure (23) which has a crown structure for Na^+ to K^+ selectivity and an aniline for the nitrogen linkage. It was reasoned that in the design of fluorescent Na^+ and K^+ indicators the site of the metal-binding cavity should result in the necessary metal-binding selectivity while the delocalization of the pair of electrons on the nitrogen atoms should result in the changes in fluorescence response. As is typical for development of new ion indicators, several prototype structures were synthesized before finding the appropriate combination of binding affinity and spectral rseponse. Compound (23) had been synthesized by Dix and Vogtle but had not been characterized for cation binding properties. It was found that its affinity for Na^+ or K^+ in water was too weak to be determined accurately. The structure was therefore modified to (24), incorporating an amine and a carboxylic acid to make it water soluble. The selectivity was 13:1 for Na^+ over K^+ but the basicity of the amine on the benzene ring (pKa of 9.21) was too high for it to be useful. The aniline was replaced with a quinoline (25) in an effort to obtain something fluorescent like Quin-2. Compound (25) had a desirable selectivity for Na^+ and a suitable pKa but the high affinity for Mg^{2+} interfered with its use. Compound (26) had suitable selectivity over Mg^{2+}; however, its fluorescence quantum yield was too low to be useful. Several modifications were made on compound (26) in an attempt to increase the fluorescence yield. This resulted in synthesis of compound (27), a Fura-22 analog. The extinction coefficient was doubled and so was the quantum yield but the affinity for Na^+ dropped to an undesirable level of 200 mM. Changing the oxazole in (27) to a guinolino-oxazole in (28) restored the affinity to an unusually high level (4.7 mM). Through a series of modifications, compound (29) (SBFP) was obtained with the right selectivity of Na^+ to K^+, affinity for Na^+, respectable quantum yield as well as a reasonable pKa. Unfortunately, the AM ester of SBFP did not hydrolyse inside the cells to the original dye. Changing the phthallate to isophthallate gave SBFI (21) which has since been successfully used in various cells.

2. Application. The loading of Na^+ and K^+ indicators is like the Ca^{2+} indicators with the following exceptions [Harootunian et al., 1989]: a) Because of the lower quantum yields of these indicators compared to Fura-2, it is necessary to incubate cells with more of the AM esters (2–10 μM). Like most AM esters, they are usually dissolved in DMSO and may be solubilized with Pluronic F-127. b) Because of the poorer solubility of the AM esters in water, cell loading may require longer incubation time (1–3 h). SBFI and PBFI undergo small excitation shifts to shorter wavelengths on metal binding. Excitation ratioing can therefore be used in ratio imaging [Harootunian et al., 1989]. Calibration of the Na^+ and K^+ indicators [Harootunian et al., 1989] is different than the Ca^{2+} indicators. It is accomplished by application of gramicidin D, amphotericin B, or a combination of monensin and nigericin. These ionophores produce transmembrane pores that are highly conductive and selective for monovalent ions. The ionophores do not, however, distinguish markedly between Na^+ and K^+. Exposure of the cells with various mixtures of Na-gluconate and K-gluconate in the presence of these ionophores permits calibration of 340/385 nm excitation ratios in terms of Na^+.

E. Magnesium

Cytoplasmic Mg^{2+} is crucial in the investigation of intracellular mechanisms, from organelle function to enzyme kinetics. It has also been implicated in various aspects of cell activation such as cell growth, differentiation and proliferation [Rubin et al., 1979]. In addition, hypomagnesia results from chronic diseases such as diabetes, essential hypertension and alcoholism. Hypermagnesia is associated with renal vascular diseases and circulatory shock [Altura and Altura, 1981]. Four fluorescent Mg^{2+} indicators have been developed since 1988 with a varying range of affinities (Table I).

1. Design. The EGTA-type indicators have about 10^6 selectivity for Ca^{2+} over Mg^{2+}. Availability of all four carboxylic acids of EGTA appears to be essential for tight binding of Ca^{2+}. To reduce Ca^{2+} binding it appears reasonable to use only one half of the molecule and to allow the remaining two carboxylic acids to envelope the smaller Mg^{2+} ion. Indeed this has been observed by Tsien (personal communications) and has been confirmed with the synthesis and use of APTRA, a non-fluorescent Mg^{2+} indicator [Levy et al., 1988]. The fluorescent Mg^{2+} indicators now available take advantage of the fluorophores used by Tsien, namely Quin-2, Fura-2, and Indo-1. The Fura-2 version [Raju et al., 1989] was called Furaptra by Levy and co-workers. Molecular Probes has made this analog and the other ones and decided to name them after their Ca^{2+} analogues—hence the names Mag-Fura-2 (30), Mag-Indo-1 (31), Mag-Quin-1 (32), and Mag-Quin-2 (33). The spectral shifts of Mag-Fura-2 and Mag-Indo-1 are very similar to their Ca^{2+} counterparts and should permit ratio measurements of intracellular Mg^{2+} flux. While Ca^{2+} binding by the Mg^{2+} indicators is generally stronger than Mg^{2+} binding (e.g., K_d of Ca^{2+} for Mag-Fura-2 is 54 μM), the selectivity for Mg^{2+} over Ca^{2+} is generally sufficient so that the indicators respond only to Mg^{2+} except perhaps in cells that undergo unusually high spikes of Ca^{2+} flux. The Mag-Quin analogs spectra undergo predominantly intensity changes on metal binding and cannot be ratioed.

2. Application. Mag-Fura-2 [Raju et al., 1989] has been used recently to determine the cytosolic free Mg^{2+} concentration for isolated rat hepatocytes. The dyes are loaded in the AM ester forms. Calibrations and calculations for the ratioable Mg^{2+} indicators are all similar to Fura-2 for Ca^{2+}, using 4-bromo-A23187 [Alvarez-Leefmans et al., 1987] as a Mg^{2+} ionophore.

F. Chloride

Chloride ion-fluxes play an important role in cellular regulatory, absorptive, and secretary processes. Only one compound SPQ (34), 6-methoxy-N-(3-sulfopropyl) quinolinium, has been used extensively as a chloride indicator in vesicles and erythrocyte ghosts [Illsley and Verkman, 1987] and in epithelial cells [Fong et al., 1988]. It differs from all the other indicators in that the chloride interacts with the excited state of SPQ causing a radiationless quenching of fluorescence of the dye with no change in the absorption spectrum.

1. Design. SPQ was selected from a series of compounds synthesized by Wolfbeis and Urbano [1982] while searching for standards for pH-independent fluorescence measurements. Unlike EGTA for Ca^{2+} and the cryptand-like structures for Na^+ and K^+, a suitable model for a Cl-specific chelator is not obvious. The current Cl indicators based on the weak tendency of Cl to quench fluorophores has been the only approach found useful to obtain equilibrium Cl indicators.

2. Application. Since SPQ is a polar zwitterion with a quartenary nitrogen cation and a sulfonate anion, cells must be loaded with high external concentrations of the dye. Once the excess external dye is removed, however, the dye tends to leak out. The quenching constants K_q of SPQ by chloride obtained in cell-free solution may be ten times higher than in intact cells in which the rate of chloride diffusion may be more limited or the dye may interact with other ions or undergo binding to cellular structures [Krapf et al., 1988]. Calibration of the response is done with ionophores such as tributyltin, a Cl^-/OH^- exchanger together with nigericin [Krapf et al., 1988].

G. Other Ions

Heavy metals tend to quench the fluorescence of most of the cation-sensitive indicators. The compound tetrakis (-2-pyridylmethyl)ethylenediamine (TPEN) (35) has been recommended as a control for intracellular heavy metal artifacts [Arslan et al., 1985]. Fluorescence quenching by Mn^{2+} has been exploited in calibration of Ca^{2+} indicators. Zinc enhances the fluorescence on binding to the Ca^{2+} indicators. Methods used in studying Zn^{2+} in nutrition still involve cell-destructive assays because there is no suitable indicator. The compound TSQ (36), N-(6-methoxy-8-quinolyl)-p-toluenesulfonamide, has been used as a zinc stain in boutons [Frederickson, 1987].

III. FUTURE PROSPECTS AND CONCLUSIONS

There are fluorescent indicators for almost all the useful ions in cell biology, but the deficiencies discussed for them make it obvious that improvements are still possible that will facilitate future progress. A long-wavelength ratioable dye for measuring Ca^{2+} with good pKa and a lower affinity (about 1 μM) should be a target for research. For pH it appears more work on the SNARF and SNAFL to lower pKa and increase retention may be the direction for synthetic research. The remaining ions need indicators that operate in the visible spectrum. Relatively few laboratories are pursuing this chemistry. Nevertheless, commercial availability of the current indicators and their importance have made their application widespread.

Instrumentation for ratio measurements appears to be advancing faster than the chemistry of the dyes. Refinement in digital imaging microscopy is proceeding quite steadily. There are already photomultipliers for detecting in the infrared should such dyes become available. The support for making better indicators, it appears, should therefore be paramount in the future of noninvasive dyes in cell biology.

ACKNOWLEDGMENTS

We thank Mrs. Lisa Wright and Ms. Elizabeth Schoen for the preparation of the manuscript including the structures and Linda Bovard and Andrew Hansen for the figures. This work has been supported by a grant from the National Institutes of Health (GM 37347-02).

IV. REFERENCES

Agard DA (1984) Optical sectioning microscopy—cellular architecture in three dimensions. Annu Rev Biophys Bioeng 13:191–219.

Altura BM, Altura BT (1981) Mg^{2+} ions and contraction of vascular smooth muscle: Relationship to some vascular diseases. Fed Proc 40:2672–2679.

Alvarez-Leefmans FJ, Giraldez F, Gamino SM (1987) Intracellular free magnesium in excitable cells: Its measurement and biologic significance. Can J Physiol Pharmacol 65:915–925.

Ammann D, Gueggi M, Pretsch E, Simon W (1975) Improved Ca^{2+} ion-selective electrode based on a neutral carrier. Anal Lett 8:709–720.

Arslan P, DiVirgilio F, Beltrame M, Tsien RY, Pozzan T (1985) Cytosolic Ca^{2+} homeostasis in Ehrlich and Yoshida carcinomas. A new membrane-permeant chelator of heavy metals reveals that these ascites tumor cell lines have normal cytosolic free Ca^{2+}. J Biol Chem 260:2719–2727.

Bright GR, Taylor DL (1986) Imaging at low light level in fluorescence microscopy. In Taylor DL, Waggoner AS, Murphy RF, Lanni F, Birge RR (eds): Applications of Fluorescence in the Biomedical Sciences. New York: Alan R. Liss, Inc.

Brown RG, Porter G (1977) Effect of pH on the emission and absorption characteristics of 2,3-dicyano-p-hydroquinone. Chem Soc Faraday Trans I 73:1281–1285.

Cobbold PH, Rink TJ (1987) Fluorescence and bioluminescence measurements of cytoplasmic free Ca^{2+}. Biochem J 248:313–328.

Dix JP, Vogtle F (1980) Ionenselektive farbstoffkronenether. Chem Berichte 113:457–470.

Fong P, Illsley NP, Widdicombe JH, Verkman AS (1988) Chloride transport in apical membrane vesicles from bovine tracheal epithelium. J Membr Biol 104:233–239.

Frederickson CJ (1987) A quinoline fluorescence method for visualising and assaying the histochemically reactive zinc (bouton zinc) in the brain. J Neurosci Methods 20:91–103.

Grynkiewicz G, Poenie M, Tsien RY (1985) A new generation of Ca^{2+} indicators with greatly improved fluorescence properties. J Biol Chem 260:3440–3450.

Harootunian A, Eckert B, Minta A, Tsien RY (1988) Ratio imaging using a newly developed fluorescent indicator in rat embryo fibroblast. FASEB J 2:A728.

Harootunian A, Kao JPY, Eckert BK, Tsien RY (1989) Fluorescent ratio imaging of cytosolic free Na^+ in individual fibroblasts and lymphocytes. J Biol Chem (in press).

Haugland RP (1989) Handbook of Fluorescent Probes and Research Chemicals. Eugene, OR: Molecular Probes, Inc.

Hesketh TR, Smith GA, Moore JP, Taylor MV, Metcalfe JC (1983) Free cytoplasmic calcium concentration and the mitogenic stimulation of lymphocytes. J Biol Chem 258:4876–4882.

Illsley NP, Verkman AS (1987) Membrane chloride transport measured using a chloride sensitive fluorescent probe. Biochemistry 26:1215–1219.

Kao JPY, Harootunian AT, Tsien TY (1989) Photochemically generated cytosolic Ca^{2+} pulses and their detection. J Biol Chem 264:8179–8184.

Krapf R, Berry CA, Verkman AS (1988) Estimation of intracellular chloride activity in isolated perfused rabbit proximal convoluted tubules using a fluorescent indicator. Biophys J 53:955–962.

Kurtz I, Balaban RS (1985) Fluorescence emission spectroscopy of 1,4-dihydroxyphthallonitrile. Biophys J 48:499–508.

Levy LA, Murphy E, Raju B, London RE (1988) Measurement of cytosolic free Mg^{2+} ion concentration by ^{19}F NMR. Biochemistry 27:4041–4048.

Marban E, Kitakaze M, Kusuoka H, Porterfield JK, Yue DT, Chacko VP (1987) Intracellular free calcium concentration measured with ^{19}F-NMR spectroscopy in intact ferret hearts. Proc Natl Acad Sci USA 84:6005–6009.

Minta A, Harootunian AT, Kao JPY, Tsien RY (1987) New fluorescent indicators for intracellular sodium and Calcium. J Cell Biol 105:89a.

Minta A, Kao JPY, Tsien RY (1989) Fluorescent indicators for cytosolic Ca^{2+} based on rhodamine and fluorescein. J Biol Chem 264:8171–8178.

Minta A, Tsien RY (1989) Fluorescent indicators for cytosolic sodium. J Biol Chem (in press).

Moore EDW, Minta A, Tsien RY, Fay FS (1988) Measurement of intracellular sodium with SBFI, a newly developed sodium sensitive fluorescent dye. FASEB J 2:A754.

Negulescu PA, Harootunian A, Minta A, Tsien RY, Machen TE (1988) Intracellular sodium regulation in rabbit gastric glands determined using a fluorescent sodium indicator. J Gen Physiol 92:26a.

Paradiso AM, Tsien RY, Machen TE (1984) Na^+-M^+ exchange in gastric glands as measured with a cytoplasmic-trapped, fluorescent pH indicator. Proc Natl Acad Sci USA 81:7436–7440.

Poenie M, Alderton JA, Steinhardt PA, Tsien RY (1986) Ca^{2+} rises abruptly and briefly throughout the cell at the onset of anaphase. Science 233:886–889.

Poenie M, Alderton JA, Tsien RY, Steinhardt PA (1985) Changes in free Ca^{2+} levels with stages of the cell division cycle. Nature 315:147–149.

Raju B, Murphy E, Levy LA, Hall RD, London RE (1989) A fluorescent indicator for measuring cytosolic free Mg^{2+}. Am J Physiol 256:C540–C548.

Rink TJ, Tsien RY, Pozzan T (1982) Cytoplasmic pH and free Mg^{2+} in lymphocytes. J Cell Biol 95:189–196.

Rubin AH, Terasaki M, Sanmi H (1979) Major intracellular cations and growth control: Correspondence among Mg^{2+} content, protein synthesis, and the onset of DNA synthesis in Balb/c3T3 cells. Proc Natl Acad Sci USA 76:3917–3921.

Smith GA, Hesketh TR, Metcalfe JC (1988) Design and properties of a fluorescent indicator of intracellular free Na^+ concentrations. Biochem J 250:227–232.

Thomas JA, Birschbaum RN, Zimniak A, Racker E (1979) Intracellular pH measurements in Ehrlich ascites tumor cells utilizing spectroscopic probes generated *in situ*. Biochemistry 18:2210–2218.

Tsien RY (1989) Fluorescent probes for cell signalling. Annu Rev Neurosci 12:227–253.

Tsien RY (1983) Intracellular measurements of ion activities. Annu Rev Biophys Bioeng 12:91–116.

Tsien RY (1981) A non-disruptive technique for loading Ca^{2+} buffers and indicators. Nature 290:527–528.

Tsien RY (1980) New Ca^{2+} indicators and buffers with high selectivity against Mg^{2+} and protons. Biochemistry 19:2396–2404.

Tsien RY, Poenie M (1986) Fluorescence ratio-imaging: A new window into intracellular ion signalling. Trends Biochem Sci 2:450–455.

Tsien RY, Pozzan T, Rink TJ (1982) Calcium homeostasis in intact lymphocytes. J Cell Biol 94:325–334.

Whitaker JE, Haugland RP, Prendergast FG (1988) SNAFLs and SNARFs: Dual emission pH indicators. Biophys J 5:197.

Wolfbeis OS, Urbano E (1982) Synthesis of fluorescent dyes XIV. Standards for fluorescence measurement in the near neutral pH range. J Heterocyclic Chem 19:841.

Noninvasive Techniques in Cell Biology: 21–51
© 1990 Wiley-Liss, Inc.

2. Quantitative Single Cell Fluorescence Imaging of Indicator Dyes

David J. Gross

Department of Biochemistry and Program in Molecular and Cellular Biology,
University of Massachusetts, Amherst, Massachusetts 01003

I. INTRODUCTION

The adage "a picture is worth a thousand words" is as relevant to modern cell biology today as it was when it was coined. With the advent of high-sensitivity imaging detectors and high-speed digital processing, cell biologists have begun to explore phenomena which were impossible to observe only a decade ago. In this chapter I aim to discuss the application of optical imaging techniques to problems in cell biology, and especially to studies employing extrinsic fluorescent indicator probes at the level of the individual cell.

A. Biological Parameters

1. Simultaneous individual cell monitoring. In an image of the field of view of the optical microscope one finds from one to hundreds of individual cells depending on the cell type, size, density, and the microscope magnification. Thus, the photons emanating from any particular cell impinge on an imaging photodetector at the same time as those from any other cell, allowing one to monitor simultaneous events in all cells in the field of view. This fact, in combination with the intrinsic ability to resolve spatial information within the image of a single cell, provides quantitative optical imaging with a powerful advantage over non-imaging detection schemes. As with any technique, the advantages of imaging must be considered along with its disadvantages which include limitations on temporal resolution, massive volumes of data and the need for formidable data processing capacity.

2. Ca^{2+} ion activity. With the advent of the Ca^{2+} chelating fluorescent probe fura-2 [Grynkiewicz et al., 1985] the measurement of $[Ca^{2+}]$ in individual living cells has become quite popular. The earliest studies were of dye-loaded cells in a bulk fluorometer or of individual cells monitored with a photomultiplier tube with sub-second time resolution [Tsien et al., 1985] and of single cells with high spatial imaging resolution but low temporal resolution [Williams et al., 1985]. More recent studies have employed rapid time-lapse imaging of fura-2-loaded cells to examine the kinetics of $[Ca^{2+}]$ transients in several cells monitored simultaneously with temporal resolution down to 1.5 s [Millard et al., 1988; Gonzalez et al., 1988a,b; Jacob et al., 1988; Prentki et al., 1988].

Perhaps most striking in the application of quantitative imaging to the measurement of $[Ca^{2+}]$ in single cells are measurements of apparent spatial gradients of $[Ca^{2+}]$ within the cytosol [Connor et al., 1988; Poenie et al., 1987; Chandra et al., 1989]. Such measurements are not possible for most cells without high-sensitivity imaging detectors, digital processing, and the appropriate fluorescent probe.

Although this chapter focuses on fluorescence imaging techniques, the application of calcium-sensitive photoproteins such as aequorin and spectral-shift absorption probes such as arsenazo III deserves mention. One of the earliest and perhaps most dramatic demonstrations of a simultaneous temporal and spatial variation of $[Ca^{2+}]$ was accomplished with aequorin [Gilkey et al., 1978]. This study demonstrated that a dramatic wave of $[Ca^{2+}]$ traverses an activated fish egg from the point of stimulation to the opposite pole of the cell. The rather smaller size of most other cells and thus the diminished intensity of photoprotein-generated light has made this approach less popular than fura-2 $[Ca^{2+}]$ imaging. Absorption probes, which have found application in neuronal cells with a considerable optical path length, are also not well suited for imaging of small cells [however, see Bolsover and Spector, 1987 for a non-imaging study].

3. Membrane potential imaging. Fluorescent indicators of transmembrane electric potential fall into two general categories: 1) charged amphiphilic molecules which accumulate within the cytosol or within intracellular organelles under an electrochemical gradient according to the Nernst equation and 2) charged or zwitterionic lipophilic probes which incorporate into membranes and which respond to the transmembrane electric field by a fluorescence shift. The former category of probe responds sluggishly to a change in transmembrane potential since a wholesale flux of probe molecules is involved. The latter category, however, is quite temporally responsive since the measured signal derives from molecular-scale motions or electronic transitions within the probe molecules. The relative properties of the two categories of probes have been discussed in detail recently [Gross and Loew, 1989].

Imaging studies employing both categories of potential-sensitive probes have been reported. The use of redistribution probes is less common, although Loew and colleagues have recently demonstrated that highly quantitative measurements are readily accomplished by the use of imaging techniques [Ehrenberg et al., 1988]. Membrane-bound probe studies employing imaging have found a wider application, particularly for neurophysiological preparations. Most of these studies have employed photodiode array detectors which provide high temporal resolution but limited spatial resolution [Cohen and Salzberg, 1978; Salzberg et al., 1983; Grinvald et al., 1984]. Recently, high spatial resolution imaging of changes in transmembrane potential in individual cells has been reported [Gross et al., 1986; Kinosita et al., 1988; Montana et al., 1989].

One major drawback of the membrane-bound probes of transmembrane electric field is their limited dynamic range. Di-4-ANEPPS, one probe that has been used extensively, has one of the highest responsivities of this class of probe at 10% fluorescence change per 100 mV change in transmembrane potential (Fluhler et al., 1985). Thus, the use of these probes is difficult and requires careful control and optimal performance of the experimental apparatus.

B. The Imaging Approach

1. Quantitation. As noted above, quantitative imaging of extrinsic fluorescent indicators incorporated within living cells provides a spatially resolved record of the quantity of interest. Both intracellular spatial and temporal variations as well as simultaneously recorded variations in several different cells are accessible to this approach. The signals one records are subject to the same sorts of calibration restrictions as are those obtained by point photometry with the added complication that the calibrations must be measured at all points of the image which contain data of interest. In general, one must measure background light and system gain at all wavelengths of light which are relevant to a particular experiment. In some cases, corrections for geometric distortion of the optical system and spatially dependent fluorochrome photobleaching effects must be taken into account [Benson et al., 1985; Jeričević et al., 1989].

2. Instrumentation. Spatially resolved photometry can be accomplished in two ways. Either an imaging detector can record photons from a stationary object or a single element detector can record from a scanned object. The latter can be accomplished by moving the object, the light source or the detector in a rastered pattern, although the first two of these modes of scanning are more common. The imaging detector has the advantage of simultaneous observation of all the points within the field of view while the single element detector is less expensive and easier to calibrate. Either method can prove to be the more useful one for any given measurement; the experimentalist must make that decision.

3. Limitations and advantages. The use of images to study cell biological phenomena clearly has the advantage of permitting one to examine the spatial characteristics of a signal. For the optical microscope operating in the standard mode of image formation, the limiting spatial resolution is of the order of the wavelength of light whch forms the image; one can expect resolution down to 0.25 μm. Thus, for a field of view measuring 125×125 μm, the array of resolvable points in the image is 500×500. This degree of resolution is difficult to obtain with many imaging detectors.

Digital processing of microscope-generated images has become the preferred method of quantitation. This mode of analysis is advantageous in that a wide variety of algorithms with which images can be analyzed are available. However, the massive volume of data associated with a digital image poses a problem for storage, retrieval, display, and processing speed. Dedicated hardware helps ease these difficulties, but equipment expense and the need for some degree of technical expertise can be offsetting factors.

II. IMAGING TECHNIQUES
A. Hardware

1. Detectors. Aside from the biological preparation itself and the microscope optics, the most important component of a quantitative imaging system is the imaging detector. The quality of the output signal from the detector directly controls the quality of the data derived from that signal. Important

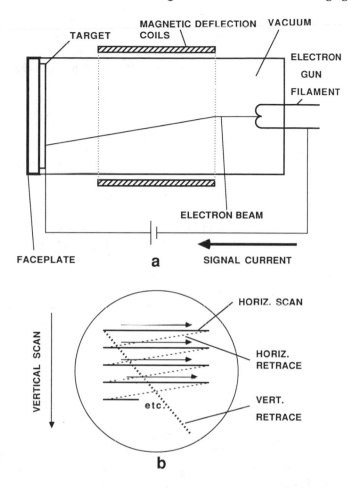

Fig. 1. Schematic diagram of a vidicon imaging device. **a:** As the electron beam scans the target, the current flowing through the beam is measured, giving rise to the video signal. **b:** Raster scan pattern for a vidicon. Horizontal lines are scanned rapidly as the beam is scanned vertically. During the horizontal and vertical retrace, the beam is shut off to prevent discharge of the target.

properties of the detector which control the quality of the output signal include sensitivity, dynamic range, resolution, geometric distortion, and noise characteristics [Bright and Taylor, 1986]. An excellent reference text devoted to this subject is that of Inoué [1986].

Video cameras have become the most widely used imaging detectors for quantitative imaging microscopy. A schematic diagram of a vidicon camera tube is shown in Figure 1a. For this type of detector, the primary transducing element is the target, which essentially stores photons which strike it as electron-hole pairs. An electron beam generated from a hot filament is accelerated toward the target by an applied voltage from target (anode) to electron gun (cathode). When the electron beam strikes a region of the target containing stored electron-hole pairs, a transient signal current flows through the target, electron beam and external circuitry. The video signal is derived from this current and is nearly proportional to the intensity of the faceplate illumination within the dynamic range of the target.

The array of electron-hole pairs produced on the target by the photon image focused on the faceplate is scanned and thus converted to the video signal by rastering the electron beam across the target as shown schematically in Figure 1b. The electron beam is scanned repetitively in the horizontal and vertical directions by magnetic deflection coils or in some cases by electrostatic deflection electrodes. A horizontal scan cycle is composed of one sweep in which the electron beam is turned off and positioned at the beginning of the next line. Simultaneous with the horizontal scan, the vertical position of the beam is being scanned at a slower rate. For US standard video cameras, 262.5 horizontal lines are scanned for one vertical sweep across the target. As with the horizontal scan, the electron beam is turned off during the vertical retrace of the beam. A second series of 262.5 horizontal lines is scanned during the next vertical sweep; this time each scan line is midway between the scan lines of the previous vertical sweep. Thus, at the completion of the second vertical retrace a frame of 525 horizontal lines has been swept across the target, the frame being composed of two interlaced fields of 262.5 lines each. For US standard video, each field is scanned every 16.7 ms, or each full frame is scanned every 33.33 ms. Thus, any given point in the image is sampled briefly at a rate of 30 samples per second, while the average time between sampling any two points in the image is 16.7 ms. Higher-resolution scanning with more lines per frame is also possible under different standards recommended by the Electronic Industries Association.

A silicon-intensified target (SIT) video camera operates slightly differently than a vidicon. Photons falling on the SIT faceplate excite photoelectrons from a phosphor deposited there. These photoelectrons are accelerated toward and imaged on a silicon target in which they generate electron-hole pairs that are detected by a scanned electron beam as for the vidicon tube. The primary advantage of the SIT over the vidicon is its increased sensitivity due to the first-stage signal amplification of the primary photoelectrons.

Image intensifiers have been employed to amplify the input photon signal to both vidicon and SIT cameras. In the image intensifier, photons strike an input phosphor screen and eject photoelectrons which are accelerated and magnetically or electrostatically focused on an output phosphor screen which emits many photons per photoelectron. Effective amplifications of 100 per intensifier stage are practical and up to four intensifier stages can be cascaded for very high gain. A second type of image intensifier, the microchannel plate, also uses a photoelectron-emitting front face phosphor, but in this case the photoelectrons impinge on an array of hollow cylinders which are coated with a material that emits secondary electrons when struck by an energetic primary electron. Thus, a cascade of secondary electrons is generated in each microchannel proportional to the number of incident photoelectrons. These secondary electrons strike an output phosphor screen which emits photons in response to the electron flux. The overall gain of the microchannel plate can be 1,000 or more.

Although the sensitivity of intensified video cameras can be quite impressive, the high degree of signal amplification leads to increased noise. Since thermal fluctuations produce random release of electrons in the intensifier stages, the intrinsic detector noise for these cameras is quite high. Also, the requisite electron optical stages which intervene between the primary image and the amplified image typically introduce considerable geometric distortion into the output image; this distortion can be a function of the degree of amplification and the brightness of the input image. In addition, the most appropriate phosphors used in SIT cameras and intensifiers can have quite long persistence times, thus making low-light cameras slow to respond to time-dependent changes in light intensity. Microchannel plate intensifiers can circumvent the lag problem and are intrinsically less susceptible to geometric distortion [see Spring and Smith, 1987].

Implicit in the above extended discussion of intensified video cameras is that these detectors are more appropriate for fluorescence imaging than nonintensified vidicon cameras. Vidicons have sensitivity appropriate to detect images of high brightness, for example, a microscope field easily seen by the non-dark-adapted eye. Intensified video cameras, on the other hand, can detect images three or more orders of magnitude dimmer, for example, a microscope field that is barely visible to the dark-adapted eye. Since many fluorescence images are intrinsically dim or are susceptible to chromophore photobleaching, intensified video cameras are preferred for this low light level work. The intrascene dynamic range of most video detectors does not exceed 100:1, and is generally poorer for the intensified cameras. The nonlinearity of response of these cameras can also be quite high, and must be carefully measured to allow proper image quantitation.

Solid-state imaging sensors, particularly charge-coupled device (CCD) sensors, are rapidly gaining popularity due to their excellent imaging characteristics [Hiraoka et al., 1987; Connor, 1988]. As schematized in Figure 2, a CCD is composed of a rectangular array of n rows and m columns of charge collection wells. The wells are typically of square geometry about 15 μm on a side. CCDs are fabricated from silicon, which produces electron-hole pairs when

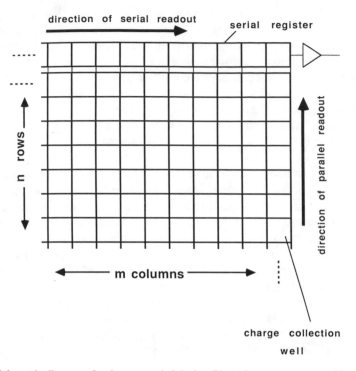

Fig. 2.   Schematic diagram of a charge-coupled device. Photoelectrons are captured in charge collection wells and are transferred via parallel and serial registers to the output amplifier.

photons strike it. A CCD is designed to trap the electrons in the region in which they were generated; thus, a photon image focused on the CCD will produce a corresponding photoelectron image in the CCD. This photoelectron image is read from the CCD by transferring photoelectrons in parallel from row to row of charge wells until all the charge is removed from the device. As each row is sequentially transferred into the first row, the charge that had been in the first row of the array is transferred into a serial register. Before the next row is transferred, the contents of the serial register are transferred well by well to an amplifier where the photoelectron packets from each potential well are converted to an analog output signal. Since CCD arrays can be very large (up to 2,048 × 2,048 elements), the efficiency of photoelectron transfer from well to well is of necessity very high.

In addition to photon-generated electrons, the CCD also accumulates thermally generated electrons. Thus, instrumentation CCD imagers are cooled to temperatures ranging from −40 to −120°C. This level of cooling allows image collection times of many seconds to hours with only a few thermal electrons generated per charge collection well.

For low-light-level CCD imaging, the dominant source of noise in the output image is the charge amplifier. The noise generated in the amplifier increases as the speed of image readout increases. Thus, instrumentation grade CCD cameras are read out at relatively slow rates, from 50,000 to 500,000 charge packets per second. A CCD array of size 512 × 512 would require 0.5–5 s to convert a full frame of stored charge to an electronic signal. Since the readout time is proportional to the number of charge packets read, reducing this number will speed image readout. This reduction can be accomplished by reading only a subarray of charge packets and/or combining, before amplification, the charge of a number of adjacent charge packets in the row and/or column directions. The former reduces the available field of view, while the latter reduces the available spatial resolution.

Since the CCD is a fixed array which directly converts a photon image to an electronic signal, it has no geometric distortion. The response of the CCD is also linear over its entire dynamic range, making it easy to calibrate. The available intrascene dynamic range on instrumentation devices approaches 50,000:1. The relative sensitivity of a CCD camera is comparable to that of an intensified video camera for most low-light-level imaging. Since the resolution of the CCD is determined by the size of the charge collection array and does not degrade with decreasing illumination levels, the CCD will, in general, produce a better image than a video camera for most low-light-level microscopy.

The above arguments apply to slow scan, instrumentation grade CCD imagers. CCD-based video cameras, that is, those solid-state cameras that produce images at 25–30 frames per second, do not share the low-noise, high-resolution and wide dynamic range properties of the slow scan cameras. They are, for the most part, not as suitable for fluorescence imaging as intensified vidicon or SIT cameras.

Both video and CCD cameras share the attribute of signal temporal integration at any given point in an image between image collection cycles. One may consider them essentially as arrays of imaging points which collect, in parallel, the signal from all points in the image at once. An alternate mode of image generation, scanned point imaging, contrasts with the imaging detectors in this respect. Since each point in an image is sampled serially by sweeping a light beam through the specimen or vice versa, the various points in the specimen are sampled only for a short time during each sweep.

When one needs to examine time-dependent changes in an image, the time course over which the input signal is changing must be considered in comparison with the above characteristics of the different imaging systems. For example, if a spatially uniform image increases in brightness during the time the image is collected, the CCD will collect an image that is spatially uniform with intensity equal to the mean intensity of the input image over the image

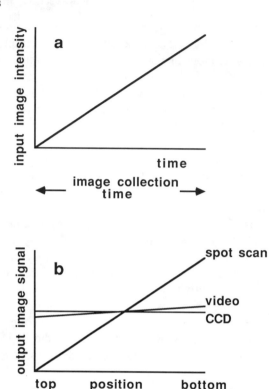

Fig. 3. **a:** Image intensity as a function of time for a hypothetical uniformly fluorescent speci-men. **b:** Spatial variation from top to bottom of the detected image of the time-dependent input image for spot-scan, video, and CCD detectors.

collection interval (see Fig. 3). A video camera will collect an image that is brighter at the end position on the camera target that the electron beam sweeps than at the beginning; the magnitude of the difference depends on the number of frames swept during the imaging interval as well as the speed of response of the camera target and/or phosphors. A spot scanned image will show an even greater difference in detected image brightness than does a video camera for any given image collection time since spot scans are typically much slower than video camera frame scan times. The advantage of the CCD in this respect is due to its purely parallel mode of imaging; one should note that the video camera, even with its built in bias from beginning to end of a field scan, can respond faster to the changing signal and that, for any particular point in an image, the spot scan mode of imaging can provide the highest possible tem-poral resolution.

2. Digital processing. Once an image has been collected by an imaging device, it then must be stored and analyzed. Although analog processing, particularly for video images, has been employed in some instances, digital processing has become the method of choice in quantitative image analysis. The analog signal from a video or CCD camera, which is a serial representation of the two-dimensional image, is digitized in a serial fashion in nearly all cases. This stream of digital data encodes the original image as an array of numbers, with array indices that correspond to position in the image. Two general formats are common for the hardware in which the bulk of the processing is performed: a dedicated image processing system or a general purpose computer.

A schematic diagram of a typical dedicated image processing system is shown in Figure 4. In this system, digital image data are stored in volatile frame buffer memory. This data can be processed with a dedicated, fast arithmetic and logic unit (ALU) and other specialized processors for such manipulations as image arithmetic, geometric corrections, fourier transform computation, region of interest processing, and the like. Display of data contained in the frame buffers is usually on a separate high resolution monitor driven by a dedicated digital to analog converter. Typically, input image data are digitized by

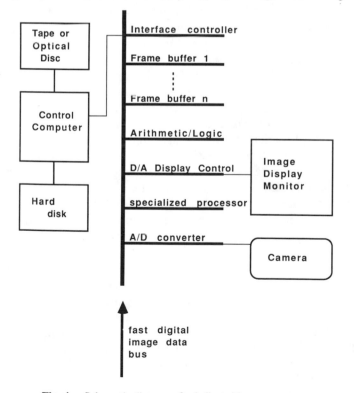

Fig. 4.   Schematic diagram of a dedicated image processor.

a dedicated analog to digital converter. All digital communication between these subsystems is via a high speed, specialized image data bus which allows rapid data manipulation. Control of the image processing system is through a computer via an intervening interface controller which communicates with the high-speed image bus and the computer backplane bus via direct memory access (DMA). Usually, temporary image data archiving is on a hard disk run by the computer while more permanent data archiving is on a magnetic tape or optical disk drive also controlled by the computer. Many variations on this general scheme have been implemented.

The exact format of the frame buffers and the numerical resolution of the ALU varies from machine to machine, although all are based on integer data. An image point is represented as a positive 8-bit or 16-bit datum, providing maximal data values of 256 or 65,536, respectively. Processing through the ALU is commonly carried out at 16-bit accuracy. The analog to digital converter of most systems is limited to 8 bits of accuracy for video input, with 12 to 16 bits of accuracy for slow-scan input. The display converter typically provides 8-bit digital to analog conversion for output video channels to drive red, green, and blue inputs to the monitor. The spatial resolution of dedicated processors varies from machine to machine, with 512×512 pixel frame buffers being the most common. In general, the price of the dedicated processor increases dramatically with resolution.

The use of a general purpose computer for image processing is another option for quantitative data analysis. This approach allows a great deal of flexibility in terms of image spatial format and data type, although it is slower in processing speed for 8-bit or 16-bit image data as compared to a dedicated processor. The ability to perform floating point or complex floating point calculations can offset the relative lack of speed for this format, while the addition of an array processor can bring the image computation speed of this hardware closer to that of the dedicated processor. A schematic diagram of one such arrangement, currently implemented in my laboratory, is shown schematically in Figure 5. In this arrangement, the central processing unit (CPU) communicates with the main system data bus (Multibus) as well as with a high-speed memory interconnect bus. Also in communication with the memory bus is the system random access memory (6 Mb) and an array processor. In communication with the system bus are a graphics processor, a hard disk controller, a magnetic tape controller, a high-speed IEEE-488 parallel interface controller, and a data acquisition controller. The last item mediates control of a real-time data acquisition bus. This system, a Masscomp 5500-PEP computer (Concurrent Computer Corp., Westford, MA), is well suited to the demands of image acquisition and processing.

One important feature of this system is its operating system, a modified version of UNIX, which allows near real-time instrument control and data

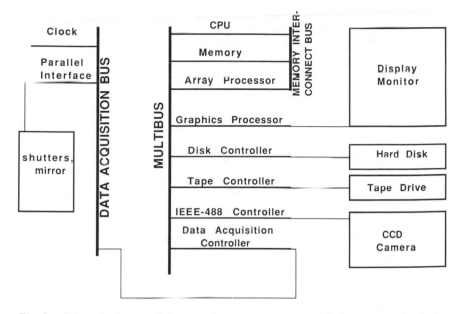

Fig. 5. Schematic diagram of the general purpose computer used for image processing in the author's laboratory.

collection. This is an important consideration for any type of image collection instrumentation since precise timing of image exposure times and time-lapse image sequences is necessary. In this system, shutter timing is independent of main CPU activity or Multibus data traffic, while time-lapse sequence timing is reasonably well controlled by the real-time modifications to UNIX coupled with the independent data acquisition bus. A second notable feature is the tightly coupled array processor. This accelerator allows fast processing of arrays stored in the system random access memory, improving processing speed for memory-resident images to near that of a dedicated image processor. The array processor performs computations on real (non-integer) image arrays at the same speed as computations on integer images.

In any general-purpose computer used for image processing, the ability to display images is an important consideration. The display system shown in Figure 5 is an integrated graphics processor/display monitor which serves both as an image display and interactive computer terminal. This particular graphics terminal has a pair of frame buffers of size $1,152 \times 910 \times 12$ bits. Of the 12 bits of display capacity, 10 bits are mapped to monochrome or pseudocolor output lookup tables while two bits are strictly overlay planes. Thus, this display system permits high lateral resolution display along with extended (beyond 8-bit) pseudocolor display, which can be quite useful.

No matter which type of image processing system one chooses, image data storage and archiving is an important concern. Since a $512 \times 512 \times 8$ bit image contains 0.25 Mb of data, a modest series of images can consume a large volume of data storage space. For those applications in which a large number of images are captured or processed, large (300 Mb +) disk storage capacity is a necessity. Most image processing systems use hard disks for intermediate image storage and magnetic tape storage for image archiving. One should note that floppy disk storage of images is wholly inadequate since only a very few images will fill any floppy disk.

One difficulty with hard disk storage of image data beyond the capacity consideration is the speed of data access. Typical image disk-to-memory transfer times are from one to several seconds, thus setting the minimum image capture rate for a sustained time lapse series or limiting the speed of processing of a sequence of images. Some dedicated image processors alleviate this image transfer bottleneck by the use of high-performance hard disks which can read or write (byte) image data at the speed of the video bus.

One's choice of hardware depends on the type of imaging to be done. In general, microcomputers have insufficient compute power to be used as general purpose image processing computers, while most mini- and supermini-computers suffer in real-time control situations. In any imaging system, more disk capacity and faster archival storage capability are always better.

B. Fluorescent Probes

1. Responsivity. No matter how accurate and powerful an imaging system is applied to any given problem, the data that result from the measurement are only as good as the fluorescent probe that is employed. Since this volume has a chapter devoted to the properties of indicator dyes [Haugland and Minta, 1990], only a brief discussion of the relation between probe properties and imaging system properties will be undertaken here.

Clearly, fluorescent probes with high selectivity for the molecule or parameter of interest are preferable to those which are less specific. Although one may be able to employ calibration techniques to correct for the effects of an interfering ion or other molecule, more correction steps lead to more potential errors and to a degradation of the signal to noise properties of the output data. Aside from the question of selectivity, probes that interact very little with the cell are preferred over those that do not. For example, a $[Ca^{2+}]$ indicator like fura-2 can become incorporated within intracellular organelles as a function of time. Even though one might be able to visualize the organelles using imaging techniques, proper quantitation of cytoplasmic vs. organelle $[Ca^{2+}]$ is very difficult due to the nonstationary nature of the dye distribution and of the cell itself.

Another consideration in the use of fluorescent probes to study living cells is the toxicity of the probes. Not only can they have intrinsic toxicity to the cell, but also free radicals formed during photoexcitation can be toxic. One must keep in mind that imaging requires that the whole cell is illuminated and thus many chromophores are simultaneously activated; additionally, the level of illumination must produce sufficient light from all parts of the cell. Thus, a non-imaging experiment can be performed at a much lower total dose of radiation as compared to an imaging experiment. This problem is exacerbated during long duration time-lapse imaging experiments.

2. Relation to instrumentation. As with any optical technique, the spectral properties of the imaging detector must be compatible with those of the chromophores in use. Intensified video cameras generally have a spectral response function which spans the visible spectrum into the infrared, with a peak around 450 nm. CCD cameras generally have a response peak around 600 nm, with a rapid drop off below 500 nm. Thus dyes with emission similar to that of rhodamine are relatively better suited to CCD imaging than are dyes similar to fluorescein or fura-2. However, even for these shorter wavelength emission dyes, the high quantum efficiency of many CCD detectors along with their other imaging characteristics make them an excellent choice for quantitative fluorescence imaging.

For imaging of time-independent cell-associated fluorescent probes, the temporal response characteristics of the imaging system are unimportant. For time-dependent cell processes, the ability of the probe to respond to a change in its environment as well as the inherent speed of the cell process must be considered in relation to the temporal response characteristics of the imaging system. As a practical example, consider the imaging of the membrane potential change which propagates along an axon during an action potential. If a slow redistribution dye is used, a vanishingly small signal would be measured no matter what the detector characteristics since the characteristic biological response time, ≈ 1 ms, is much faster than the dye equilibration time, of the order of tens of seconds. A fast, membrane-bound probe of transmembrane electric field strength has the requisite speed of response, $\approx 1 \mu s$, so that the temporal resolution of the imaging system is limiting and must be somewhat better than the 1 ms duration of the action potential. High temporal resolution, low spatial resolution imaging of action potential propagation has been accomplished using fluorescent probes and photodiode arrays [Cohen and Salzberg, 1978; Grinvald et al., 1984], while single-frame high-resolution imaging of changes in membrane potential has been accomplished using fluorescent probes, a video detector, and stroboscopic illumination [Kinosita et al., 1988]. The continuous monitoring of very high speed cellular phenomena measured with fluorescent probes at high spatial resolution presents technical difficulties which are beyond the capabilities of current instrumentation.

C. Noise and Resolution

1. Instrumentation noise vs. indicator noise. The uncertainty in the detected level of fluorescence at any point in an image determines the degree of confidence one can place in the data derived from the image. Sources of noise within a pixel include imaging system noise, photon shot noise, and chromophore fluctuations. Included in imaging system noise are fluctuations from thermal excitation in the detector, electronic noise, and external interference noise. Included in chromophore noise are fluctuations in the number of chromophores within the area imaged in a pixel, fluctuations in the fluorescence properties of the chromophores, and other cell-associated fluctuations.

In many cases, particularly for indicator dyes, a chromophore can be considered to exist as two separate species, f and b, which exhibit different fluorescence properties. In the case of indicator dyes such as fura-2 and BCECF, species f is the unbound dye molecule and species b is the chelated dye molecule; the difference in fluorescence properties is a shift in the excitation spectrum between free dye and bound dye. Assuming that dye molecule D and ligand L are in equilibrium and that L and D form a 1:1 complex, then the amount of free (D) and chelated (DL) dye is

$$D = D_{TOT} K/(K+L)$$

$$DL = D_{TOT} L/(K+L) \tag{1}$$

where K is the dissociation constant of the DL complex and D_{TOT} is the total amount of dye present. If the mean fluorescence associated with a free dye molecule is F_O and that associated with a chelated dye molecule is rF_O, then the total fluorescence F detected as a function of ligand concentration is

$$F = F_O(D+rDL) = F_O D_{TOT}(K+rL)/(K+L) \tag{2}$$

Thus, so long as the fluorescence properties of species D differ from species DL (i.e., if $r \neq 1$), then the measured fluorescence will change with changes in L. It is clear from the form of Equation 2 that variations in the total amount of dye present (by cell shape changes or dye flux) or in F_O (by nonspecific shifts in dye fluorescence or photobleaching) make the quantitation of F, and thereby the measurement of L, difficult. For chromophores that have a spectral shift associated with the free to bound transition of the dye molecule, one can measure F at two convenient points in the dye spectrum and thus generate a self-referencing measurement. In Equation 2, if F_O and r vary in a known way with wavelength, then one can unambiguously determine L and D_{TOT} since two equations with two unknowns can be solved.

The most common approach is to measure F at two separate wavelengths, λ_n and λ_d. These may be either two excitation wavelengths, two emission wavelengths, or a combination of both. The two separate fluorescence measurements, F_n and F_d, are then combined as a ratio $R = F_n/F_d$. From Equation 2 one can readily see that

$$R = (F_{O,n}/F_{O,d}) (K + r_n L)/(K + r_d L) \tag{3}$$

since the total amount of dye is the same in both cases. With the definitions that $R = R_{min}$ for $L = 0$, $R = R_{max}$ for L very large and $\beta - F_d(L=0)/F_d(L$ very large), one can show (Grynkeiwicz et al., 1985) that

$$R = R_{min}(K + R_{max}L/\beta R_{min})/(K + L/\beta)$$

or

$$\tag{4}$$

$$L = K\beta(R - R_{min})/(R_{max} - R)$$

Thus, for nonzero values of β, L can be determined from the value of R after proper system calibration. This ratiometric determination ignores potential complications such as wavelength-dependent light absorption, which appear to be small in most cases [Connor, 1988].

Equation 4 can be applied to signals from both bulk fluorometry and imaging. For imaging, it is important to consider the effects of fluctuations on the measured signal. From Equation 2, one can see that fluctuations in the total amount of dye or in the free ligand concentration within a point of the image can lead to fluctuations in the detected fluorescence. Large fluctuations in the small number of molecules sampled within a pixel can be a concern. Using fura-2 imaging as an example, if a pixel in an image samples an area 0.25 μm on a side, and the depth of the sampled region in the cell is 2 μm, the total volume sampled is 1.25×10^{-13} cm^3. For [Ca^{2+}] $- 100$ nM, only 1.25×10^{-23} mole of free calcium is present in this volume, which amounts to only about 8 Ca^{2+} ions! One would then expect large fluctuations, of order of 3 ions, about the mean number of 8 within this volume. These fluctuations occur over times of order of the diffusion time of an ion, calculated from the apparent diffusion coefficient of calcium in the cytoplasm [D $\approx 10^{-8}$ cm^2/s; Nasi and Tillotson, 1985], in this case ≈ 30 ms. If [Ca^{2+}] is monitored with fura-2, then in addition one must consider fluctuations of the number of chromophores in the sampled volume as well as the kinetics of Ca^{2+} binding to the dye. Since intracellular fura-2 concentration is usually used at levels greater than 10 μM, relative fluctuations in the number of dye molecules in the sampled volume are at least an order of magnitude less than those of Ca^{2+} fluctuations. The kinetic response of Ca^{2+} binding to fura-2 is in the 10 ms range [Cannell et al., 1987; Jackson et al., 1987]; thus one would expect fluctuations due to the indicator or to Ca^{2+} to be due mainly to $[Ca^{2+}]$ fluctuations. These fluctuations should be observed for measurement times less than 30 ms. Most imaging experiments employ longer image collection times, typically to reduce the noise in the images. Additionally, spatial averaging over many pixels is used to reduce noise. One should exercise caution in interpreting ratiometric "images" of a ligand L (such as Ca^{2+} imaged with fura-2) in which the image pair was collected over times comparable to the diffusion time, since large pixel-to-pixel fluctuations in L can occur during the imaging.

Another source of noise is due to the imaging process. The emission and detection of photons both lead to fluctuations in the number of detected photons. The size of these fluctuations is proportional to the square root of the mean number of detected photons. Fluctuations associated with the detector also contribute to the noise of the detected signal. These two components add in quadrature, i.e.,

$$\sigma_o = (\sigma_S{}^2 + \sigma_D{}^2)^{1/2} \qquad (5)$$

where σ_o is the noise in the output image, σ_S and σ_D are, respectively, the photon shot noise and detector noise. For low-light-level imaging typical of fluorescent specimens, detector noise dominates shot noise, so $\sigma_o \approx \sigma_D$.

As can be seen from Equation 5 and the fact that detector noise is independent of input light intensity for fixed detector gain, the signal to noise ratio (S/N) improves with increasing signal strength, at first linearly and then as the square root of the input signal strength. Thus, a brighter image will have a better S/N, all other conditions being the same. For the case of indicator dyes which change image intensity with changes in ligand concentration, the S/N will also vary with ligand concentration.

For the two-wavelength ratiometric method of fluorescence quantitation, the fractional noise in the ratio $R = F_N/F_D$ is

$$\sigma_R/R = (\sigma_N{}^2/F_N{}^2 + \sigma_D{}^2/F_D{}^2)^{1/2} \qquad (6)$$

where σ_R, σ_N, and σ_D are the standard deviations of fluctuations about the mean values of R, F_N, and F_D, respectively [Parratt, 1961]. If we presume that the numerator and denominator images are taken under low-light-level conditions, then one can apply Equation 5 to Equation 6, giving

$$\sigma_R/R = \sigma_D (1/F_N{}^2 + 1/F_D{}^2)^{1/2} \qquad (7)$$

Combining this expression with Equations 2 and 4 gives

$$\sigma_R/R = [\sigma_D(K + L)/F_0 D_{TOT}] \, [1/(K + r_N L)^2 + 1/(K + r_D L)^2]^{1/2}$$

or   (8)

$$\sigma_R/R = [\sigma_D(K + L)/F_0 D_{TOT}] \, [K + R_{max}L/\beta R_{min})^{-2} + (K + L/\beta)^{-2}]^{1/2}$$

Since R_{min}, R_{max}, and β vary with the wavelength values selected for the numerator and denominator images, the dependence of the image relative noise level on the ligand activity L will also depend on the choice of wavelengths.

More relevant than the noise in the ratio image is the noise in the computed ligand activity. From Equation 4 we see that L depends on R, R_{min}, R_{max}, and β, all of which are quantities derived from measurement and all of which contribute to the uncertainty in L. As for Equation 8, the relative standard deviation of L about its mean can be computed:

$$\sigma_L/L = \{\sigma_\beta^2/\beta^2 + \sigma_R^2[(R - R_{min})^{-1} + (R_{max} - R)^{-1}]^2 +$$
$$\sigma_{R_{min}}^2(R - R_{min})^{-2} + \sigma_{R_{max}}^2(R_{max} - R)^{-2}\}^{1/2} \qquad (9)$$

where σ_β, σ_{Rmin}, and σ_{Rmax} are the standard deviations of β, R_{min}, and R_{max} about their means. It is easy to see from Equation 9 that σ_L/L increases rapidly as R approaches R_{min} or R_{max}, as expected.

2. Image oversampling. Another important consideration in quantitative fluorescence imaging is the limit on spatial resolution set by instrumentation. Aside from the intrinsic diffraction-limited resolution of image features, the spatial resolution of the imaging detector as well as of digital representations of the image can limit resolvability of small objects. The general rule of thumb, the Nyquist criterion, is that the imaging system must be able to sample with spatial resolution half that of the smallest object to be resolved [Inoué, 1986]. Thus, if the image being considered has features of interest of dimension $\approx$ 0.25 µm, a field of view 64 × 64 µm in a 512 × 512 image array would just satisfy the Nyquist criterion.

In some instances, spatial resolution is not a critical parameter. In these cases, one might choose to image a larger field of view to include more cells. Alternately, one might choose to limit digital spatial resolution to enhance temporal resolution since the total throughput of data in a digital imaging system is a time-limiting parameter. Illustrations of the use of both of these compromises are given in the next section.

III. APPLICATIONS

A. Fura-2 Imaging

1. Spatial resolution. Numerous studies of the spatial and temporal distribution of $[Ca^{2+}]$ have been reported. A wide variety of imaging systems have been employed for these studies. I have used two different systems for fura-2 imaging, one based on an intensified video camera and an independent image processor [Millard et al., 1988; Gonzalez et al., 1988a,b; Chandra et al., 1989] and one based on a CCD imager and a general purpose computer [Linderman et al., 1990; this chapter]. Although a direct comparison between these two types of systems is difficult, I prefer the CCD-based imaging system primarily for its ability to produce photometrically accurate images with a wide dynamic range, although the intrinsic flexibility of image format is also a useful feature.

The details of the imaging system have been described previously [Linderman et al., 1990]; thus only a brief outline of the hardware is given here. The system is based on a Zeiss IM-35 microscope (Carl Zeiss, Inc. Thornwood, NY) equipped with a Nikon 40 × /1.3 NA CF Fluor UV lens (Nikon Inc. Instrument Division, Garden City, NJ). A 75 W Hg/Xe arc lamp (Hammamatsu Corp, Bridgewater, NJ) illuminates a pair of monochromators (Instruments SA, Inc., Metuchen, NJ) via a home-built sliding mirror system. The monochromators are coupled to the epi-illumination port of the microscope via a bifurcated quartz fiber optic bundle (Volpi Manufacturing USA, Auburn, NY). Images are collected on a CCD instrumentation camera (Photometrics, Ltd., Tucson, AZ) equipped with a Texas Instruments 4849 CCD chip. Computer control of the CCD camera, system timing, image storage, image processing, and image display is by a Masscomp 5500 computer (Concurrent Computer Corp., Oceanport, NJ; see Fig. 5).

The field of view of the microscope is adjusted so that it overfills the 390 × 584 pixel area of the chip; a pixel on the CCD maps to a square region on the specimen ≈ 0.2 μm on a side. Thus, by the Nyquist criterion, this system can resolve image detail down to 0.4 μm. This is larger than the resolution limit of the microscope, but it allows a larger field of view (and thus more cells) to be imaged at any one time. For most experiments, the resolution of the image is further reduced by a factor of 2 in both the horizontal and vertical directions by combining or binning, before readout, the photoelectron charge in four adjacent pixels. This reduces the size of the data array needed to store the image and thereby reduces the time needed for data transfer. Additionally, binning improves S/N in the image since photoelectrons are combined before they pass through the readout amplifier.

An image pair of a clump of human epidermoid carcinoma A-431 cells loaded with fura-2 and illuminated at both 365 and 334 nm is shown in Figure 6a. The cells are exposed to 2 μM Fura-2/AM (the acetoxymethyl ester of fura-2,

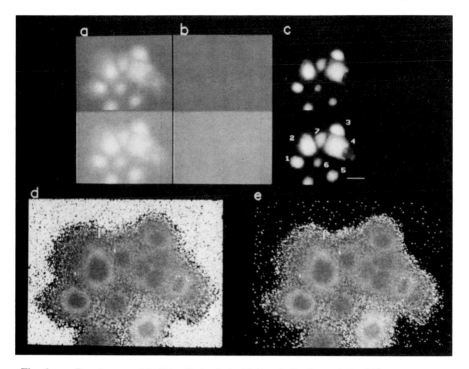

Fig. 6.   **a:** Raw images of A-431 cells loaded with fura-2. Top image is for 365 nm excitation, bottom is for 334 nm excitation. **b:** Background (fluorescence plus camera dark current) for a blank region of the coverslip. **c:** Background-subtracted fluorescence of the cells in panel a, with gray levels expanded linearly for display clarity. Bar is 20 μm. **d:** Ratio image computed by dividing the bottom of panel c by the top of panel c. **e:** Same as panel d except pixels corresponding to gray levels below a value of 1 in the denominator pixel have been set to zero. Both d and e are magnified 2× compared to a–c.

Molecular Probes, Eugene, OR) in a 37°C 5% CO_2/95% air humidified incubator for 30 min in the medium in which they are cultured (Dulbecco's modified Eagle's medium [DMEM] plus 10% fetal calf serum). After two washes in 140 mM NaCl, 5 mM KCl, 1 mM $CaCl_2$, 10 mM HEPES, pH 7.4 (HBS), the cells are mounted on the microscope on a heated stage in a flow chamber. The 365 nm image is captured on the bottom half of the CCD chip (the top half is masked by a partial field stop), then the 365 image is transferred electronically to the top half of the CCD, and finally the 334 image is captured on the now-blank bottom half of the CCD. This frame-transfer procedure is fast, allowing rapid image-pair collection [Linderman et al., 1990].

Figure 6b,c shows, respectively, the background image pair and the background-subtracted fura-2 image pair of the cells. Background images are obtained by imaging a portion of the coverslip, free of cells or debris, under conditions identical to those for which the cell image is obtained. If the background image does not correctly represent the true cell background, the cell fluorescence image will be systematically biased. Autofluorescence in these cells is less than 5% of the fura-2 signal [Gonzalez et al., 1988b].

Ratio images are computed by dividing the bottom half of Figure 6c by the top half of Figure 6c. Results of this division are shown in Figure 6d. Since the numerator and denominator fluorescence value are approximately equal [Fig. 6c], ratio values are close to unity. Thus, in order to display the ratio image on a 0 to 255 grey scale, the calculated ratios are multiplied by 100 to produce Figure 6d. The non-cell regions of Figure 6d are all rather bright since these regions are calculated from background areas, thus very small values divided by very small values multiplied by 100 produce a bright, noisy background. Figure 6e is the result of the same ratio calculation, except that pixels in the ratio image which correspond to pixels in the denominator image (Fig. 6c, top) below a value of 1 have been set to zero. This procedure removes much of the obtrusive background information, but it also arbitrarily selects the region of the image to display. Note that areas of very dim fluorescence, not easily seen in Figure 6a,c, contribute to the ratio image of panel e.

A second mode of display is shown in Figure 7. Panel a is a pseudocolor

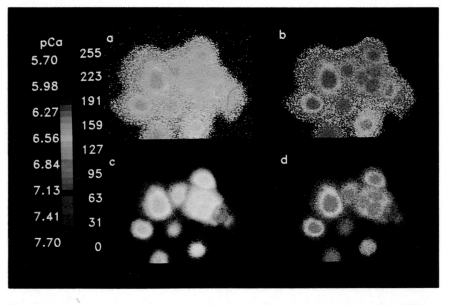

Fig. 7.   Pseudocolor display of the data in Figure 6. **a:** The ratio as shown in panel e of Figure 6. **b:** Same as panel a except conversion to pCa has been completed on a pixel-by-pixel basis. Note scale calibration. **c:** Intensity-modulated display of the data in panel a. **d:** Intensity-modulated display of the data in panel b.

version of Figure 6e in which the gray scale is converted to 31 bands of color as noted on the left scale of the color bar. Panel b is a similar pseudocolor display of the calibrated ratio image. Panel b was produced by computing $[Ca^{2+}]$ on a pixel-by-pixel basis from Equation 4 with $K = 225$ nM and R_{min}, R_{max} and β derived from calibration images of 10 μM fura-2-free acid in high Ca^{2+} and Ca^{2+}-free HBS. The calibration values were corrected by constant factors for the effect of the intracellular milieu on the fura-2 excitation spectrum [Konishi et al., 1988]. This correction procedure has been described elsewhere [Linderman et al., 1990]. Panel b displays pCa = $-\log[Ca^{2+}]$ rather than $[Ca^{2+}]$ since pCa is nearly linear with fura-2 ratio for the 334/365 ratio pair [Millard et al., 1988; Linderman et al, 1990]. The pseudocolor scheme for this set of cells maps zero gray level to pCa = 7.7 ($[Ca^{2+}]$ = 20 nM) and 255 gray level to pCa = 5.7 ($[Ca^{2+}]$ = 2,000 nM). Panels c and d display the same data as panels a and b, respectively, except that the brightness of the displayed image is controlled by the brightness of the denominator image [top of Fig. 6c]. In this display scheme, 5 bits of ratio or pCa image data map to the 31 colors representing ratio or pCa values, while 5 bits of the denominator image map to intensity information. This type of display allows one to correlate the brightness of the image (i.e., regions of high dye concentration or of greatest cell thickness) with ratio or pCa values.

2. Temporal resolution. The cells of Figures 6 and 7 were stimulated with a large dose of epidermal growth factor (EGF, Collaborative Research, Bedford, MA) during a time-lapse imaging sequence. Computations of pCa as in Figure 7 were carried out for all images, and the mean value of pCa was measured in each of the cells. To measure mean pCa, a mask corresponding to the area of each cell was generated, and the average of the pixels in the pCa image which corresponded to mask pixels was computed for every image in the sequence. Pixels which corresponded to division by zero were rejected. All calculations were done on floating point real pixel values. The mean pCa in each of the cells of Figures 6 and 7 are plotted in Figure 8. Note the strong rise in $[Ca^{2+}]$ subsequent to EGF stimulation in most cells.

Interpretation of fura-2 data as shown in Figures 6–8 is colored by several potential artifacts. Aside from the intrinsic difficulty with accurate subtraction of background, the dye itself can pose problems for absolute calibration. Photobleaching artifacts [Becker and Fay, 1987], viscosity and protein binding spectral shifts [Tsien and Poenie, 1986; Konishi et al., 1988], and the presence of uncleaved fura-2/AM [Scanlon et al., 1987] all make absolute calibration difficult. The intracellular compartmentalization of fura-2 as well as potential toxic effects of the dye, photoinduced free radicals, or effects of the UV excitation light all can change with time, thus introducing a further difficulty. It seems safe to say that EGF stimulates a rise in $[Ca^{2+}]$ in these cells; the magnitude of that rise is more difficult to quantitate. The use of

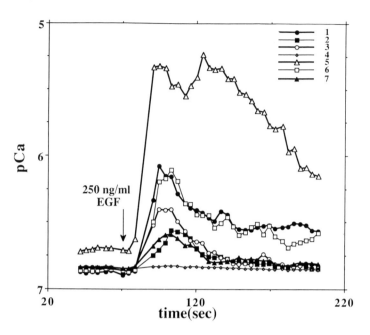

Fig. 8. Time dependence of the computed pCa levels in the cells of Figures 6 and 7. EGF addition was at the arrow.

simultaneous multiple cell recording lends credence to the data, although absolute calibration is still questionable. In fact, the strikingly larger response of cell 5 as well as its elevated resting $[Ca^{2+}]$ may be related to differences in dye-cell interactions in this cell.

B. Membrane Potential Imaging

1. Spatial resolution. Fluorescent indicators of membrane potential (V_m) have been used in a small number of studies to image changes in V_m with spatial resolution approaching that of the optical microscope [Gross et al., 1986; Kinosita et al., 1988; Montana et al., 1989]. All of these studies were performed with video camera detectors and independent digital image processors. As with fura-2 imaging described above, CCD detection of the fluorescence of the V_m-dependent dye is described here.

The indicator used here and in a previous study, di-4-ANEPPS [Fluhler et al., 1985], responds to changes in the electric field strength directed along the axis of the molecule in a manner consistent with an electrochromic mechanism [Loew, 1982]. Specifically, the energy of electronic transitions of the chromophore are related to the electric field strength when electronic redistribution from the ground state to excited states of the chromophore is along the field axis. The shift in the transition energy can be written as

$$h\Delta\upsilon = -q\,\vec{r}\cdot\vec{E} + \Delta\vec{\mu}\cdot\vec{E} + \Delta\alpha E^2 \qquad (10)$$

where h = Planck's constant; $\Delta\upsilon$ = transition frequency shift; q = effective charge redistributed; r = distance through which the charge moves; E = field strength; $\Delta\mu$ = change in the dipole moment from ground to excited state, and $\Delta\alpha$ = change in polarizability from ground to excited state [Loew, 1982]. The third term in Equation 10 is not significant for values of E found in biological membranes, so the spectral shift of the dye is linearly related to the applied field strength, i.e.,

$$F = D_o F_o(1 + KE) \qquad (11)$$

where F = fluorescence; D_o = amount of dye present; F_o = zero-field fluorescence of 1 molecule of dye and K = constant. F_o and K are functions of the dye characteristics, the excitation and emission wavelengths used, and the local membrane environment, and thus must be calibrated for each experiment. D_o depends on the local dye concentration and on the amount of membrane stained with the dye. As with the chelating indicator dyes, the latter variable can make absolute calibration difficult [Gross et al., 1986]. The ratio of fluorescence excited at two separate wavelengths (or imaged at two separate emission wavelengths) removes this factor from the data:

$$R = [F_{o,N}(1 + K_N E)]/[F_{o,D}(1 + K_D E)] \qquad (12)$$

For di-4-ANEPPS the values of K_N and K_D are small, so that one can approximate Equation 12 as

$$R = R_O[1 + (K_N - K_D)E] \qquad (13)$$

for values of E encountered in cell membranes.

Although measurements made with membrane-bound dyes such as di-4-ANEPPS are often termed membrane potential measurements, they are actually measurements of the electric field strength in the membrane. These two quantities are intimately related, although other factors beside the electric field strength within the lipid portion of the membrane bilayer can affect the transmembrane potential V_m [Gross and Loew, 1989].

Two images of the fluorescence of a clump of A-431 cells stained with di-4-ANEPPS are shown in Figure 9a,b. The cells are exposed to 10 μM di-4-ANEPPS in HBS on ice for 10 min, are washed in cold HBS, and are mounted on the microscope to minimize dye internalization; experiments are carried out at 10°C. Panel a is the image for excitation at 436 ± 5 nm, while panel b is for 546 ± 10 nm. These two wavelengths were chosen since they correspond to points in the dye excitation spectrum at which the relative change in fluorescence with change in electric field strength is large [Fluhler et al., 1985] and they correspond to lines in the Hg arc spectrum. The images were collected as described for fura-2 imaging above except a Zeiss $63 \times /1.4$ NA objective was used, the excitation wavelengths were as noted, and a different dichroic mirror and barrier filter were employed. In this case, emission above 610 nm is collected. For these parameters, K_N and K_D in Equation 12 correspond to $+7\%$ and -2% for a change in E equivalent to a 100 mV hyperpolarization of V_m [Fluhler et al., 1985].

Figure 9a,b has been compressed in gray scale to a displayable 8 bits (0–255) from the raw CCD data, which is collected at 12-bit (0–4,095) resolution. The photography has not reproduced the full dynamic range of the data. This higher resolution is important if small changes in the transmembrane electric field are to be detected. For example, a change in V_m of 10 mV will change the fluorescence in panel a by $+0.7\%$ and in panel b by -0.2%. The latter change is about 0.5 gray levels out of 256, i.e. it is not resolveable. For a signal of 3000 (presuming less than full saturation of the detector), the signal would be $+6$ gray levels, which is somewhat larger than the detector noise. Thus, much smaller changes in the transmembrane electric field (or V_m) are detectable on a pixel-by-pixel basis with the 12-bit data of the CCD than for 8-bit (or less) data from a video camera.

Panels c and d of Figure 9 display the 546/436 ratio similar to panels a and c of Figure 7. Figure 9c is the ratio displayed in the 31 pseudocolor levels shown by the color bar, while Figure 9d is the 10-bit display in which the ratio value of a pixel controls its color and the denominator fluorescence intensity of a pixel [Fig. 9b] controls the brightness. The advantage of the latter display is clear since the important features, the plasma membranes which are sharply in focus, are highlighted. The cells near the bottom of the image contain internalized dye, a condition which precludes accurate calibration. The apparent high values of fluorescence ratio in these cells is uninterpretable.

In a separate series of images, ratio images of this clump of cells were collected to calibrate the 546/436 ratio. Calibrations were accomplished by exposing the cells to a series of HBS solutions with KCl replacing NaCl for final $[K^+]$ of 5 mM to 150 mM. Valinomycin (500 nM) was added to the calibration solutions to facilitate K^+ diffusion across the cell membranes. Apparent V_m was calculated from the Nernst equation, assuming that $[K^+] = 150$ mM inside the cells. The resulting plot of R vs. V_m was linear with slope 3% per 100 mV hyperpolarization. The difference between the expected 9%/100 mV and the measured value could be due to dye flip-flop across the bilayer [Gross et al., 1986] or for other reasons not yet well understood.

The data of Figure 9d indicate that the di-4-ANEPPS fluorescence ratio varies by a few percent laterally across the plasma membrane of an individual

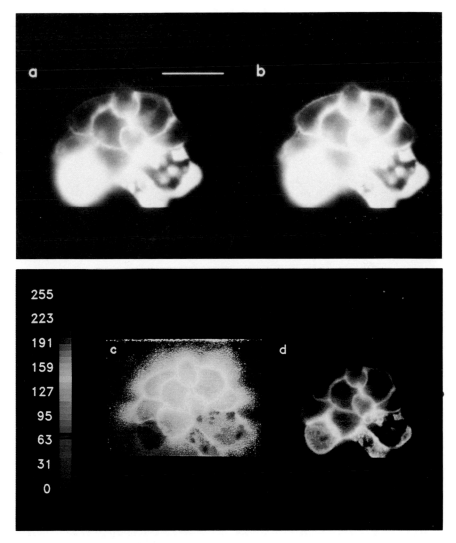

Fig. 9.   **a:** Background-subtracted fluorescence of di-4-ANEPPS-stained A-431 cells excited at 436 nm. Bar is 20 μm. **b:** 546 nm-excited di-4-ANEPPS fluorescence. **c:** Pseudocolor display of the ratio of panel a to panel b. **d:** Intensity-modulated ratio similar to that in Figure 7c,d.

cell. The variation is such that portions of plasma membrane facing other plasma membranes are depolarized relative to those facing medium. As this technique is new and all artifacts may not yet be understood, one should be cautious in interpreting these data. One possible interpretation of these data, if they are found to be artifact-free, is that membrane surface charge varies widely

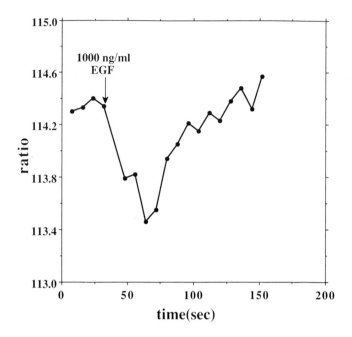

Fig. 10. Time course of changes in di-4-ANEPPS fluorescence ratio in A-431 cells stimulated with EGF at the time indicated by the arrow.

between regions of the plasma membrane and that this surface charge variation translates to variation in the transmembrane electric field [see Gross, 1988, and Gross and Loew, 1989, for a further discussion of such effects].

A separate clump of A-431 cells was similarly labeled and imaged, but at 15°C. This clump was stimulated with 1,000 ng/ml EGF, a stimulus that is known to produce a rise in $[Ca^{2+}]$ [see Fig. 8]. The time course of the change in the di-4-ANEPPS ratio is shown in Figure 10. The data of Figure 10 were taken from plasma membrane regions of the cells by a selective masking procedure similar to that for fura-2 data collection, except data were collected from all cells in the image rather than individual cells. Calibrations performed on the same cell clump indicated that the size of the measured hyperpolarization after EGF stimulation was $\approx$ 90 mV. As the time course of this response is similar to the $[Ca^{2+}]$ response time course, the validity of this data is supported. One should note that hyperpolarization of the plasma membrane is counter to a presumed depolarization due to Ca^{2+} influx. This EGF-stimulated hyperpolarization has also recently been reported by Pandiella et al. [1989]. They suggest that a Ca^{2+}-activated K^+ channel is responsible for the hyperpolarization.

IV. DISCUSSION

The recent advances in both imaging hardware and computers have stimulated a burgeoning growth in the use of quantitative fluorescence imaging in cell biology. The concomitant development of fluorescent indicator probes has produced an explosive expansion in knowledge of the details of cellular processes measurable by these probes—knowledge that was unobtainable only a few years ago. The future of such studies seems bright indeed since much effort is being brought to bear on the development of both hardware and probes.

The examples of quantitative imaging shown in this chapter illustrate both the strengths of and the caveats concerning quantitative fluorescence imaging. The ability to resolve spatial information is a very powerful one. This ability makes possible the characterization of variations in response within a single cell and from cell to cell. Particularly for the multiple cell response, the strength of the data is amplified by the built-in control that all cells provide for any individual cell. It should also be emphasized that the data must be interpreted cautiously. As with any new technique, unforeseen complications can bias results in surprising ways. Thus, absolute calibrations of any probe must be subject to meticulous scrutiny before new results become scientific dogma. A healthy dose of skepticism is in order for any new phenomenon reported using these new tools.

In my view, the strongest case for the application of quantitative imaging of cells can be made when changes in a particular parameter can be measured over time as a function of stimulation. In such experiments, the possibility of regional variation of probe response or cell to cell variability can be minimized.

Although I have focused on two specific fluorescent probes, fura-2 and di-4-ANEPPS, many more indicators exist or are being developed for a wide variety of applications. As the interactions of these probes with their local environment become better understood, the application of quantitative imaging to cell biological systems will greatly expand our knowledge of the details of a wide variety of dynamic processes in the living cell.

ACKNOWLEDGMENTS

I would like to thank Tamara Cheyette for her assistance with the EGF stimulation experiments. I also want to thank Jennifer Linderman, Catherine Randall, Lisa Harris, and Noah Sciaky for contributions to hardware and software development. I acknowledge support from the National Science Foundation (DMB-8803826) and the National Institutes of Health (RR07048).

V. REFERENCES

Becker PL, Fay FS (1987) Photobleaching of fura-2 and its effect on determination of calcium concentrations. Am J Physiol 253:C613–C618.

Benson DM, Bryan J, Plant AL, Gotto Jr. AM, Smith LC (1985) Digital imaging fluorescence microscopy: Spatial heterogeneity of photobleaching rate constants in individual cells. J Cell Biol 100:1309–1323.

Bolsover SR, Spector I (1987) Measurements of calcium transients in the soma, neurite, and growth cone of single cultured neurons. J Neurosci 6:1934–1940.

Bright GR, Taylor DL (1986) Imaging at low light level in fluorescence microscopy. In Taylor DL, Waggoner AS, Murphy RF, Lanni F, Birge R (eds): Applications of Fluorescence in the Biomedical Sciences. New York: Alan R. Liss, Inc., pp 257–288.

Cannell MB, Berlin JR, Lederer WJ (1987) Effect of membrane potential changes on the calcium transient in single rat cardiac muscle cells. Science 238:1419–1423.

Chandra S, Gross D, Ling Y-C, Morrison GH (1989) Quantitative imaging of free and total intracellular calcium in cultured cells. Proc Natl Acad Sci USA 86:1870–1874.

Cohen LB, Salzberg BM (1978) Optical measurement of membrane potential. Rev Physiol Biochem Pharmacol 83:35–88.

Connor JA (1988) Fluorescence imaging applied to the measurement of Ca^{2+} in mammalian neurons. In Grinnell AD, Armstrong D, Jackson MB (eds): Calcium and Ion Channel Modulation. New York: Plenum Publishing Corp., pp 395–406.

Connor JA, Wadman WJ, Hockberger PE, Wong RKS (1988) Sustained dendritic gradients of $Ca^{2}+$ induced by excitatory amino acids in CA1 hippocampal neurons. Science 240:649–653.

Ehrenberg B, Montana V, Wei M-D, Wuskell JP, Loew LM (1988) Membrane potential can be determined in individual cells from the Nernstian distribution of cationic dyes. Biophys J 53:785–794.

Fluhler E, Burnham VG, Loew LM (1985) Spectra, membrane binding, and potentiometric responses of new charge shift probes. Biochemistry 24:5749–5755.

Gilkey JC, Jaffe LF, Ridgway EB, Reynolds GT (1978): A free calcium wave traverses the activating egg of the medaka, *Oryzias latipes*. J Cell Biol 76:448–466.

Gonzalez FA, Heppel LA, Gross DJ, Webb WW, Parries G (1988a) The rapid desensitization of receptors for platelet derived growth factor, bradykinin and ATP: Studies on individual cells using quantitative digital video microscopy. Biochem Biophys Res Commun 151:1205–1212.

Gonzalez FA, Gross DJ, Heppel LA, Webb WW (1988b) Studies on the increase in cytosolic free calcium induced by epidermal growth factor, serum and nucleotides in individual A431 cells. J Cell Physiol 135:269–276.

Grinvald A, Anglister L, Freeman JA, Hildesheim R, Manker A (1984) Real-time optical imaging of naturally evoked electrical activity in intact frog brain. Nature (Lond) 308:848–850.

Gross D (1988) Electromobile surface charge alters membrane potential changes induced by applied electric fields. Biophys J 54:879–884.

Gross D, Loew LM (1989) Fluorescent indicators of membrane potential: Microspectrofluorometry and imaging. In Taylor DL, Wang Y-L (eds): Methods in Cell Biology, Volume 30. San Diego: Academic Press, Inc., pp 193–218.

Gross D, Loew LM, Webb WW (1986) Optical imaging of cell membrane potential changes induced by applied electric fields. Biophys J 50:339–348.

Grynkiewicz G, Poenie M, Tsien RY (1985) A new generation of Ca^{2+} indicators with greatly improved fluorescence properties. J Biol Chem 260:3440–3450.

Haugland R, Minta A (1990) Design and application of indicator dyes. In Foskett JK, Grinstein S (eds): Noninvasive Techniques in Cell Biology. New York: Wiley-Liss, Inc. pp. 1–20.

Hiraoka Y, Sedat JW, Agard DA (1987) The use of a charge-coupled device for quantitative optical microscopy of biological structures. Science 238:36–41.

Inoué S (1986) Video Microscopy. New York: Plenum Publishing Corp.

Jackson AP, Timmerman MP, Bagshaw CR, Ashley CC (1987) The kinetics of calcium binding to fura-2 and indo-1. FEBS Lett 216:35–39.

Jacob R, Merritt JE, Hallam TJ, Rink TJ (1988): Repetitive spikes in cytoplasmic calcium evoked by histamine in human endothelial cells. Nature (Lond) 335:40–45.

Jeričivič Z, Wiese B, Bryan J, Smith LC (1989) Validation of an imaging system. In Taylor DL, Wang Y-L (eds): Methods in Cell Biology, Volume 30. San Diego: Academic Press, Inc., pp 48–83.

Kinosita Jr. K, Aohikowa I, Saita N, Yoshimura H, Itoh H, Nagayama K, Ikegami A (1988) Electroporation of cell membrane visualized under a pulsed-laser fluorescence microscope. Biophys J 53:1015–1019.

Konishi M, Olson A, Hollingworth S, Baylor SM (1988) Myoplasmic binding of fura-2 investigated by steady-state fluorescence and absorbance measurements. Biophys J 54:1089–1104.

Linderman JJ, Harris LJ, Slakey LL, Gross DJ (1990) Charge-coupled device imaging of rapid calcium transients in cultured arterial smooth muscle cells. Cell Calcium 11:131–144.

Loew LM (1982): Design and characterization of electrochromic membrane probes. J Biochem Biophys Methods 6:243–260.

Millard PJ, Gross D, Webb WW, Fewtrell C (1988) Imaging asynchronous changes in intracellular Ca^{2+} in individual stimulated tumor mast cells. Proc Natl Acad Sci USA 85:1854–1858.

Montana V, Farkas DL, Loew LM (1989) Dual wavelength ratiometric fluorescence measurements of membrane potential. Biochemistry 28:4536–4539.

Nasi E, Tillotson D (1985) The rate of diffusion of Ca^{2+} and Ba^{2+} in a nerve cell body. Biophys J 47:735–738.

Pandiella A, Magni M, Lovisolo D, Meldolesi J (1989) The effects of epidermal growth factor on membrane potential: Rapid hyperpolarization followed by persistent fluctuations. J Biol Chem 264:12914–12921.

Parratt LG (1961) Probability and Experimental Errors in Science. New York: John Wiley and Sons, Inc.

Poenie M, Tsien RY, Schmitt-Verhulst A-M (1987) Sequential activation and lethal hit measured by $[Ca^{2+}]_i$ in individual cytolytic T cells and targets. EMBO J 6:2223–2232.

Prentki M, Glennon MC, Thomas AP, Morris RL, Matschinsky FM, Corkey BE (1988) Cell-specific patterns of oscillating free Ca^{2+} in carbamylcholine-stimulated insulinoma cells. J Biol Chem 263:11044–11047.

Salzberg BM, Obaid AL, Senseman DM, Gainer H (1983) Optical recording of action potentials from vertebrate nerve terminals using potentiometric probes provides evidence for sodium and calcium components. Nature (Lond) 306:36–40.

Scanlon M, Williams DA, Fay FS (1987) A Ca^{2+}-insensitive form of fura-2 associated with polymorphonuclear leukocytes. J Biol Chem 262:6308–6312.

Spring KR, Smith PD (1987) Illumination and detection systems for quantitative fluorescence microscopy. J Microsc 147:265–278.

Tsien RY, Poenie M (1986) Fluorescence ratio imaging: A new window into intracellular ionic signalling. Trends Biochem Sci 11:450–455.

Tsien RY, Rink TJ, Poenie M (1985) Measurement of cytosolic free Ca^{2+} in individual small cells using fluorescence microscopy with dual excitation wavelengths. Cell Calcium 6:145–157.

Williams DA, Fogarty KE, Tsien RY, Fay FS (1985) Calcium gradients in single smooth muscle cells revealed by the digital imaging microscope using Fura-2. Nature (Lond) 318:558–561.

Noninvasive Techniques in Cell Biology: 53–72
© 1990 Wiley-Liss, Inc.

3. 3D Fluorescence Imaging of Single Cells Using Image Restoration

W.A. Carrington, K.E. Fogarty, and F.S. Fay

Program in Molecular Medicine, Physiology Department, University of
Massachusetts Medical School, Worcester, Massachusetts 01605

I. INTRODUCTION

Changes in intracellular molecular and ionic distribution are believed to underlie a wide range of cell functions, such as cell motility, cell polarization,
morphogenesis, and cellular response to hormones. Highly specific bright fluorescent probes can now be used to view distributions of specific ions, proteins, and nucleic acids in a light microscope. The ability to see a molecular
distribution in a fixed cell and changes in a molecular distribution in a living
cell is an invaluable tool in elucidating the biological processes underlying
those molecular distributions.

A cell is a complex three-dimensional (3D) structure. A single two-dimensional view frequently cannot provide insight into the spatial relationship between parts of the cell. Two superimposed features might be either close to each other or far apart in the axial direction. Therefore, a full 3 D view of the object will often be necessary to understand the true relationships within the cell. Optical sectioning—acquiring a series of images at different optical depths—provides a 3D view of the cell using a digital imaging microscope which uses a digital camera and motorized focus under computerized control. Once stored in a computer in digitized form a number of computer manipulations become possible.

In fact, all light microscopy is 3D. Even when a single 2D image of a cell contains all the needed information, it is degraded by light originating above and below the plane of interest. Details that should be well within the theoretical resolving power of the light microscope are obscured by this out of focus haze. Two methods to reduce this problem, confocal microscopy and image restoration, are under investigation by a number of workers. In this chapter we discuss image restoration (also called deconvolution) as a means to remove this out-of-focus haze. Image restoration is a process that combines quantitative knowledge of the 3D imaging process with a set of optical sections of a cell to partially reverse the blurring process in the computer. The result is a 3D image of the cell that has most of the out of focus haze removed, a better signal to noise ratio, greater dynamic range, higher resolution, and greater numerical accuracy than the original image [Carrington and Fogarty, 1987; Fay et al., 1986, 1989; Castleman, 1979; Agard, 1984]. Figure 1 shows the results of applying image restoration to an image of microtubules in a newt eosinophil.

Scanning confocal microscopes [Wilson and Sheppard, 1984] also provide some of the same benefits. In fact, image restoration applied to 3D confocal images would theoretically result in images with the highest resolution. Currently available confocal microscopes are unfortunately significantly less efficient at collecting emitted fluorescent light than a conventional microscope with a state of the art CCD camera. This lower efficiency limits the suitability of current confocal instruments for high resolution 3D imaging of single cells where photobleaching or photodamage are a factor or where the fluorescence is not bright. This is discussed further in section II and in Carrington et al. [1989].

II. COMPARISON WITH CONFOCAL MICROSCOPY

Confocal microscopes can also provide optical discrimination against out-of-focus light. This discrimination is accomplished by passing excitation and emission light through a pinhole which is focused into the sample; this point is then scanned relative to the sample. The main benefit of this microscope is that much of the emitted out-of-focus light is excluded by the pinhole before it

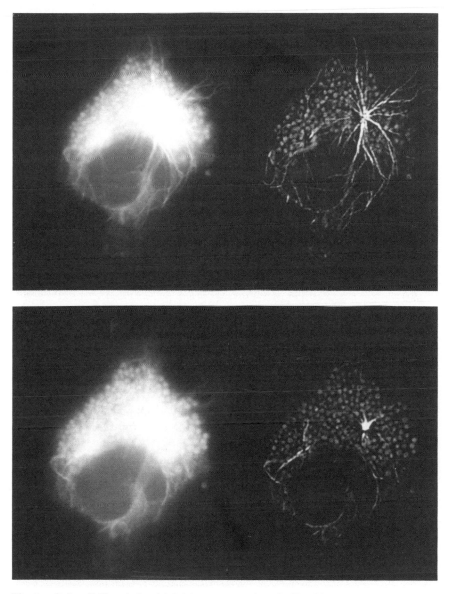

Fig. 1. Before **(left)** and after **(right)** image restoration of a 3D wide-field microscope image. Two optical sections 0.75 μm apart from a series of 46 optical sections at 0.25 μm axial increments of a Newt eosinophil with rhodamine-labeled tubulin, using a 63 × , 1.4NA lens on a Zeiss IM35 with a Photometrics cooled CCD. Voxel size is 200 × 200 × 250 nm. Note the detail revealed in the microtubule organizing center that is obscured by out-of-focus flare in the original data on the left that is revealed in the image restoration on the right.

reaches the detector, so the general background of out-of-focus light present in unprocessed wide-field images is eliminated. Confocal microscopes have both advantages and disadvantages compared to a wide-field microscope combined with image restoration. The main advantage of the confocal microscope is that it supplies clear images almost immediately without lengthy computer processing. It is able to do this using a single optical section, unlike the image restoration approach, which requires multiple optical sections.

Confocal microscopes suffer from a number of disadvantages, some of which are inherent in the concept. Image acquisition time is slower than for a wide-field microscope, and will likely remain so in future designs. In a wide-field microscope the whole image is acquired simultaneously while in a scanning confocal microscope each pixel (or in a tandem scanning microscope, an array of spaced pixels) is acquired sequentially. This greater parallelism of image acquisition in a wide-field microscope means that less time is required to acquire an image. If fluorescence saturation is a limiting factor in acquisition speed, the wide-field microscope has a considerable advantage because its peak excitation intensity is equal to its average intensity; in a confocal microscope, peak intensity of excitation illumination is related to the average by the ratio of time to acquire a whole 2D image to the time spent at one scan position. When only one pinhole is scanned, this ratio is equal to the number of pixels in the image, which is typically on the order of 10^5. A less obvious difficulty for quantitative fluorescence measurements that is inherent to confocal microscopes is also its main advantage—the ability to discriminate against out-of-focus light; this makes quantative analysis of fluorescent images more difficult because the total fluorescence received from each plane is dependent on how far out of focus it is. If a fluorescent structure is located between two optical sections, the percentage of fluorescence detected from it is lower than if it had been located in the plane of one of the optical sections. This can be corrected in postprocessing, but the process is not nearly as simple as for a wide-field microscope. A further difficulty of confocal microscopy is a greater sensitivity to chromatic aberration than in conventional wide-field microscopy; chromatic aberration causes the pinhole to be focused to different depths for the excitation and emission wavelengths. This has the effect of an axial misalignment and transverse off-axis misalignment of the pinhole for excitation and emission, which both decreases the efficiency of light collection and decreases resolution. For designs, such as the current commercial instruments, that use a single objective lens and pinhole for both excitation and emission, this effect will make it difficult to use dyes, such as the calcium indicator fura-2, that emit visible light but are excited in the ultraviolet.

There are also a number of ways that currently available confocal microscopes can fail to achieve the level of performance that is theoretically possible. The major problem with these current designs is light collection and detection efficiency, which can be surprisingly low. Wells et al. [1989] have measured the light collection efficiency of an early commercial instrument to be as low as 0.01%–0.2% (the manufacturer has since improved this instrument). The best of the commerically available instruments will still have losses from optical elements not present in a conventional microscope and losses from the confocal pinhole. Steltzer [1989] reports losses from these optical elements for his design of an efficient confocal microscope built onto a conventional microscope that result in the collection of 24%–80% of the light exiting from the conventional microscope. In addition, many confocal microscope designs must use a photo-multiplier tube rather than the higher quantum efficiency CCD cameras. These CCD cameras can require 1/3 or less of the light to produce the same signal to noise ratio in the acquired image as the most efficient photomultiplier tubes at wavelengths typical of rhodamine or fluorescein. We recommend testing samples typical of your experiments on the instrument before purchase, especially for living cells or for cells where fluorescent bleaching may be a problem.

The image restoration we describe below can be used on 3D images from a confocal microscope as well as those from a wide field microscope. In fact, the combination of confocal microscope with image restoration provides, in computer simulations, the highest resolution images. We expect these theoretical results from this combination to be achieved as the available instruments improve in sensitivity.

III. IMAGE RESTORATION

A. How Does Image Restoration Work?

A 3D image of a cell is acquired by optically sectioning; i.e., a series of images are recorded by focussing on a series of different planes in the cell. Image restoration combines this 3D data with knowledge of the imaging process and any other information available about the cell in order to invert the imaging process mathematically. Because the image formation process is inherently 3D when a 3D object is viewed, image restoration is an estimate of the distribution of the fluorescent stain within the cell that is closer to the true distribution than is the unprocessed image. The blurring caused by the microscope optics is partially reversed. The resulting image also has better resolution, contrast, dynamic range, and depth discrimination and better accuracy for numerical measurements than the original data set. We are able to obtain 3D images from a conventional fluorescence microscope and a CCD camera that are comparable to (and sometimes superior to) images from a confocal microscope.

The key to understanding the process is that image restoration uses more information than just the measured 3D image; an integral part of the process is that it incorporates quantitative knowledge of the way the microscope blurs objects. Our approach to image restoration can be so effective because it uses quantitative information on the microscope's 3D blurring; it operates on the full 3D image simultaneously; it can incorporate additional information on the object (e.g., that a fluorescent dye density is non-negative).

These three together can be very powerful; noisy images are mathematically transformed to a 3D reconstruction that fits *all* the information at our disposal. For example, some noisy images are not consistent with the physics of optical systems; the pixel to pixel variation caused by noise may be greater than is possible for any optical system because a real optical system blurs out such rapid variations. We improve images by bringing more information to bear on the problem and using mathematical techniques that effectively combine all of that information.

Image restoration is able to use knowledge of the manner in which the optics spreads and blurs light to reassign blurred or out of focus light closer to its correct place of origin. The 3D blurring of the microscope is characterized by a 3D image of a point source (the point spread function). Figure 2 shows several optical sections of a point spread function. It is easy to see that if we had a 2D image of an out-of-focus point, we could find the best match in figure 2 to estimate how far out of focus it is. To distinguish it from an in-focus diffuse ring-shaped object, we would need to compare at least two planes of focus. We can even look at Figure 3 and see that the images on top are optical sections of a point source which lies to the right outside the field of view. While

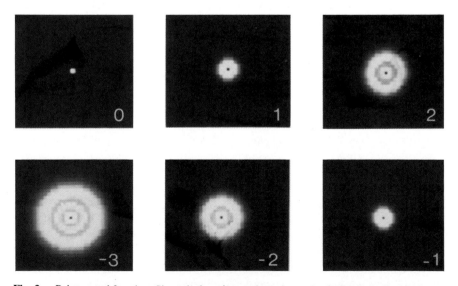

Fig. 2.   Point spread function. Six optical sections at 1 μm intervals of a theoretically calculated point spread function of a 1.4NA oil immersion lens. Voxel size is $250 \times 250 \times 250$ nm.

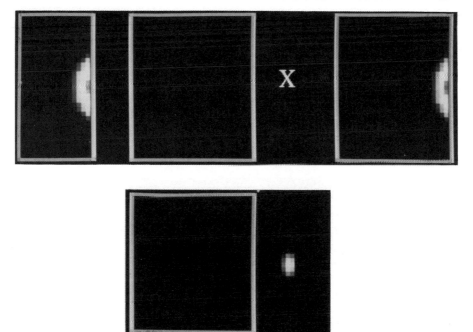

Fig. 3.   Top: Three optical sections at 2.5 μm intervals of a point source that lies 1 μ to the right of the field of view, marked by X. The borders enclose the field of view used; the center image is in the same plane of focus as the point source. **Bottom:** Image restoration applied to the top image. The restored point source is shown superimposed on the frame of the field of view; the data used was just that inside the field of view, but at 0.25 μm intervals. *This is not a recommended way to acquire data,* but illustrates the ability of the image restoration procedure to explain light originating from outside the field of view and place it in its correct position. Resolution using this data is less than is possible if data placing the point source inside the field of view were used.

it is possible visually to match overlays of a point spread function to simple and very sparse images of a small number of point sources, this is too difficult to do visually on more complex images, as is illustrated in Figure 4, which is an out-of-focus image of three closely spaced point sources.

In the next section we note that the human visual system is very good at discerning patterns. However, it is not so good at correlating the multiple images of an optically sectioned image. An advantage of computer processing is the possibility of operating on all the 2D slices in the 3D image simultaneously.

B. Limitations of the Human Visual System and Direct Viewing

The limitations of directly viewing cells through a microscope are in a way more the limitations of the human visual system than of the microscope. The human visual system is a very sophisticated image processing system which

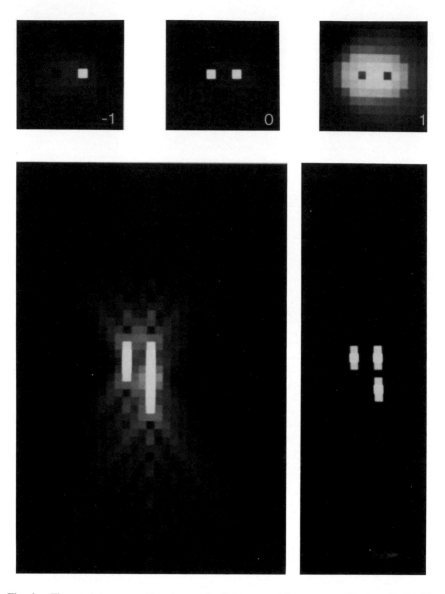

Fig. 4. Three point sources: Two at same depth 1 μm apart in transverse direcion, the third is μm below the first. **Top:** Three optical sections at 1 μm intervals of these three point sources blurred with the point spread function in figure 2. **Bottom left:** A series of optical sections at 0.25 μm intervals of the three blurred points displayed as a section parallel to the optical axis through the three points. **Bottom right:** Image restoration of a series of 64 optical sections of the blurred image of the three points displayed as the same axial slice as the blurred image on the left.

has evolved to perform well on scenes different from 3D microscope images. 3D fluorescent images are 3D densities, which have the appearance of translucent clouds or fog. On the other hand, natural scenes are mostly composed of opaque objects which are viewed by light reflected from their surfaces. We perceive three-dimensionality by stereovision, by variations in intensity for different angles between a visual surface and a light source, and by motion cues. In each case, we solve a computational problem—obtaining a 3D representation of a scene from limited and indirect information—that is conceptually the same as in image restoration but with very different data. Solving these visual computations that we do automatically is an active area of research in computer vision [Marr, 1982; Horn, 1986].

Our visual system cannot use a stack of optical sections to visualize three dimensionality with the same efficiency that it can from surface representations. Nor can it effectively use knowledge of a microscope's 3D blurring to correct that 3D blurring. It is the task of image restoration to perform that correction. It is the task of other image processing techniques and computer graphics to present those corrected images in ways that can be understood visually.

C. Visual Versus Numerical Improvement in Images

There are a number of approaches used in image processing and computer graphics whose purpose is to extract information that is present numerically in a digitized image and improve it visually. Methods such as thresholding or autoscaling add no new information beyond that present in the image itself, but do allow details that are present in the image to be *seen* more clearly. Image processing methods that operate on an image without additional information about the image formation process differ from image restoration which does use additional information of both the imaging process and the object itself. Image restoration uses quantitative knowledge of the imaging process to improve the accuracy of the restored image compared to the original image. The restored image has higher resolution than the unprocessed image and is closer visually to the true object; more important, the relative estimates of dye density are more accurate at each point than the unprocessed image provides. This improvement in accuracy and improvement in resolution is not possible with image processing methods that do not incorporate knowledge of the imaging process; they are not able to reassign out of focus light to its correct place of origin.

Methods that do not use information on the imaging process are able to extract information that is present numerically in the digitized image but is not apparent visually. Simple methods are very useful in examining a digitized image; these include autoscaling, thresholding, pseudo-coloring, the ability to slice or project the image from any angle and to display it in stereo. These operations can all be performed on inexpensive personal computers. More sophisticated computer graphics methods can be used to render 3D fluorescent images in a variety of ways. The question of rendering volume densities in a way that can be understood visually also occurs in medical imaging and is a difficult problem that is an area of active research. Motion cues, such as might be provided by a rotating stereo display of the cell, are also useful in conveying visual information. Volume rendering and display of motion, especially under interactive control, require the use of computer workstations that are more powerful than personal computers. These computer graphics methods are very useful in interpreting and analyzing images because they translate information contained in the data to a visually meaningful form. They do not increase resolution, remove blurring, or increase the numerical accuracy of estimates of dye density. The information they provide is present numerically in the unprocessed image and in principle could be extracted by direct examination of the numbers in the digitized image. That is a very slow process; what these methods do provide is a much higher bandwidth connection to the data by using the human visual system. To actually improve upon the data requires the use of more information about either the imaging or the dye distribution itself; the image restoration process uses both types of additional information.

There are image processing algorithms that make use of qualitative information on the imaging process. For example, out-of-focus structures are blurred and can contribute no sharp detail to an image. High-pass filtering attenuates the low spatial frequencies (slow variations from voxel to voxel). High-pass filtering tends to enhance sharp details and attenuate out-of-focus haze. It can render fine details more visible. However, it suppresses any slowly varying component of the underlying molecular distribution, and so is entirely unsuitable on diffuse distributions, such as ions, and can provide confusing images of structures such as filaments that vary slowly in one direction. The major drawback of high-pass filtering is that it also enhances noise, which always has substantial high spatial frequency components. It does not increase the numerical accuracy of estimates of dye density in a compartment as image restoration can. Edge detection techniques have advantages and disadvantages similar to high-pass filtering. These methods use fairly weak additional information to obtain—sometimes—a modest increase in the ability to visualize localized structures. They usually *decrease* our ability to visualize less localized or diffuse dye distributions.

IV. THE FORWARD PROBLEM: CALIBRATION OF THE MICROSCOPE

A. Definition of the Forward Problem

There are two parts to the image restoration process, which mathematicians call the forward problem and the inverse problem. The forward problem is to describe mathematically the 3D imaging process; i.e., the solution to the forward problem is a mathematical equation that calculates the 3D optically sectioned image that results from any given 3D fluorescent dye distribution. The forward problem can be thought of as a mathematical modelling problem—given any simulated cell we must calculate a simulated 3D microscope image of that cell. The inverse problem is to calculate the 3D fluorescent density from the 3D optically sectioned image. We can express these problems symbolically by letting f = the 3D dye density in the cell, and g = the 3D optically sectioned image of the cell. The microscope transforms the dye density, f, into an array of numbers, g, which is the digitized image of the cell. We write this as: Kf = g. The forward problem is to describe K mathematically. The inverse problem is to solve for f using our measurements, g, and knowledge of K. The inverse problem will be discussed in section V.

There are several useful facts about optical systems we will need. The microscope's optical system is linear; that is, if the dye density is multiplied by a constant then the image intensities are multiplied by the same constant and if two densities are added then the image intensities also are added. Mathematically we write

$K(af) = aK(f)$, where a is a constant

$K(f_1 + f_2) = K(f_1) + K(f_2)$, where f_1 and f_2 are two dye distributions

Digital cameras sample the image at only a finite number of points (pixels) and we can only acquire images at a finite number of planes of focus. So, we have only a finite, though large, number of volume elements (voxels) in our 3D image—between 1 and 10 million voxels is typical. An ideal, but physically impossible, microscope would collect light only from a neat cubical volume of the cell. But the blurring of the microscope causes each voxel to collect light from a large volume of the cell. Thus, the intensity of voxel number i, g_i, is related to the fluorescent dye density by

$$g_i = \iiint k_i(x,y,z)\, f(x,y,z)\, dx dy dz \qquad (1)$$

where k_i is a function that describes the response of voxel number i to a point source at each point (x,y,z) in space. In other words, each voxel's intensity is a weighted average of the light emitted by the cell.

When well-corrected microscope optics are used, if an object is moved, its image is unchanged except for a corresponding change in position in the field of view. A system with this property is called spatially invariant, and the operator, K, describing the system is derived from the mathematical operation of convolution. Each of the functions, $k_i(x,y,z)$, is to a good approximation the same as every other except for a translation to a different position. The optical system is then described by any one of these functions, which is called the point spread function and is described in the next section.

Finally, any physical measurement contains noise or error, so the noisy imaging process is described by

$$\tilde{g}_i = \iiint k_i(x,y,z) \, f(x,y,z) \, dxdydz + \epsilon_i, \ i = 1,2,3,\ldots,N \qquad (2)$$

where $\tilde{g}_i$ is the noisy measurement of intensity at voxel i, ϵ_i is random noise, and N is the number of voxels. Random noise will be composed of photon noise, i.e., the random variation of photons which is described by the Poisson distribution, plus camera readout noise, dark current, and thermal noise of the electronic circuitry.

B. Equipment Description

We have developed the digital imaging microscope to collect digitized images of fluorescent probes in cells in 2D and in 3D. The system is based on an inverted microscope equipped for epifluorescence and modified to collect images under computer control. Images are recorded with a cooled slow scan CCD camera, which is the camera of choice for low-light-level fluorescence microscopy. The CCD cameras we use have quantum efficiencies between 0.4 and 0.8 at 500 nm wavelength, with a readout noise of 50 photons (rms); more recent cameras have quantum efficiencies of about 0.4 with readout noise of 6 photons (rms). The microscope focus is motorized under computer control using feedback on position from an eddy current sensor. Typically, series of optical sections are acquired at 0.2–0.5 μm intervals through focus. The microscope control computer (a PDP11/73), a VAX-11/750, Silicon Graphics Personal Iris workstations and a Silicon Graphics 4 processor 240/GTX computer server are connected by Ethernet. A Biorad MRC500 confocal microscope is also available and connected to our other systems by Ethernet.

C. Calibration: Measurement of the Point Spread Function

Image restoration depends on quantitative knowledge of the 3D image formation process in the microscope. This is characterized by the 3D image of a point source, i.e., the so-called point spread function. In principle, the point spread function can be calculated using the methods of Fourier optics; however, these theoretically calculated point spread functions have not yet incorporated all important factors and are not yet accurate enough to use routinely [Tella, 1984]. Empirical determination of the point spread function for each optical setup is then the best course. It is measured by optically sectioning a subresolution fluorescent bead (200 nm diameter or less). The point spread function plays the role in image restoration that calibration of an instrument does in other measurements. Differences in the optical setup between the measurement of the point spread function and the cell image degrade the final results. The objective lens must be the same for both. Axial spacing and magnification could differ but then interpolation must be used to obtain an image and cell with matching voxel sizes, with some loss of accuracy. Emission wavelengths of the fluorophores used in the point spread function and cell image should be similar. (In a confocal microscope, excitation wavelengths should also be similar). If there is any significant change in the microscope alignment, the point spread function should preferably be remeasured.

The bead preparation used should be appropriate to the experiment. The bead should be mounted in the same media as the cell to avoid mismatches caused by different indices of refraction. Since a small fluorescent bead contains only a small amount of fluorophore, low signal to noise ratio is a problem. Every effort should be made to maximise the labelling of the bead and to minimise the fluorescent background.

There are several important points to emphasize about the point spread function. Though there is some variation in excitation intensity in epi-illumination, the fraction of emitted light from a bead collected by the camera is the same no matter what plane is focussed on. If a light source of constant intensity moves out of focus, there is no fall off in total light collected from it. The peak intensity from a point source will fall off rapidly as it goes out of focus, but the same total amount of light will be spread out over a larger area as the point goes out of focus. The wide field microscope, unlike the confocal microscope, does not optically reject out of focus light. Also, as a point goes out of focus it spreads transversely, so in a given region, some of the light detected will come from out of focus points located outside of the region in the transverse direction.

V. THE INVERSE PROBLEM
A. Description of the Inverse Problem

In the previous section we described the forward problem of microscopy; its solution allows us to simulate the microscope imaging process. The inverse problem is to find a good estimate of the dye density in the cell that fits all the available information. When we use the forward problem to simulate the image resulting from this estimate of dye density it should match the actual image to within the noise level. We will later describe algorithmic approaches to finding this good estimate.

The solution of the forward problem can be incorporated into a computer program that would allow us to simulate the 3D images that arise from arbitrary dye distributions. In principle, we could hypothesize a dye distribution and test it by simulating the microscope blurring and comparing the simulated image to the actual measured 3D optically sectioned image. We can predict certain results. If we use the actual 3D data as our estimate of the density, we will be blurring an already blurred image, and it will not match the actual data; a point source in the cell will appear in its simulated image broader than in the real image. We might then guess that where there is a small blob in the image there is a point source in the real object. On simulating the blurring of this point source, we might find that it matches the image well, which would argue that there is in fact a point source there, or we might find that the blurred point source is narrower than the blob in the image which would argue that the source is either more than a single point source or is a broader object. We could theoretically vary our hypothesized dye distribution till it matches the measured data. Of course, this kind of interactive process is impractical because it would take forever; 8 bit images and 1 million voxels involves $256^{1,000,000}$ choices. We need a more focused and algorithmic method for exploring the choice of dye density estimate and numerical criteria for deciding which estimate to use from the huge number possible.

Furthermore, we can see the possibilities for noise reduction in this process. A noise peak might appear to be the image of a point source in the cell, but it is not consistent with the surrounding voxels; a bright point in the cell produces a copy of the point spread function added to the rest of the image and this will not be the case with a noise peak. If a bright point in an image corresponds to a bright point in the cell, not to noise, the pattern of fall off of intensity as it goes out of focus must match the point spread function. Most voxel to voxel variation in the image is not consistent with the image of any physical object. So, an approach that properly uses the full set of data and the point spread function will be able to reject much of the noise as inconsistent with any possible cell dye density. This is especially true when we add the fact that any density must be non-negative.

B. Solving the Inverse Problem

The mathematical problem we must solve is to find an estimate of the dye density in the cell that is a good approximation to the actual unknown dye density. The information that is the data for this problem is a description of the imaging process:

$$g_i = \iiint k_i(x,y,z) \, f(x,y,z) \, dxdydz, \, i=1,2,...,N \tag{3}$$

In addition, prior information may be available to constrain the solution. A simple but powerful piece of information is that any density is non-negative, so we have also

$$f(x,y,z) \geq 0 \text{ at every } (x,y,z)$$

Also, there is noise present in the measurements, so the measured data is

$$\tilde{g}_i = g_i + \epsilon_i \tag{4}$$

There are some mathematical difficulties that must be solved. First, the finite amount of measured data usually cannot uniquely determine the dye distribution even in the absence of noise. Second, the microscope is insensitive to rapid spatial oscillations in the dye density; it averages out rapid variations. Two dye distributions can produce images whose difference is very small, yet the dye distributions can differ by the presence of wild spatial oscillations in one distribution, but not the other. These difficulties are characteristic of a class of problems called ill-posed problems in the mathematical literature [Tichonov and Arsenin, 1977; Morozov, 1984; Twomey, 1977]. A well-posed problem is one whose solution exists, is unique, and depends continuously on the data; a problem that is not well-posed is called an ill-posed problem. There is a substantial mathematical literature on the solution of ill-posed problems. Inverse problems are frequently ill-posed; perhaps the best known ill-posed problem is the tomographic reconstruction of a cross section from its projections that is the basis of medical CAT scans.

The inverse problem of 3D microscopy is also ill-posed. The solution could fail to exist because noise renders it inconsistent with any possible solution. If the image could be sampled continuously, the exact solution of the inverse problem would depend discontinuously on the image; arbitrarily small errors in the image would correspond to large errors in the solution. When the data is sampled, this discontinuity in the continuously sampled problem becomes a magnified sensitivity to noise in the measurements when the most direct solution methods are attempted. Finally, when the image is sampled at only a finite number of voxels, the continuously defined dye density cannot be uniquely determined from only a finite number of measurements.

The most direct solution is to use the method of least squares to find the f that minimizes the residuals between the data and the image corresponding to f:

$$\text{MIN } \sum |\tilde{g}_i - \iiint k_i(x,y,z)\, f(x,y,z)\, dxdydz|^2 \tag{5}$$

This least squares solution suffers from sensitivity to any noise present in the data. This is so since the microscope blurs or averages out any spatial oscillations in the solution. Since an oscillatory solution and a smooth solution can yield images that differ by only a very small amount, the least squares solution will tend to have artifacts consisting of wild oscillations. Solution methods that ignore the ill-posed nature of the problem will usually result in such oscillatory artifacts. The intuitive solution to this problem is to enforce some sort of smoothing on the solution to smooth out these oscillations. This smoothing has been done in a variety of ways. One common but unreliable method is to use some iterative method such as the Jansson/van Cittert method [Jansson, 1970] to solve Equation 1 and to stop before the solution is reached; typically the stopping point is determined by the time one is willing to wait for the calculations to finish. As computer speeds increase and more iterations are used, the result does not improve, but becomes worse. This is because these methods attempt to fit the data exactly, and the exact solution magnifies the effect of noise; early iterations are far enough away from the exact solution that they are smooth but further iterations begin to fit the noise in the data. Linear smoothing filters such as Wiener filters are also frequently used in similar problems to obtain a smoothed version of the solution [Castleman, 1979]; they have the advantage of shorter computational times and disadvantages we will discuss later.

For biological applications, there are several requirements imposed on an image restoration method. It must be reliable on the whole range of fluorescent images encountered in biology. It must be flexible in its use of data; data collection should be determined as much as possible by the needs of the biological experiment rather than the needs of an image restoration algorithm. It is desirable that the image restoration be possible on computer workstations rather than supercomputers. This also is a requirement on flexibility in use of data. Because biological 3D images are so large, unless the image restoration can be done separately on pieces of the image, very large amounts of computer memory are needed. In both cases, the difficulty is that the data must be truncated. In the axial direction, no matter how far out of focus the cell is, the total amount of light collected does not become small. Therefore, it becomes necessary to either truncate the image before intensities become small or to artificially roll off the intensities (apodization). In both cases, if an algorithm requires the full data, as most deconvolution algorithms do, an error is introduced. If the data is truncated and zeros inserted for the missing data, this error is sizable; the image cannot drop to zero from one plane of focus to the next. If apodization is used to roll off the intensities more gradually in the axial direction, error is still introduced; something has artifically been introduced into the data that cannot be accounted for by the point spread function. Similarly, if we subdivide the image for processing in pieces, at the edge of each piece light enters from the adjacent piece that must be accounted for or an error is introduced. This border effect is less important than in the axial direction because the artifacts introduced occur near the border and can be discarded. However, the region that needs to be discarded can be a large percentage of the piece; when we used apodization, only one quarter of the volume of each piece was useable with $64 \times 64 \times 64$ pieces. In defining the forward problem, $f(x,y,z)$ represents a function with a continuous domain; we do not discretize (or pixelize) the cell when we define the problem. The measured image is discrete—a finite number of voxels—but the cell is continuous. We also do not specify how the image is sampled. Our cameras give us a regular rectangular array of pixels, but even if they were distributed non-uniformly the algorithm we present would work with some increase in computational time. More important is the fact that we do not specify the number or spacing of the planes of focus in the formulation of the problem or in the resulting algorithm. The data can be truncated in the axial direction or planes omitted, and the algorithm still works. We have even successfully performed restorations where no in focus information was used, only out of focus planes above and below the plane of interest were used. In figure 3b we show the results of a restoration where the only source of light was a point source 1 μm outside the field of view. This is clearly not the best way to acquire data for a restoration, but shows how the algorithm we use is able to account correctly for light that originates outside the field of view, rather than allowing it to introduce artifacts.

There are also mathematical properties that are desirable. We want the method we use to have the property that as the quality and quantity of our data improves, the solution converges to the true dye density, i.e., we can make our estimate as close to the exact distribution as we wish if we obtain a large enough number of voxels with a large enough signal to noise ratio. Such a method is said to regularize the ill-posed problem. While it may seem easy to do arbitrarily well by obtaining arbitrarily good data, many methods fail this modest test. A least squares solution does not regularize the problem; in fact, as the number of measurements increases, the sensitivity to noise increases sharply and the least square solution becomes further from the true distribution. Many methods used for deconvolution are intuitive; they smooth their solution estimate, but do so in an ad hoc manner. Frequently such methods have been tuned to do well on a narrow class of image but require a major research project to modify for a different class of image. In fluorescent microscopy where there is a broad range of image types, there is a great advantage to a method that has been rigorously analyzed and *proven* to regularize the problem. Once such a mathematical method is decided upon, a reliable numerical algorithm, preferably one that is efficient and also proven to converge to the same estimate as the mathematical method expects. The numerical analysis required for successful implementation of a solution to an ill-posed problem is non-trivial; the ill-posed nature of the problem is frequently reflected in the ill-conditioned nature of some of the calculations in an algorithm.

The term regularization is also used for a particular method for regularizing the ill-posed problem that is closely related to least squares. It consists of adding a term onto the least squares expression that forces the solution to be smooth. In regularization we choose the f that minimizes:

$$\text{MIN } \sum |\tilde{g}_i - \iiint k_i(x,y,z)\, dxdydz|^2 + \alpha\, S(f) \qquad (6)$$

where S(f) is a term that measures the smoothness of f; the larger S(f) is the less smooth f is. In its original form it is called Tichonov regularization and S(f) is an integral of a term involving derivatives of f. In most situations in fluorescent microscopy this forces the solution to be too smooth; it does not allow the discontinuity present in a stained filament structure such as the microtubule images shown. A better choice is to let

$$S(f) = \iiint |f(x,y,z)|^2 dxdydz$$

This is a measure of the smoothness of $f(x,y,z)$. Since f is squared, the peaks in an oscillation cause this measure to be larger than if the peaks were lowered; thus by penalizing the peaks in oscillations it enforces smoothness. This is actually equivalent to using a linear smoothing filter when there is no data truncation. Its advantage over a linear smoothing filter is the greater flexibility in data requirements.

A surprisingly powerful piece of information that can be incorporated into Equation 6 is the simple fact that a dye density must be non-negative. This non-negativity cannot be used with linear smoothing filters, which is their second major disadvantage. To incorporate it into Equation 5 we simply perform the minimization over just the non-negative functions:
Find the $f(x,y,z) \geq 0$ that minimizes

$$\underset{f \geq 0}{\text{MIN}} \; \sum |\tilde{g}_i - \iiint k_i(x,y,z) f(x,y,z) \, dxdydz|^2 + \alpha \iiint |f(x,y,z)|^2 dxdydz \quad (7)$$

In Equation 7 we have two terms: the first measures how well f fits the data; the second measures how smooth the solution estimate is. If the data are fitted exactly, the first term is small, and the solution is too sensitive to noise. If we make the second term very small, f will be very smooth and insensitive to noise, but will just be a broad flat distribution no matter what the cell is like. We need to balance between the two extremes; that is done by the proper choice of α. One way of choosing α is to choose it so that the first term, the residuals, is larger than an estimate of the noise variance. Carrington [1982] has shown that this choice of α regularizes the problem. The solution is not extremely sensitive to α, and $\alpha = 0.001$ works adequately on most problems in 3D microscopy. Generally, the noisier the data, the larger α should be.

We now need an efficient and reliable numerical algorithm for performing the minimization in Equation 7. Carrington and Fogarty [1987] have presented one such algorithm which we can prove mathematically converges to the minimum in Equation 7. The function we must minimize has infinite dimensional domain because f is a function defined on a continuous domain and it is nondifferentiable because of the non-negativity constraint. The approach we take is to transform it into a related differentiable, finite dimensional problem using methods from convex analysis. This related problem is then minimized using calculus methods. Generally, ill-posed problems require great attention to the numerical analysis of the minimization algorithm and its implementation; the ill-posed nature of the problem causes some calculations that appear to be straight forward to be very sensitive to numerical roundoff errors. The reader is referred to other discussions [Carrington and Fogarty, 1987; Butler et al., 1982; Carrington, 1990] for the technical details of our approach to the numerical algorithm.

VI. RESULTS

We have applied this image restoration to 3D images of a variety of fluorescently labeled cells, both living and fixed. We have also tested it on computer simulations. It has proven robust and reliable. As can be seen in the restoration of labeled microtubules in figure 1, out-of-focus haze is substantially eliminated and details not clear in the original data are visible in the restoration. Dynamic range, contrast, and numerical accuracy are also increased in the restored 3D image. Axial resolution is also improved. In figure 1, the full width at half intensity in the axial direction of a single microtubule is 850 nm; for images containing a fluorescently labeled bead, this axial resolution if 590 nm. This compares favorably to 800 nm measured in a confocal microscope by Brakenhoff et al. [1989]. Computer simulations show that image restoration combined with a confocal microscope will have an axial resolution of 350 nm, compared to the theoretical value of 450 nm for an unprocessed confocal image.

Restoration of a $100 \times 100 \times 40$ voxel section of a 3D image takes 70 min on a single processor Silicon Graphics 240/GTX and will take about 18 min when we implement a parallel version of the algorithm. Inexpensive RISC-based computer workstations are becoming available that will perform such a restoration in less than an hour; inexpensive array processors are becoming available that will perform it in less than 15 min.

VII. CONCLUSIONS

Image restoration is applied to 3D data sets from a conventional wide-field microscope provides images that have increased resolution, contrast, and dynamic range, and reduced out-of-focus haze. When high quantum efficiency CCD cameras on a conventional microscope are used, images can be obtained from dim fluorescent samples that are comparable to those obtained on confocal microscopes from brighter samples. The ultimate resolution will be obtained by applying image restoration to confocal images.

ACKNOWLEDGMENTS

This work was supported by NSF grant DIR-8720188.

VIII. REFERENCES

Agard DA (1984) Annu Rev Biophys Bioeng 13:191–219.

Brakenhoff GJ, van der Voort HTM, van Spronsen EA, Nanninga N (1989) Three-dimensional imaging in fluorescence by confocal scanning microscopy. J Microsc 153 (pt. 2):151–159.

Butler JP, Reeds JA, Dawson SV (1982) Estimating solutions of first kind integral equations with non-negative constraints and optimal smoothing. SIAM J Numer Anal 19, No. 3:381–397.

Carrington W (1982) Moment problems and ill-posed operator equations with convex constraints. Ph.D. Dissertation, Washington Univ., St. Louis.

Carrington W, Fogarty KE (1987) 3-D Molecular distribution in living cells by deconvolution of optical sections using light microscopy. In Foster K (ed): Proc of the 13th Annual Northeast Bioengineering Conference. IEEE, pp 108–111.

Carrington WA, Fogarty KE, Fay DS (1989) Three-dimensional imaging on confocal and wide-field microscopes. In Pawley J. (ed): The Handbook of Biological Confocal Microscopy. New York and London: Plenum Press.

Carrington WA (1980) Image Restoration in 3D Microscopy: Regularization with Convex Constraints, to appear in S.P.I.E. Proceedings.

Castleman KR (1979) Digital Imaging Processing. Englewood Cliffs, NJ: Prentice-Hall, Inc.

Fay FS, Carrington W, Fogarty KE (1989) Three-dimensional molecular distribution in single cells analysed using the digital imaging microscope. J Microsc 153 pt. 2:133–149.

Fay FS, Fogarty KE, Coggins JM (1986) Analysis of molecular distribution in single cells using a digital imaging microscope. In DeWeer P, Salzberg B (eds): Optical Methods in Cell Physiology. New York: John Wiley & Sons, pp 51–62.

Horn BKP (1986) Robot Vision. Cambridge, MA: The MIT Press.

Jannson PA, Hunt RH, Plyler EK (1970) Resolution enhancement of spectra. J Opt Soc Am 60: pp 596–599.

Marr D (1982) Vision: A Computational Investigation into the Human Representation and Processing of Visual Information. San Francisco: W.H. Freeman and Co.

Morozov VA (1984) Methods for Solving Incorrectly Posed Problems. New York: Springer-Verlag.

Steltzer EHK (1989) Considerations on the intermediate optical system in confocal microscopes. In Pawley J (ed): The Handbook of Biological Confocal Microscopy. Madison, Wisconsin: IMR Press.

Tella LL (1985) The determination of a microscope's three-dimensional transfer function for use in image restoration. Master's Thesis, Worcester Polytechnic Institute, Worcester, MA.

Tichonov AN, Arsenin VY (1977) Solutions of Ill-Posed Problems. Washington, D.C.: Winston and Sons.

Twomey S (1977) Introduction to the Mathematics of Inversion in Remote Sensing and Indirect Measurements. Amsterdam: Elsevier Scientific Publishing Co.

Wells KS, Sandison DR, Strickler J, Webb WW (1989) Quantative fluorescence imaging with laser scanning confocal microscopy. In Pawley J (ed): The Handbook of Biological Confocal Microscopy. Madison, Wisconsin: IMR Press.

Wilson T, Sheppard C (1984) Theory and Practice of Scanning Optical Microscopy. Academic Press, publishers: Harcourt Brace Jovanovich.

Noninvasive Techniques in Cell Biology: 73–92
© 1990 Wiley-Liss, Inc.

4. Confocal Fluorescence Microscopy in Cytology

Ernst H.K. Stelzer

Confocal Light Microscopy Group, Physical Instrumentation Programme,
European Molecular Biology Laboratory (EMBL), D-6900 Heidelberg,
Federal Republic of Germany

I. INTRODUCTION: APPLICATIONS OF CONFOCAL FLUORESCENCE MICROSCOPES

Epithelial Madin-Darby Canine Kidney (MDCK) cells have been intensively studied in the Cell Biology Programme at the European Molecular Biology Laboratory (EMBL) for several years [Simons and Fuller, 1985]. Grown on permeable polycarbonate supports, MDCK cells form a monolayer with a high transepithelial resistance that comes close to the organization

Abbreviations used: BHK = baby hamster kidney; C_6-NBD-Ceramide = N-6[7-nitro-2,1,3-benzoxadiazol-4yl]aminocaproyl sphingosine galactoside; CFM = confocal fluorescence microscope/y; CBSLM = confocal beam scanning laser microscope; FWHM = full width half-maximum; MCM = modular confocal microscope; MDCK = Madin-Darby Canine kidney; N-Rh-PE = N-rhodamine-phosphatidyl-ethanolamine.

observed in vivo [Richardson et al., 1981; Simons and Fuller, 1985]. Under such conditions the MDCK cell line provides a model system to study the establishment and maintenance of cell polarity, endocytosis, cytoskeletal organization, functional relationships, and many other interesting aspects of polarized cell lines.

Due to their large depth (15–18 μm), the contributions from out-of-focus planes observed in a conventional fluorescence microscope after labeling the sample with a fluorescent antibody are, in most cases, relatively strong and do not allow a detailed observation. Until recently these cells have been observed only in the electron microscope or were the subject of experimental studies applying biochemical approaches [Van-Meer anad Simons, 1982; Hansson et al., 1986]. During many investigations, in which several groups at the EMBL participated since 1984, confocal fluorescence microscopy has proven to be an ideal method to perform morphological studies applying, among others [Van-Meer et al., 1987], the immunocytochemical tools of modern cell biology to investigate these cells [Wijnaendts-van-Resandt et al., 1985; Stelzer and Wijnaendts-van-Resandt, 1986; Stelzer and Wijnaendts-van-Resandt, 1987; Stelzer et al., 1989].

Confocal fluorescence microscopy has since been successfully applied in a number of investigations of, e.g., the distribution of F-actin in MDCK cells [Stelzer and Wijnaendts-van-Resandt, 1989], the distribution of microtubules in polarizing MDCK cells [Bacallao et al., 1989], the transcytosis of fluid phase markers applied from the apical and the basolateral plasma membranes to MDCK cells [Bomsel et al., 1989], the distribution of secretogranin I containing granules [Rosa et al., 1989], and the distribution of desmin and villin in chicken gizzard cells [Draeger et al., 1989], to name a few.

In all successful applications of confocal fluorescence microscopy the labeling was very dense and when observed in conventional fluorescence microscopy, the details were obscured by an out-of-focus haze. The depth discrimination properties of the confocal microscope were essential to achieve a detailed view of the target distribution. All applications relied heavily on the colocalization of different targets. The conclusion is nevertheless that conventional fluorescence microscopy should be applied first and the confocal fluorescence microscope should be used only if it is absolutely necessary. A development cycle is outlined in Figure 1 [Bacallao et al., 1989].

II. OVERVIEW

A. Principles of Confocal Fluorescence Microscopy

The principal optical layout of a confocal fluorescence microscope is shown in Figure 2. The light of a laser beam is focused into the source pinhole. The spatially filtered light is deflected by a dichroic mirror in the direction of a microscope objective and focused into the specimen. In the specimen, the

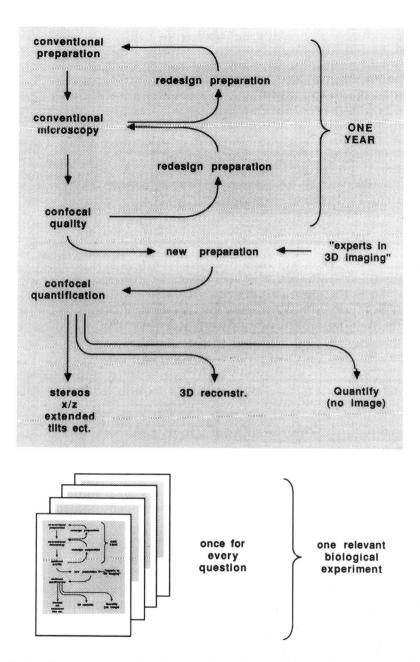

Fig. 1. The development of sample preparation techniques that preserve the three-dimensional structure. Most applications will start with a technique that had been developed to look at the sample in a conventional fluorescence microscope. Observing this sample in a confocal fluorescence microscope will start a new development cycle since, e.g., the preparation is essentially flat or only partially labeled. The new technique is then applied to get the the three-dimensional data set that is then further processed until the desired result is achieved.

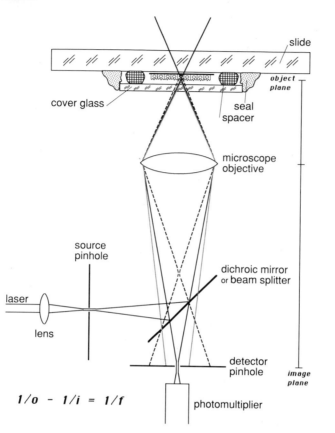

slide

object plane

cover glass

seal
spacer

microscope
objective

source
pinhole

dichroic mirror
or beam splitter

laser

lens

detector
pinhole

image
plane

$$1/o - 1/i = 1/f$$

photomultiplier

Fig. 2.   A simplified view of the optical system of a confocal fluorescence microscope. The images of the source and the detector pinhole are formed by the microscope objective in the object. Both pinholes are centered on the optical axis in an image plane of the objective lens. The microscope is confocal if the images of the two pinholes are diffraction limited and overlap in the object. Although there are some differences, this graph may also be used to explain the principle of discrimination in confocal reflection microscopes.

fluorescent light is emitted in all directions; only a part of it is collected by the microscope objective and focused into a pinhole in front of the detector. This light passes the pinhole and is converted to an electrical signal in the detector. Since the difference between object and image distances is fixed, light that is emitted from locations either in front of or behind the focal point in the object is focused to points either in front of or behind the detector pinhole. These beams are expanded in the plane of the pinhole and therefore only a fraction of this light enters the detector. The detector pinhole hence discriminates against the out-of-focus contributions of the object.

Another way of describing a confocal microscope and emphasizing symmetry considerations is, *The microscope objective forms an image of the source and detector pinhole in the object. The instrumental setup is confocal, if the diffraction limited images overlap.*

If the detector pinhole is removed, all the light collected by the microscope objective will penetrate the detector and the instrument will no longer discriminate against out-of-focus contributions. All the light that is emitted along a common optical axis is centered in the detector plane. The intensities add and hence cause a neighborhood-dependent background. The confocal reflection microscope will also discriminate against out-of-focus contributions. Reflected or back-scattered light is, however, not isotropic, and since it preserves its coherence, the amplitudes must be added, and not the intensities. As only one point is (or in the case of disk scanners a few distant points are) observed at a given time point, a confocal fluorescence microscope requires that either the object is moved through the beam [Minsky, 1961; Brakenhoff et al., 1979; Marsman et al., 1983; Stelzer and Wijnaendts-van-Resandt, 1986], the beam is moved through the sample [Carlsson et al., 1985; Stelzer et al., 1988] or an aperture is scanning the image plane [Petran and Hadravsky, 1968; Xiao et al., 1988].

A conventional fluorescence microscope can be easily made by removing the pinhole and collecting all the light that is penetrating the image plane. Hence, by changing the size of the pinhole one can "tune" a scanning microscope to have the properties of a confocal or of a conventional fluorescence microscope.

B. A Simplified Theory

Figure 3 outlines a confocal microscope and defines all the variables that are used in this paragraph. If the instrument is well aligned the part of the fluorescent object in the focal point will receive an intensity I_1 that is proportional to the incident intensity I_0. The object will emit an intensity that is proportional to I_1 and the detector will receive an intensity I_2 that is also proportional to I_1. If the object is now moved a distance z, the image is moved a distance $z' = M^2 z$. M is the magnification of the microscope objective. To a first approximation a shift of the object along the optical axis by the distance z reduces the intensity that is incident in the object by $1/z^2$. Since the pinhole will also receive an intensity that is reduced by $1/z'^2$, the effect of moving an object by a distance z out of the focal point is that the measured intensity relative to the incident intensity falls off with a factor that is proportional to $1/z^4$.

The following two formulas show an improved approximation for fluorescence and reflection, respectively, and Figure 4 graphically displays the relationship between integrated intensity and the distance from the geometrical focus z (z is the distance along the optical axis; z_r^{ex}, is the Rayleigh length).

78   **Stelzer**

$$I(z) = \frac{1}{\left(1 + \left(\frac{Z}{Z_r^{ex}}\right)^2\right)^2} \tag{1}$$

$$I(z) = \frac{1}{\left(1 + \left(\frac{Z}{Z_r^{ex}}\right)^2\right)\left(1 + \left(2\frac{Z}{Z_r^{ex}}\right)^2\right)} \tag{2}$$

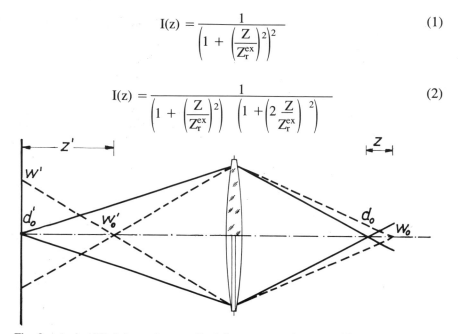

Fig. 3.   A simplified theory for a confocal fluorescence microscope. The source pinhole is omitted for simplicity. The detector pinhole is on the left-hand side, the object is on the right-hand side. The lens focuses the excitation light into the object and collects the light emitted by the object. Moving the object by a distance z away from the focal plane results in an image that is a distance M^{2}z away from the image plane. This results in a decreased excitation and a decrease in the detectable emitted light.

The results from this intuitive and (hopefully) easy to understand theory compare very well with those derived by more straightforward theories [Wilson and Sheppard 1984, p 70]. Expressed in terms of the full width half-maximum (FWHM), it is possible to calculate the performance of a microscope objective once its numerical aperture is known. The assumption in the case of fluorescence is that emission and excitation wavelengths are very close.(NA is the numerical aperture of the objective lens, λ is incident wavelength and n is the refractive index of the immersion medium.)

$$FWHM = 1.5\frac{b\lambda_0}{(NA)^2} \tag{3}$$

$$FWHM = 1.0\frac{b\lambda_0}{(NA)^2} \tag{4}$$

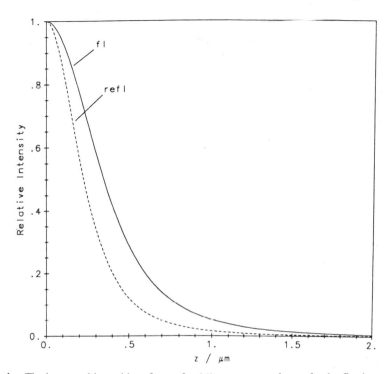

Fig. 4. The integrated intensities of a confocal fluorescence and a confocal reflection microscope as a function of z. The assumptions are that an oil immersion lens of numerical aperture 1.3 is used. In the case of fluorescence microscopy, the excitation light has a wavelength of 514 nm, the fluorescent light is detected above 560 nm (typical case for rhodamine). In the case of reflection microscopy the wavelength is 514 nm. Similar graphs have been published elsewhere [Wilson and Sheppard, 1984, p 70] using the normalized axial coordinate u and assuming a numerical aperture of 1.

It is very important to understand the term "integrated intensity." As outlined above it is assumed that a fluorescent object is placed in the plane of focus. The above formulas and the graphs in Figure 4 return the intensity that is measured in the detector as a function of the distance of the center of the sampling light spot from the fluorescing object. If the detector pinhole is removed, no discrimination takes place. The intensity will not vary as a function of $1/z^4$. If we place a point object in the plane of focus, the image intensity will spread in the detector plane and decrease with $1/z^2$ in the center, since the excitation intensity varies.

C. What to Expect From a Confocal Fluorescence Microscope

The lateral resolution of a confocal fluorescence microscope is more-or-less equivalent to that of a conventional fluorescence microscope [Cox, 1984;

Wilson and Sheppard, 1984, p 48]. Every scanning digital recording system (i.e., confocal microscopes, video microscopes) is ultimately limited by the number of pixels per line and the number of lines per image. Assuming an incident wavelength of 0.5 μm a lens of 1.3/100× has a lateral resolution on the order of 230 nm and hence a maximum field size of 512*200 nm/2 = 60 μm if every pixel is to be resolved. The conventional fluorescence microscope will allow a field size of at least 200 μm. The discrimination against out-of-focus contributions is important. The confocal fluorescence microscope will *not* discriminate against a uniform background that comes from the object. It may, however, discriminate against the autofluorescence of filters, lenses, mirrors, or sample-supporting parts such as dishes or synthetic and paper filters.

The novice user often compares the confocal images seen on the screen with photographic prints taken with conventional fluorescence microscopy. The latter very often have a small dynamic range and a high contrast. Structures are either visible or invisible. Recording data in this manner is a sensible approach since in most cases only the morphological data is of interest and varying grey levels would confuse the message. The confocal fluorescence microscope equipped with a photomultiplier or a CCD camera is a quantitative device. A dynamic range of 4 bits is easily achieved. There is no need to operate with a negative background or a gain that saturates the amplifiers [Bacallao and Stelzer, 1989]. "Nice images" may be easily generated once the data set is recorded. If the incident power is low, the device will most probably operate in the Poisson limit: the signal to noise ratio in the image is dominated by the small number of detectable photons.

When comparing confocal recording to conventional recording using CCD cameras with subsequent deconvolution of the data set [Agard and Sedat, 1983; Fay et al., 1989; Carrington et al., 1989; Chen et al., 1989], it should be realized that a high dynamic range is not necessary for confocal images. Deconvolution of conventional images on the other hand must start with a high dynamic range of the data set (12 or more bits) since every step in the computation adds to the noise and hence decreases the dynamic range.

From the formulas presented above, it is straightforward to calculate the performance of a confocal fluorescence microscope. Assuming an incident wavelength of 0.5 μm a lens with a numerical aperture of 1.3/100× should achieve an axial resolution of 700 nm (expressed in full width half-maximum) if the pinhole has the optimal size equivalent to 46 μm in the primary image plane. If the pinhole is twice as large the axial resolution will be in the order of 1,000 nm. This is demonstrated in Figure 5.

The axial resolution depends very much on the structure that is observed [Wilson and Carlini, 1987; Wilson, 1989; Van-der-Voort and Brakenhoff, 1989]. This is not different from the lateral resolution in conventional microscopy. Resolution is always object-dependent [Goodman, 1968, p 129]! Observing a sphere along the optical axis will produce the image of an ellipsoid. Observing a ring along the optical axis will produce sharp lateral sides, but diffuse axial sides.

Figure 6 shows an x/z image of microtubules in the periphery of flat MDCK cells grown on glass coverslips for 1 day. The impression is that some microtubules are closer to the apical membrane and others are closer to the basal membrane. The cell periphery is usually regarded as a flat region in a cell since there is only about 1 μm between the two bilayers. Now, what is the definition of flat in a confocal fluorescence microscope?

III. TIME-RESOLVED CONFOCAL FLUORESCENCE MICROSCOPY

The future of cell biology is, in part, the determination of dynamic processes. These must be investigated in biological systems in vivo. In the simple case this will mean that a flat (two-dimensional) system is investigated as a function of time, but in a real case an extended (three-dimensional) system must be observed at different time points. Light microscopic techniques are the system of choice to obtain morphological data and, in an adequate scientific environment, are well complemented by biochemical and immunocytochemical approaches.

A. Fundamental Limits

Using living cells for three-dimensional microscopy is done, for example, to avoid fixing the cells [Van-Meer et al., 1987] and definitely avoiding the production of artefacts, or to gain information on their three-dimensional activities. In order to observe in vivo systems severe obstacles must be overcome:

- the recording speed of the instrument must be sufficient to observe *this* process;
- the object must "survive" repeated observations;
- the cells must be kept in a vital environment;
- adequate means must be established to determine if the cells are still vital;
- the dyes must not be toxic or become toxic during the observation period;
- the dyes must "survive" repeated observations; and
- enormous amounts of data must be processed as quickly as possible.

B. Instrumental Requirements

A prime requirement is a very efficient device, an instrumental arrangement that produces low noise and makes use of the emission signal as well as possible. A high dynamic range will very often not be necessary. It is sufficient to determine the spatial distribution of a fluorophore and hence the spatial distribution of the target in the cell. This requires the answer to a mere yes or no question equivalent to a dynamic range of 1 bit. Such a result may be

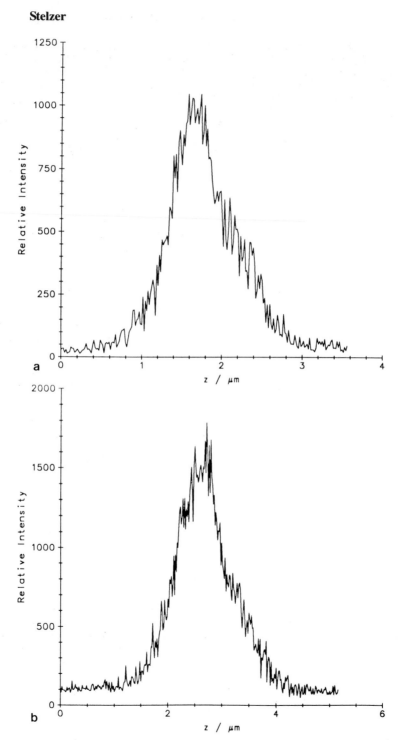

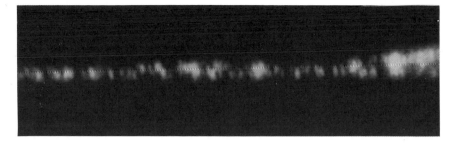

Fig. 6. An x/z view of microtubules in the peripheral region of flat MDCK cells. The MDCK cell has been cultured on a glass coverslip [Bacallao et al., 1989] for 1 day. An antibody pair was used to label the microtubules. The field size is 30 μm, the distance between the apical and the basal plasma membrane is 1 μm in the cell periphery (this picture was produced in collaboration with Robert Bacallao and Eric Karsenti on the CBSLM at the EMBL, Heidelberg).

achieved by instrumental means and by the sophisticated combination of instrumentation and data processing. The principal limit of the Poisson noise cannot be overcome on the single pixel level. Neighbouring pixels, a well known spatial distribution, a priori knowledge concerning the properties of the target, etc., may be used to improve the signal to noise ratio.

An instrument that produces 8,000 lines per second will generate roughly 15 images of 512 lines per image per second. If the line consists of 512 picture elements (pixels) the average data rate will be in the order of 4 megapixel per second and the peak performance (considering back sweep, linear movement of the scanner, etc.) will be in the order of 12 mega-pixel per second. The time per pixel is then in the order of 80 ns. A sample of 20 μm thickness sampled at 0.4 μm per section would be recorded within about 4 s, on average. The storage space would be in the order of 13 mega-pixel per three-dimensional data set per recorded channel. If the life of a cell is followed for 5 min at an interval of 15 s at least 250 mega-pixel must be stored and processed. These numbers give a rough idea of what is required:

- a fast recording, data processing, and storage device;
- a large mass storage and backup device;

Fig. 5. The integrated intensity of fluorescent latex spheres recorded with different pinhole sizes as a function of z. A fluorescently labeled latex sphere (Polysciences Fluoresbrite #16661, diameter 60 ± 3 nm) was observed with an optimal pinhole size of 50 μm **(a)** and a pinhole that was twice the size **(b)**. The lens was a ZEISS Neofluar 1.3/100×. The wavelength of the excitation was 476.5 nm, the emission was recorded above 545 nm. The expectation value for the full width half-maximum is 700 nm, a value of 720 nm has been measured (this data has been recorded on the modular confocal microscope at the EMBL, Heidelberg.)

• sufficient computer power to post-process the data;
• fluorophores that decay within 80 ns;
• a reasonable approach to the problem and a rough idea of how much data is
  absolutely necessary to solve *this* problem.

Recording speed can be enhanced by reducing the number of lines per image. Processing speed can be improved by reducing the number of pixels per line. It may also be debatable if the section pitch must be so small. A very important question that must be answered before the experiments are performed is, "What should be done with the data once it is recorded?" All the numbers given above are reasonable and, at least in principle, within the reach of current technology. To my knowledge, however, no device has been set up that will record, store, and process data at that speed. Probably the fastest confocal microscope is that described by Xiao, Corle, and Kino [Xiao et al., 1988] but this instrument is not yet known to work well for fluorescence.

C. Lipid Metabolism in MDCK Cells

The metabolism of C_6-NBD-ceramide (N-6[7-nitro-2,1,3-benzoxadiazol-4yl] aminocaproyl sphingosine galactoside) was studied in living MDCK cells [Van-Meer et al., 1987] to avoid any artefacts due to a fixation of the sample. To study the intracellular transport of newly synthesized lipids in MDCK cells, a fluorescent ceramide analog was used to serve as a probe. The same probe had been previously used to study a similar transport in fibroblasts [Lipsky and Pagano, 1983]. The ceramide analog was easily taken up by the MDCK cells from liposomes at a temperature of 0°C and having penetrated the cell the fluorescent ceramide analog accumulated in the Golgi area, as long as the temperature did not exceed 20°C. The confocal fluorescence microscope was used to study the 3D intensity distribution as a function of the internalization state. An analysis of the fluorescence pattern after the cells had been kept at 20°C for 1 h, revealed that the fluorescent marker had concentrated in the Golgi complex. Little fluorescence was observed at the plasma membrane or in the cytosol. After raising the temperature to 37°C for 1 h, intense plasma membrane staining occurred and there was an overall loss of fluorescence from the Golgi complex (Fig. 7). Quantitation was performed biochemically and correlated well with the data generated by the confocal fluorescence microscope. The investigation suggested that the fluorescent analog is sorted in MDCK cells in a way similar to their natural counterparts.

The observation in the confocal fluorescence microscope was performed on samples that had first been kept under the above described conditions and were then stored on ice. Small parts were cut out of the filter, placed between slide and coverslip, and observed without further delay for at most 5 min. Drops of nail polish served as spacers to avoid mechanical damage of the sam-

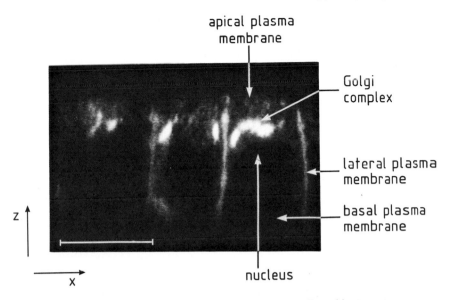

apical plasma
membrane

Golgi
complex

lateral plasma
membrane

basal plasma
membrane

z

x

nucleus

Fig. 7. An x/z view of MDCK cells having incorporated C_6-NBD-Ceramide. The cells were kept at 20°C for 1 h, at 37°C for another hour, and then observed in the confocal fluorescence microscope. The field size is 30μm, the height of the cells is about 15μm (this picture was produced in collaboration with Robert Bacallao and Eric Karsenti on the CBSLM at the EMBL, Heidelberg).

ple. Keeping the cells for 2 h under the microscope, cooling and heating it would have been interesting but was too complicated at that time.

D. Endocytosis in BHK Cells

Fluorescently labeled phospholipids are easily incorporated into the plasma membrane of baby hamster kidney (BHK) cells. Keeping the cells at 0°C blocks the metabolism of the cell, inhibits endocytosis and the transport of lipids to other cytoplasmic compartments [Van-Meer and Simons, 1983]. Increasing the temperature to 37°C releases the block and allows the cells to continue the endocytic process. The following example illustrates the problems (a detailed description of the underlying biology is in preparation; Davoust et al.).

Figures 8 and 9 show two series of images that were recorded with the confocal beam scanning laser microscope (CBSLM) at the EMBL [Stelzer et al., 1988] in collaboration with Jean Davoust. The lipid analogue N-Rh-PE (N-rhodamine-phosphatidyl-ethanolamine) was incorporated into BHK cells, a series was recorded initially and a second series was recorded 50 min later. While each image was recorded within 8 s, recording the series of 38 sections took about 10 min (in contrast to 38 · 8 s = 5 min) since 8 s per image were needed to store the image. The gain was purposely high (overexposed large white dots) in order to be able to easily detect the smaller particles and, if possible, the staining of the nucleus. This series was the result of a first test, after which the

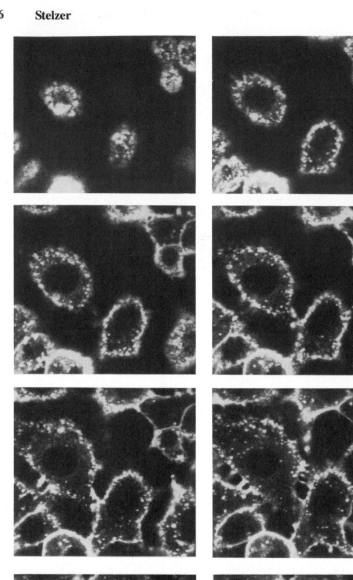

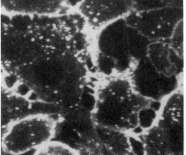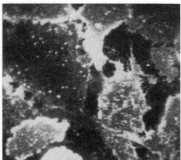

power falling into the sample was heavily reduced and an average of three scans was chosen. The reason was not that bleaching was detected, but the nucleus exhibited a very strong staining that non-illuminated cells in the neighborhood did not show, indicating the non-vital conditions of the experiment.

The pitch of 0.5 μm along the optical axis was obviously too small. One micron would have been completely sufficient since the particles were overilluminated. The field size of 100 μm is too large to resolve all detail but reduces the power per unit area. A smaller field size, fewer pixels per line, and fewer lines per image would have reduced the storage overhead without any loss in detail. An obvious problem was that the microscope stage did not remain in its place, but rather moved by about 5/19/2 μm along the x-/y-/z-axes, respectively. The environment, however, and the location of the coverglass are easily recognized so movement may be corrected for. Further tuning of the experiment may have been the reduction of the incorporated lipid, the addition of a vital anti-bleaching reagent, the avoidance of overillumination, the decrease in axial resolution by choosing a larger pinhole for the 1 μm pitch, a better synchronization of the experiment, an improved plan that determines when a series of images should be recorded, and special techniques that determine the condition of the cells.

IV. THE FUTURE OF CONFOCAL FLUORESCENCE MICROSCOPY

The 1990s will see an enormous improvement in the 3D imaging capabilities of light microscopes, and within a few more years, sufficiently powerful computers to process the 3D images as they are acquired at different time points. Appropriate biological systems must be developed until the information of interest is retrievable from an experiment of any biological relevance. Instrumentation will play a very important role but must be supported from the biological side with correctly prepared samples. The systems with the highest potential for 3D imaging are confocal fluorescence and confocal reflection microscopes. Using area detectors such as CCD cameras to gather conventional 3D data sets requires an enormous computing and storage power, but is definitely an alternative for many applications.

A number of groups will develop and implement a confocal device that gathers a complete 3D data sets within seconds. This effort combines sophisti-

---

Fig. 8.   BHK cells having incorporated N-Rh-PE shortly after the temperature block is released. The figure shows eight images out of a series of 38 images that were originally recorded at a pitch of 0.5μm along the optical axis. The full height of the cells was in the order of 15μm. The lipid accumulates in, or close to the plasma membrane of the cells. Only very few particles containing the fluorescent lipid are found inside the cell. The field size is 100μm; three averages were taken per section, resulting in a recording time of 8 s per image (this series was recorded in collaboration with Jean Davoust on the CBSLM at the EMBL, Heidelberg).

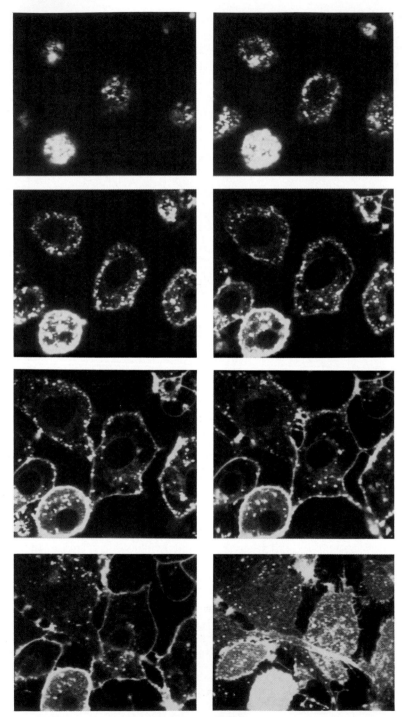

cated optics and fast electronics with parallel computing and special algorithms so that a minimal signal is sufficient to detect the object of interest. These instruments will become a main tool in cell biology within a few years. A standard part in a modern confocal microscope will be a microspectrometer to measure fluorescence spectra so that intracellular ion concentrations and pH values can be determined as a function of three dimensions. Area detectors can be used to measure the complete point spread function in the image plane. This information can be used to improve the resolution of the confocal microscope [Bertero et al., 1989]. The data recorded with a confocal microscope can be deconvoluted further to improve the lateral and depth discriminations (Kevin E. Fogarty, personal communication). Software that performs this task can be used with data from any instrument. There is no doubt that other techniques (not yet published or even known) applied to improve the axial resolution will be investigated and brought into common use. Three-dimensional light microscopy will be able to take avdantage of a lot of techniques, as well as software and hardware that are now available to the electron microscopists. The scan drivers of the microscopes will become more sophisticated and allow the implementation of scanning algorithms that avoid aliasing and reduce the photodamage imposed on the object.

In summary the following tasks will be accomplished in order to make 3D microscopy a reasonable tool for the scientific environment:

Software:

• 3D image editing and manipulation;
• 3D image visualization;
• 3D deconvolution of conventional and confocal fluorescence images;
• image exchange standards;
• combination of fluorescence, reflection and transmission images;
• combination with data recorded with other imaging techniques such as electron microscopy.

Hardware:

• fast scanning devices that operate in the diffraction limit;
• confocal fluorescence microspectrometer;
• extension of excitation spectrum by applying either conventional light sources or lasers other than the Argon ion laser;
• optimized filter sets;

Fig. 9. BHK cells having incorporated N-Rh-PE 50 min after the temperature block is released. The same cells as shown in Figure 8, but recorded 50 min later. Practically no large particles are found in the cell periphery which is visible due to a faint plasma membrane labeling. The recording conditions are the same as in Figure 8 (this series was recorded in collaboration with Jean Davoust on the CBSLM at the EMBL, Heidelberg).

Fig. 10.   An x/z view of a BHK cell after it incorporated the N-Rh-PE. The flat region shows the contact with the coverglass. The fluorophore accumulates close to the nucleus, but is definitely inside the cell. The black area on the left-hand side is the nucleus of the cell. The field size is 30 μm, the recording time was 2 s (this picture was recorded in collaboration with Jean Davoust on the CBSLM at the EMBL, Heidelberg).

- multiple detector setups;
- use of cameras (CCD and Newvicon);
- other confocal scanners;
- large storage and display devices for three-dimensional animation.

These efforts must be combined into a number of different projects:

- building a beam scanning confocal microscope;
- building microscopes that make use of all the light emitted by a point object, are easily adapted to different experiments, and do not require any adjustment;
- generating graphical information from images;
- deconvoluting confocal and conventional series of images;
- applying the instruments in cell biological research;
- investigating the theory of three-dimensional image formation and understanding its impact on cell biological research;
- improving every part that is necessary to work on living samples;
- developing fluorophores and incorporation techniques that are vital and may be combined with antibleaching reagents.

Having built precise mechanical and optical arrangements and sophisticated electronics and programmed fast computers for several years, we have realized that the application determines the success or failure of any instrument. New instruments will be of use only if there is a strong interaction between the instrument developers and the people who are supposed to use the instrument. The development of the new microscopes is viable only if appropriate efforts are made to supply biological systems that are *observable. The general direction of all efforts is towards time-resolved three-dimensional microscopy.*

ACKNOWLEDGMENTS

Andreas Merdes, Clemens Storz, and René Müller were kind enough to review the paper critically. A lot of thanks go to K. Simons and L. Philipson for their support of our projects. The applications were performed by M. Bomsel, R. Bacallao, J. Davoust, and G. Van-Meer.

V. REFERENCES

Agard DA, Sedat (1983) Three-dimensional architecture of the polytene nucleus. Nature 302:676–681.

Bacallao R, Stelzer EHK (1989) Preservation of biological specimens for observation in a confocal fluorescence microscope and operational principals of confocal fluorescence microscopy. In Tartakoff AM (ed): Methods in Cell Biology, San Diego: Academic Press, 31:437–452.

Bacallao R, Antony C, Dotti C, Karasenti E, Stelzer EHK, Simons K (1989) The subcellular organization of Madin-Darby Canine Kidney Cells during the formation of a polarized epithelium. J Cell Biol 109:2817–2832.

Bacallao R, Bomsel M, Stelzer EHK, De Mey J (1989) Guiding principles of specimen preservation for confocal fluorescence microscopy. In Pawley J (ed): The Handbook of Biological Confocal Microscopy. Madison: IMR Press, University of Wisconsin, pp 181–187.

Bertero M, Boccacci P, Defrise M, De Mol C, Pike ER (1989) Topical Meeting of the Optical Society of America on Signal Recovery and Synthesis, Cape Cod, MA: North Falmouth, pp 52–55.

Bomsel M, Prydz K, Parton RG, Gruenberg J, Simons K (1989) Functional and topological organisation of apical and basolateral endocytic pathways in MDCK cells. J Cell Biol 109:3243–3258.

Brakenhoff GJ, Blom P, Barends P (1979) Confocal scanning light microscopy with high aperture immersion lenses. J Microsc 117:219–232.

Carlsson K, Danielsson PE, Lenz R, Liljeborg A, Majlof L, Aslund N (1985) Three-dimensional microscopy using a confocal laser scanning microscope. Opt Lett 10:53–55.

Carrington WA, Fogarty KE, Lifschitz L, Fay FS (1989) Three-dimensional imaging on confocal and wide-field microscopes. In Pawley J (ed): The Handbook of Biological Confocal Microscopy. Madison: IMR Press, University of Wisconsin, pp 137–146.

Chen H, Sedat JW, Agard DA (1989) Manipulation, display, and analysis of three-dimensional biological images. In Pawley J (ed): The Handbook of Biological Confocal Microscopy. Madison: IMR Press, University of Wisconsin, pp 127–136.

Cox IJ (1984) Scanning optical fluorescence microscopy. J Microsc 133:149–154.

Draeger A, Stelzer EHK, Herzog M, Small JV (1989) Unique geometry of actin-membrane anchorage sites in avian gizzard smooth muscle cells. J Cell Sci 94:703–711.

Fay FS, Carrington WA, Fogarty KE (1989) Three-dimensional molecular distribution in single cells analysed using the digital imaging microscope. J Microsc 153:133–149.

Goodman JW (1968) Introduction to Fourier Optics. San Francisco: McGraw-Hill.

Hansson GC, Simons K, van-Meer G (1986) Two strains of the madin darbey canine kidney (MDCK) cell line have distinct glycolipid compositions. EMBO J 5:483–489.

Lipsky NG, Pagano RE (1985). Intracellular translocation of fluorescent sphingolipids in cultured fibroblasts: Endogenously synthesized sphingomyelin and glucocerebroside analogues pass through the Golgi apparatus en route to the plasma membrane. J Cell Biol 100:27–34.

Marsman HJB, Brakenhoff GJ, Blom P, Stricker R, Wijnaendts-van-Resandt RW (1983) Mechanical scan system for microscopic applications. Rev Sci Instrum 54:1047–1052.

Minsky M (1961) Microscopy Apparatus. US Patent No. 3,013,467.

Petran M, Hadravsky (1968) Tandem-scanning reflected-light microscope. J Opt Soc Am 58:661–664.

Rosa P, Ursula W, Pepperkok R, Ansorge W, Niehrs C, Stelzer EHK, Huttner W (1989) An antibody against secretogranin I (chromogranin B) is packaged into secretory granules. J Cell Biol 109:17–34.

Richardson JCW, Scalera V, Simmons NL (1981) Identification of two strains of MDCK cells which resemble separate nephron tubule segments. Biochim Biophys Acta 673:26–36.

Simons K, Fuller S (1985) Cell surface polarity in epithelia. Annu Rev Cell Biol 1:243–288.

Stelzer EHK, Wijnaendts-van-Resandt RW (1986) Applications of fluorescence microscopy in three dimensions: Microtomoscopy. SPIE 602:63–70.

Stelzer EHK, Wijnaendts-van-Resandt RW (1987) Nondestructive sectioning of fixed and living specimens using a confocal scanning laser fluorescence microscope: Microtomoscopy. SPIE 809:130–137.

Stelzer EHK, Stricker R, Pick R, Storz C (1988) Confocal fluorescence microscopes for biological research. SPIE 1028:146–151.

Stelzer EHK, Stricker R, Pick R, Storz C, Wijnaendts-van-Resandt RW (1988) Serial sectioning of cells in three dimensions with confocal scanning laser fluorescence microscopy: Microtomoscopy. SPIE 909:312–318.

Stelzer EHK, Wijnaendts-van-Resandt RW (1989) Fluorescence microscopy in three dimensions: Microtomoscopy. In Cohen E (ed): Cell Structure and Function by Microspectrofluorometry. San Diego; Academic Press, pp 131–143.

Stelzer EHK, Bomsel M, Bacallao R (1989) Confocal fluorescence microscopy of epithelial cells. In Herman B, Jacobson K (eds): Optical Microscopy for Biology. New York: Wiley-Liss, 1990, pp 45–57.

Van-der-Voort HTM, Brakenhoff GJ (1989) Three-dimensional image formation in high aperture fluorescence confocal microscopy: a numerical analysis. J Microsc, in press.

Van-Meer G, Simons K (1982) Viruses budding from either the apical or the basolateral plasma membrane domain of MDCK cells have unique phospholipid compositions. EMBO J 1:847–852.

Van-Meer G, Simons K (1983) An efficient method for introducing lipids into the plasma membrane of mammalian cells. J Cell Biol 97:1365–1374.

Van-Meer G, Stelzer EHK, Wijnaendts-van-Resandt RW, Simons K (1987) Sorting of sphingolipids in epithelial (MDCK) cells. J Cell Biol 105:1623–1635.

Wilson T, Sheppard CJR (1984) Theory and Practice of Scanning Optical Microscopy. London: Academic Press.

Wilson T, Carlini AR (1987) Size of the detector in confocal imaging systems. Opt Lett 12:227–229.

Wilson T (1989) The role of the pinhole in confocal imaging systems. In Pawley J (ed): The Handbook of Biological Confocal Microscopy. Madison: IMR Press, University of Wisconsin, pp 181–187.

Wijnaendts-van-Resandt RW, Marsmann HJB, Kaplan R, Davoust J, Stelzer EHK, Stricker R (1985) Optical fluorescence microscopy in three dimensions: Microtomoscopy. J Microsc 138:29–34.

Xiao GQ, Corle TR, Kino GS (1988) Real-time confocal scanning optical microscope. Appl Phys Lett 53:716–718.

Noninvasive Techniques in Cell Biology: 93–127
© 1990 Wiley-Liss, Inc.

5. Total Internal Reflection Fluorescence at Biological Surfaces

Daniel Axelrod

Biophysics Research Division and Department of Physics, University of Michigan, Ann Arbor, Michigan 48109

I. INTRODUCTION

The distribution and motion of molecules at or near surfaces is central to numerous phenomena in biology: e.g., binding of hormones, neurotransmitters, or antigens to receptors in the plasma membrane; surface-triggered blood coagulation; electron transport on the mitochondrial inner membrane; adhesion of cells to artificial surfaces and to other cells; possible enhancement of reaction rates by surface adsorption and two-dimensional diffusion; and the effects upon cell shape, motility, and mechanoelastic properties of submembrane cytoplasmic filament dynamics.

In many of these phenomena, certain important biomolecules are found simultaneously in both a surface-associated and nonassociated state. If the distribution and motion of such molecules is studied by fluorescence microscopy with standard epi-illumination optics, the fluorescence from the surface-associated molecules may easily be overwhelmed by the fluorescence from nearby nonassociated ones, and only a nearly featureless bright blur will be visualized. Total internal reflection fluorescence (TIRF) provides a means to overcome this problem by selectively exciting just those fluorophores in an aqueous environment very near a glass (or plastic) interface.

As applied to biological cell cultures, TIRF allows selective visualization of cell/substrate contact regions, even in samples in which fluorescence elsewhere would otherwise obscure the fluorescent pattern in contact regions. TIRF can be used to observe the position, extent, composition, topography, and motion of these contact regions. Figure 1 shows a photographic comparison of a single fluorescent labeled cell in culture as viewed by either TIRF or standard epi-illumination.

As applied to in vitro studies in surface biochemistry, TIRF can be used quantitatively to measure concentrations of fluorophores as a function of distance from the substrate, or to measure binding/unbinding equilibria and kinetic rates at a biological surface. Although much of the TIRF literature so far involves model membranes and artificially functionalized surfaces, the techniques used in such studies may be applicable to living cells. Another increasingly important but very different method of optical sectioning is confocal fluorescence microscopy; it is compared to TIRF in the Conclusions.

TIRF was first introduced by Hirschfeld [1965] for examining the fluorescence of dye solutions near a surface. Harrick and Loeb [1973] extended its use to the study of labeled protein adsorption kinetics. Axelrod [1981] further

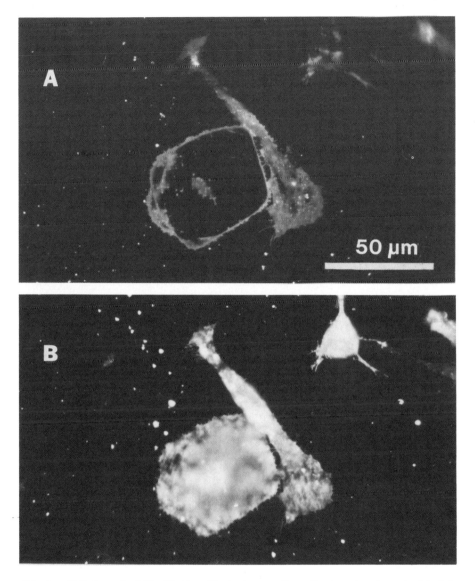

50 µm

Fig. 1. Cells in a mixed fibroblast/myoblast primary culture of embryonic rat muscle, labeled with the membrane soluble dye 3,3′-dioctadecylindocarbocyanine (diI) excited with the 514 nm line of an argon laser on an inverted microscope. **A:** TIRF; **B:** epi-illumination. TIRF shows that the large round cell is making contact with the substrate only around its periphery and in one small region near its center. A 40 × , 0.75 NA water immersion objective was used.

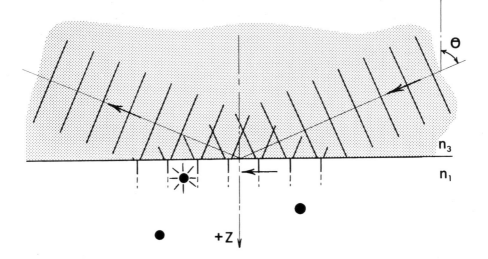

Fig. 2. Schematic drawing of the wavefronts of a laser beam incident upon a glass (refractive index n_3)/water (n_1) interface at $\theta > \theta_c$ and the evanescent wave in the water, along with selective excitation of a fluorophore (shown as a dot) near the surface. The incident, reflected, and evanescent waves are all traveling, not standing, waves.

extended its application to the examination of cell-substrate contacts in a laser/microscope configuration.

TIRF is conceptually simple. An excitation light beam traveling in a solid (e.g., a glass coverslip or tissue culture plastic) is incident at a high angle θ, measured from the normal, upon the solid/liquid interface to which the cells adhere. That angle θ must be large enough for the beam to totally internally reflect rather than refract through the interface; this condition occurs above some "critical angle." TIR generates a very thin (generally less than 200 nm) electromagnetic field in the liquid with the same frequency as the incident light and exponentially decaying in intensity with distance from the surface. The field is called the "evanescent wave" and is capable of exciting fluorophores in the liquid that are located near the surface while leaving unexcited the numerous fluorophores farther out in the liquid (see Fig. 2).

TIRF is easy to set up on a conventional upright or inverted microscope with a laser light source. It can also be set up with a conventional Xe or Hg arc or filament light source, albeit with somewhat more difficulty and reduced intensity. TIRF is completely compatible with standard epifluorescence, brightfield, darkfield, or phase contrast illumination so that these methods of illumination can be switched back and forth readily. This chapter is intended as a guide to setting up TIRF with either a laser or a conventional light source combined with a fluorescence microscope.

Section II describes optical configurations appropriate for TIRF microscopy. Section III is a brief summary of the theory of TIR fluorescence excitation. A closely related subject also discussed is how the emission light (rather than the excitation light) behaves near the TIR interface. Section IV presents a fairly complete review of recent applications of TIRF to biology and biochemistry. This last section includes nonmicroscopic TIRF studies of solution/surface biochemistry because of their possible relevance to future TIRF applications in noninvasive cell biology.

II. OPTICAL CONFIGURATIONS FOR A MICROSCOPE
A. Inverted Microscope

Figure 3A shows a schematic drawing of the sample chamber region in a possible TIRF setup for an inverted microscope. In general, a fixed-stage microscope (in which the objective moves up and down during focusing) is more convenient than a nonfixed-stage microscope (in which the stage moves up and down during focusing). A fixed stage ensures that the alignment of the beam with respect to the sample remains fixed during focusing. However, nonfixed stage microscopes are more common and either type will suffice. The laser beam first enters a focusing lens positioned obliquely above the microscope stage. The purpose of that lens is to concentrate the illumination in much the same manner as does the objective in epi-illumination and also to narrow the beam width for easier alignment. The lens' focal length is not critical (but approximately 50 mm will suffice). The focusing lens should be mounted on its own x-y-z translator with one of the axes along the direction of the beam. The translator itself may be fixed either to the stage or to the table, preferably the former for a nonfixed stage microscope.

The beam then enters a glass prism. The only purpose of the prism is to ensure that the beam will ultimately strike the TIR surface below the prism with an incidence angle $\theta > \theta_c$. Neither the size nor the shape of the prism is critical; a cube is depicted here, but it could be rectangular or 45°–45°–90° triangular or equilateral triangular in cross section. The latter two are standard commercial items. (A hemicylinder or hemisphere is also usable; the latter will be discussed in the subsection on intersecting beams.) However, a prism with a flat top such as a cube or a truncated triangle allows placement of a tungsten lamp and condenser above the prism, thereby permitting conventional illumination techniques such as brightfield, darkfield, and phase contrast.

The prism should be mounted on a single-axis translator for some limited motion in the vertical direction to allow the prism to be smoothly lowered over the sample. That translator *must* be mounted onto the stage (rather than onto the table) if a nonfixed stage microscope is used, so that the act of focus-

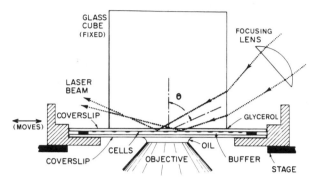

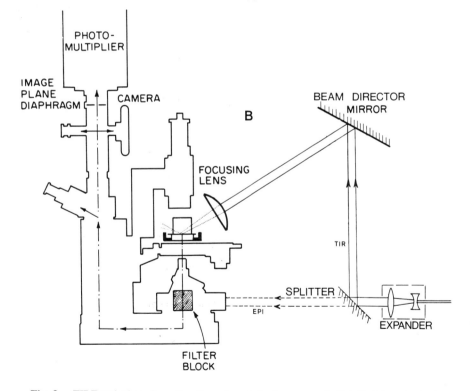

Fig. 3.   TIRF optical configuration for an inverted microscope. **A:** Detail of the sample chamber region; **B:** Incident and emission light paths.

ing does not separate the prism from the sample below it. It is perhaps most convenient to mount the prism on the carrier that normally supports the microscope condenser.

The sample, perhaps a monolayer of living cells, adheres facing down to a glass substrate plate, either a coverslip or microscope slide. On the upper side of this substrate is placed a droplet of pure glycerol, and then the prism is lowered onto it, thereby spreading the droplet into a thin layer. The glycerol serves two purposes: to ensure optical contact between the prism and the sample substrate, and to lubricate mechanically the region between the two glass surfaces. Note that as the sample below is laterally scanned by the standard stage translators, the prism remains laterally fixed. The incident laser beam should propagate obliquely through the bottom of the prism, through the glycerol, and through the substrate, totally reflecting at the bottom surface of the substrate directly over the optical axis of the microscope. The substrate thickness is not critical, except that thick substrates (e.g., those approaching 1 mm) may constrain the beam from meeting the bottom surface over the microscope's optical axis.

Glycerol should be used sparingly, because an excess will bead up at the bottom edges of the prism and interfere with the incident beam as it enters the prism. The clearance between the glycerol bead and the incident beam increases with larger prisms. On the other hand, a large prism may inhibit lateral motion of the sample below. In practice, a prism with a bottom face area of about 1 cm^2 seems suitable.

The sample chamber contains the buffer needed to maintain the sample (e.g., the cells' viability). The buffer is sandwiched between the substrate surface and a glass coverslip below by a neoprene or teflon ring spacer. Teflon rings are commercially available with thicknesses down to less than 50 μm. The depth of solution must be quite thin if high-power, high-aperture, short working distance objectives are to be used. The downward pressure of the prism upon the substrate plate may be adequate to seal the sample chamber without additional clamping. Both the substrate plate and the coverslip below should be the same size and placed in a holder that fits both of them snugly and which can be clamped by the microscope stage translator.

In the optical configuration described here, the illuminated region will remain stationary in the center of the field of view as the sample is scanned laterally. However, with a nonfixed microscope stage only, the illuminated region may move laterally as one focuses up and down, depending on whether the focusing lens and the beam director preceding it (described below) were mounted on the table or the stage.

The beam path 'upstream' from the microscope consists of three elements in addition to those normally found in a laser/microscope system: a beam expander, a beam splitter, and a beam director (see Fig. 3B).

The beam expander (commercially available) is used to control the width of the laser beam at the focusing lens. The wider the beam at the focusing lens, the higher the angle of ray convergence onto the sample and the thinner the width of the TIRF illuminated area. This optical configuration can produce an illuminated area half-width (measured from the center to the e^{-2} intensity) as small as about 2.5 μm.

The beam splitter is a mirror that diverts the horizontal beam into the vertical direction up past the level of the microscope stage for TIRF. The mirror should be mounted on a slide so that simple removal enables the beam to enter the microscope's field diaphragm port for quick and reversible conversion to epi-illumination. By using a partially silvered mirror or even a plain glass slide as the beam splitter, both epi- and TIR can be viewed simultaneously.

The beam director is a mirror that angles the beam from the vertical direction obliquely down toward the focusing lens. The height of the mirror determines the incidence angle θ at the TIRF surface. The mirror should be mounted on a biaxial rotational mount to direct the "raw" beam properly as described above. If the mirror mount is attached to the stage of a nonfixed stage microscope, rather than to the table, then the position of the TIRF illuminated region can be rendered insensitive to changes in focusing.

Other TIRF configurations for inverted microscopes have been employed. Figure 4 shows an alternative system [Weis et al., 1982]. Instead of a prism fixed with respect to the beam as above, the prism is fixed with respect to the sample. The substrate glass slide propagates the incident beam toward the microscope's optical axis via multiple internal reflections. The illuminated TIRF area will move with translation of the sample in this system.

B. Upright Microscope

A very simple setup for an upright microscope requiring few external mounts and mirrors is shown in Figure 5. The beam-expanded laser beam passes through a long focal length (~250 mm) converging lens and enters the base of the microscope stand through the port normally reserved for the transmitted illumination light source. The converging lens is mounted on an x-y translator so that the position of the beam can be finely adjusted laterally in both directions.

The beam reflects up at the internal mirror or prism already installed in the microscope base. It then enters a specially design prism made of flint glass ($n_3 = 1.62$) mounted on the holder that would normally carry the microscope condenser. The prism has a trapezoidal cross section and is manufactured by truncating and polishing the top of a commercial equilateral triangle prism. The "raw" beam position (i.e., without the focusing lens) is adjusted to enter the bottom of this prism off-center, so that it totally internally reflects at the sloping side and heads toward the upper surface at an incidence angle of 60°. With the focusing lens in place, the beam follows the same course but is much thinner and more intense.

TOP VIEW

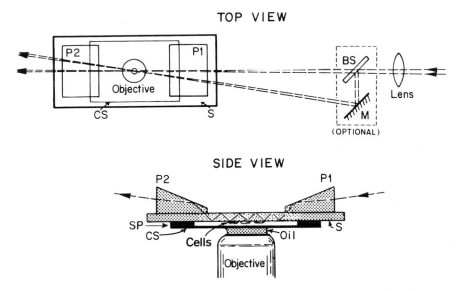

Fig. 4. TIRF microscopy with prisms fixed to substrate plate, adapted from Weis et al. [1982]. An optional second beam to create interference fringes is shown. BS, beamsplitter (optional); CS, coverslip; M, mirror (optional); P1, entrance prism; P2, exitt prism; S, glass slide; SP, spacer.

The sample substrate plate is simply placed on the microscope stage. Then the prism, with a spot of glycerol on its top, is brought up into optical contact with the lower surface of the substrate. The laser beam position is then adjusted by the focusing lens translator to totally internally reflect at the upper surface of the substrate and follow a symmetrical course back down toward the microscope base on the opposite side of the prism. This setup is suitable for any sort of substrate (microscope slides, coverslips, etc.), but is particularly well adapted for viewing cells growing on the bottom of plastic culture dishes. The cells may thereby be viewed in the same dish in which they incubated without any remounting.

The fluorescence may be viewed from above through a microscope coverslip which is held off from the cell monolayer surface by a spacer. Alternatively, a water immersion objective can be placed directly in the buffer solution bathing the cells. This upright microscope arrangement is extremely convenient, since samples may be switched as easily as they would with any standard illumination system. The position of the TIRF illumination area is also stationary and unaffected by focusing, even on a nonfixed stage microscope. The disadvantage is that the incidence angle is not adjustable without switching to another prism with a different slope angle.

With a 60° incidence angle, the refractive index of the prism must be at least 1.53 in order for TIR to take place at the substrate/water interface (regard-

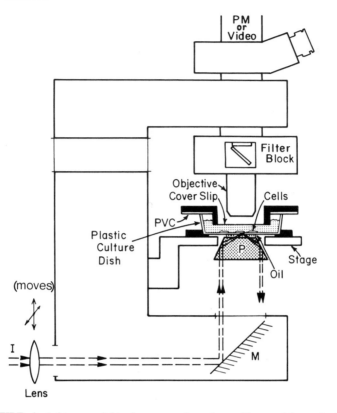

Fig. 5. TIRF adapted to an upright microscope, shown here with a special sample chamber for long term viewing of cells in a plastic tissue culture dish. The prism is a truncated equilateral triangle. The region of the sample chamber is shown enlarged relative to the rest of the microscope for pictorial clarity. I, incident light; M, mirror or prism already installed in microscope base; P, flint glass prism; PM, photomultiplier; PVC, polyvinylidene chloride (Saran Wrap™ used to seal in a 10% CO_2 atmosphere over culture medium for long term viewing).

less of the substrate material). This is why flint glass, rather than ordinary optical glass or fused silica, is the prism material in the example here.

C. Prismless TIRF

It is possible to utilize a form of epi-illumination to obtain TIRF without any prism at all, but the scheme requires a very high numerical aperture (NA) objective (generally at least 1.4), as depicted in Figure 6A. The laser beam, collimated to a narrow width by an appropriate focusing lens placed before the field diaphragm plane of the epi-illuminator, is positioned to propagate along the very edge of the objective's aperture. It emerges into the immersion oil ($n_3 = 1.52$) at a maximum angle θ derived from the definition of NA:

$$NA = n_3 \sin \theta \tag{1}$$

The critical angle for total internal reflection must satisfy the relationship

$$n = n_{water} = 1.33 = n_3 \sin \theta_c \tag{2}$$

From the TIR requirement that $\theta > \theta_c$, Equations 1 and 2 indicate that NA must be greater than 1.33, preferably by a substantial margin.

In practice, the system of Figure 6A works but the TIR illumination occurs only over part of the field of view.

A more uniform illumination would be attainable by expanding the source to cover the entire objective aperture but blocking those subcritical angle rays (those that emerge with $\theta < \theta_c$) by means of a concentric opaque disk inserted into the objective at its back focal plane. But insertion of an opaque disk of exactly the right diameter at exactly the right position inside an objective is a rather unworkable approach. Fortunately, it is also unnecessary.

To produce a hollow supercritical angle cone, an opaque disk of appropriate size can be placed in the illumination path *external* to the microscope so as to cast a sharp, real-image shadow at the objective's back focal plane [Stout and Axelrod, 1989]. This shadow allows a hollow cone of epi-illumination rays traveling at only super-critical angles to reach the glass/water interface at the sample plane. The basic approach is depicted schematically in Figure 6B as a ray diagram through a standard inverted microscope epi-illuminator.

Three kinds of TIRF illumination patterns described in more detail below, can be produced by variations of this scheme: 1) a small spot of illumination, of radius 1.5 µm, by use of a laser light source; 2) a large region of illumination, by use of a laser-illuminated diffusing screen located upbeam from the opaque disk; 3) a large region of illumination by use of a conventional mercury arc.

1. Laser spot. In this simplest configuration, the laser beam is passed through a beam expander and then through a focusing lens (shown as FL in Fig. 7a). A slightly more elaborate configuration, but one quickly interchangeable with the setup for the wide area illumination discussed below, passes an expanded beam through a short focal length lens (shown as OL in Fig. 7b). Lens OL serves to converge the beam to a focus and then diverge it past that focus to a radius again sufficient to allow FL to converge the beam at high angle to a focus at the field diaphragm plane (FDP) of the microscope.

2. Laser wide area. If a diffusing screen (e.g,. a thin sandwich of submicroscopic latex beads in suspension) is placed at the focal plane of OL, then the single point source of the laser is effective converted into an extended source. The size of the extended source, and hence the size of the illuminated region on the sample, can be adjusted simply by moving OL forward or back-

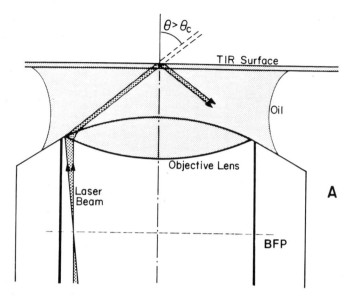

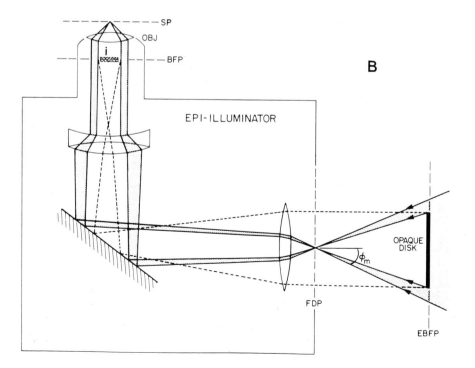

ward. An additional major advantage of a latex bead suspension is that the coherent light interference pattern produced on the sample from the superposition of scattered light waves from each bead flickers so rapidly due to bead Brownian motion that the TIR illumination appears to be uniform over much of the field of view.

A significant amount of excitation power is lost by scattering at the diffusing screen with this wide-area TIR illumination system (apart from the loss of all subcritical rays), but a standard 3W argon laser is still bright enough to produce satisfactory images.

3. Mercury arc wide area. 100 W mercury arc lamps are much more commonly used than lasers for microscope fluorescence illumination. Unfortunately, with standard prism-based TIRF, the power of a mercury arc light that can be directed through a TIRF prism at supercritical angles is insufficient to produce visible images. The prismless method, used as in section (1.) above with an arc instead of a laser, can direct more supercritical light into a smaller sample area. Nevertheless, the light available is still marginal because of the blockage of the subcritical light. Much of this light can be utilized, however, by employing a conical lens at the output of the mercury arc housing (see Fig. 7d). The conical lens serves to direct much of the arc light into a somewhat fuzzy ring. The FL (the same lens as used in the laser illumination systems in approximately the same position) then converges this ring past the opaque disk situated at the usual position toward a focus at the field diaphragm plane. The result is a wide area TIR illumination entirely free of laser interference fringes. Despite the virtues of the conical lens, the TIR intensity attainable from an Hg arc is still considerably less than that from an argon laser. Figure 8 shows fluorescent (diI) labeled erythrocyte ghosts adhered to a coverslip and photographed with this kind of illumination.

D. Intersecting Beams

Because of the geometrical contradiction between producing a highly convergent incident beam while ensuring that no rays propagate toward the TIR surface at less than the critical angle, it is difficult to produce a very small, submicron spot of TIR illumination as can be done with standard epi-illumination. However, for some types of TIRF experiments, in particular those combining TIR

Fig. 6. Prismless TIR illumination through the periphery of a high aperture objective. **A:** Beam propagating near edge of objective. The beam focuses at the back focal plane (BFP) of the objective so that it emerges collimated (at angle θ from the optical axis). **B:** Beam propagating in a ring past the image of an externally placed opaque disk. The disk's position along the optical axis is adjusted to the "equivalent back focal plane" (EBFP), defined so that the disk's image (marked as "i" and shown crosshatched) occurs at the BFP of the objective. The vertical dimension is somewhat stretched here for pictorial clarity.

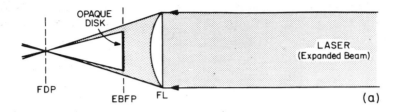

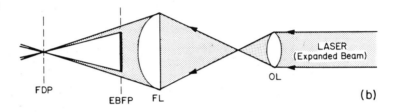

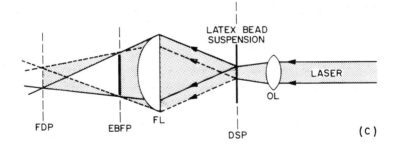

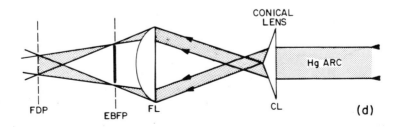

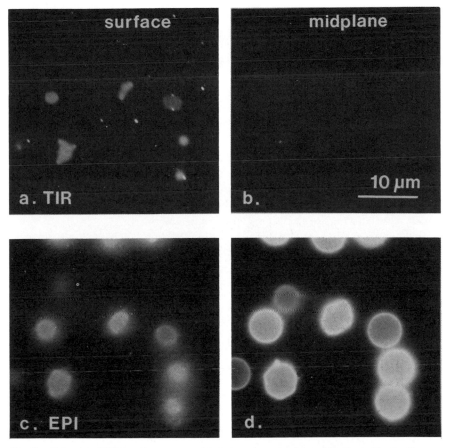

Fig. 8. Prismless TIRF on fluorescence-labeled erythrocyte ghosts, illuminated by the Hg-arc configuration depicted in Figure 7d. **a:** TIRF, focused at coverslip surface. Contact regions are clearly visible. **b:** TIRF focused at midplane of spherical ghosts. Contact regions are out of focus and nearly invisible, and the midplane is barely excited by the evanescent field. **c:** Epi-illumination, focused at coverslip surface. Contact regions are visible, but obscured in a haze of out-of-focus fluorescence. **d:** Epi-illumination focused at midplane of spherical ghosts. Note the TIRF exclusion of fluorescence from this plane evident by comparison of b and d.

Fig. 7. Optical arrangements before the epi-illuminator for four prismless TIRF configurations: **a)** laser source focused to a spot with a single focusing lens; **b)** laser source focused to a spot with two lenses, OL and FL; **c)** laser source converted to an extended source by a latex bead suspension for illuminating a large area. The size of the illuminated region on the DSP is exaggerated for pictoral clarity; **d)** mercury arc source, including the use of a conical lens to increase the power of supercritical light at the expense of subcritical light.

with fluorescence recovery after photobleaching (FRAP or FPR) for measuring surface diffusion rates, it may be desirable to produce a variegated intensity pattern with a very small characteristic distance in the evanescent wave. (See section IV.D for details on this application.)

A convenient variegated pattern compatible with TIRF with a very small characteristic distance is an interference fringe pattern created by two coherent TIR laser beams intersecting at the same region on the surface. The internode spacing d in the pattern is given by:

$$d = \lambda_o/(2\, n_3\, sin\, \theta\, sin\, \phi/2) \qquad (3)$$

where ϕ is the angle between the planes of incidence of the intersecting beams and λ_o is the vacuum wavelength of the incident light. In principle, d can be made as small as $\lambda_o/2n_3$ for oppositely directed beams.

One experimental setup for intersecting TIR beams is based on the prism/chamber arrangement shown in Figure 4. Another arrangement, which allows for a full range of incidence angles θ, intersecting beam angles ϕ, and a TIRF illumination area which is fixed in position as the sample is moved, is shown in Figure 9. This arrangement successfully produces very distinct and closely spaced interference fringes, suitable for use even on small biological cells (see Fig. 10).

E. Waveguide Fluorosensors

The last TIRF optical configuration discussed here is not based around a microscope at all, but it has great potential uses in fluorescence microscopy. A single fiber optic cylinder of small diameter (tens or hundreds of microns) can be mounted to channel excitation light down its length. Fluorescence excited within the evanescent wave along the sides or within the cone of propagating light released from the fiber optic's distal end will be captured by the fiber and be channeled back up the fiber to a detector. The optical theory and some preliminary tests of such "waveguide fluorosensors" has been presented by Glass et al. [1987]. In the future, thin fluorosensors might be employed to optically probe local exterior regions of cells, much as micropipettes electrically probe the interior.

F. General Experimental Suggestions

Regardless of the optical configuration chosen, the following suggestions may be helpful.

1. The prism used to couple the light into the system and the (usually disposable) slide or coverslip in which TIR takes place need not be matched exactly in refractive index.

2. The prism and slide may be optically coupled with glycerol, cyclohexanol, or microscope immersion oil, among other liquids. Immersion oil has a higher

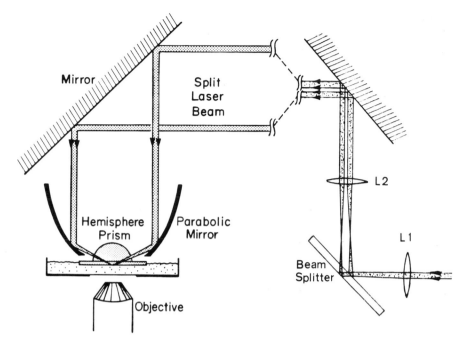

Fig. 9. Intersecting TIR beams split from the same laser to produce interference fringes for viewing in an inverted microscope. The TIR surface is positioned vertically to be at the focus of the parabola. The positions and focal lengths of lenses L1 and L2 determine the widths and separation of the two beams. The beams need not reflect at opposite sides of the paraboloid as depicted, but rather at any azimuthal angle, thereby allowing adjustment of the intersection angle ϕ.

refractive index than glycerol, thereby avoiding possible TIR at the prism/coupling liquid interface at low incidence angles), but it tends to be more autofluorescent (even the "extremely low" fluorescence types).

3. The prism and slide can both be made of ordinary optical glass for many applications, unless shorter penetration depths arising from higher refractive indices are desired. Optical glass does not transmit light below about 310 nm and also has a dim autoluminescence with a long (several hundred microseconds) delay time, which can be a problem in some photobleaching experiments. The autoluminescence of high-quality fused silica (often called "quartz") is much lower. Tissue culture dish plastic (particularly convenient as a substrate in the upright microscope setup) is also suitable, but tends to have a significant autofluorescence compared to ordinary glass. More exotic high n_3 materials such as sapphire, titanium dioxide, and strontium titanate can yield exponential decay depths d as low as $\lambda_o/20$.

4. The TIR surface need not be specially polished: the smoothness of a standard commercial microscope slide is adequate.

5. Illumination of surface adsorbed proteins can lead to apparent photo-

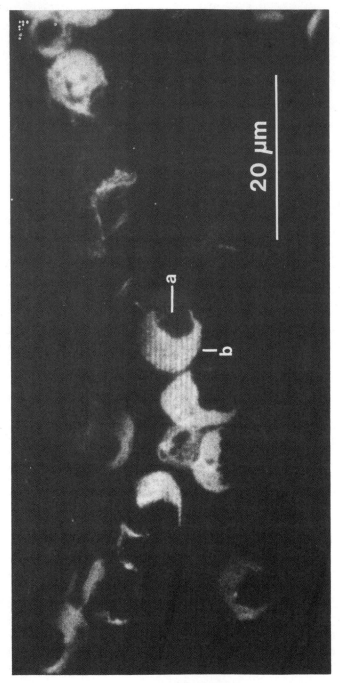

Fig. 10. Fringe pattern created by the intersecting beam configuration of Figure 9. The sample consists of flattened human erythrocyte ghosts on an aluminum film/poly-l-lysine substrate as described in the text; the TIR-excited fluorescence is from a mixture of fluorescein-epidermal growth factor (F-EGF) and free fluorescein that is nonspecifically adsorbed to the ghost membrane in equilibrium with bulk-dissolved fluorescent species. The distinctive membrane tear creating a "bite" (a) and "crescent" (b) appearance has been confirmed by other labels to expose the cytoplasmic surface in the "bite" and the outer surface in the "crescent."

chemically induced cross-linking. This effect is observed as a slow, continual, illumination-dependent increase in the observed fluorescence. It can be inhibited by deoxygenation (aided by the use of an O_2-consuming enzyme/substrate system such as protocatachuic deoxygenase/protocatachuic acid or glucose deoxygenase/glucose) or by 0.05 M cysteamine.

6. Virtually any laser with a total visible output in the 0.5 W or greater range should be adequate. The most popular laser for cell biological work with a microscope appears to be a 3 W continuous-wave argon laser.

7. The only absolute requirement for the microscope objective is that it have a long enough working distance to see the TIR surface across a thin water gap defined by the spacer thickness. Since the spacer can be made thinner than any standard working distance while still appearing infinitely deep relative to the evanescent wave, this is not normally a restriction. However, the aberration corrections of some high-magnification objectives are crucially dependent upon the sample being *immediately* on the distal side of a coverslip rather than across an additional thin water layer. An intervening water layer can shroud the image in a diffuse haze which lowers contrast. No general rule for which objectives will work best can be provided, but water-immersion objectives generally work well and success has been achieved with certain objectives of every commercially available magnification.

8. For incidence angles near θ_c, it may be difficult to determine by direct viewing whether or not TIR has been achieved. However, TIRF observed through the microscope eyepieces is quite distinct from fluorescence excited by a subcritical angle incident beam propagating at a skimming angle through the aqueous medium. TIRF originates from what appears to be a single plane, since the evanescent wave (unlike a propagating beam) is shallower than the microscope's depth of focus. Epi-illumination, on the other hand, usually excites fluorophores in a range of focal planes. At subcritical angles near θ_c, the cells appear to cast a shadow so that the whole field of view appears streaked.

III. THEORY
A. Single Interface

The critical angle θ_c for TIR is given by

$$\theta_c = sin^{-1}(n_1/n_3) \tag{4}$$

where n_1 and n_3 are the refractive indicies of the liquid and the solid respectively, where $n_1 < n_3$. For incidence angle θ less than θ_c, much of the light propagates through the interface with a refraction angle (also measured from the normal) given by Snell's law. (Some of the incident light internally reflects back into the solid.) For $\theta > \theta_c$, all of the light reflects back into the solid.

However, even with TIR, some of the incident energy penetrates through the interface and propagates parallel to the surface in the plane of incidence. (For a finite-width beam, this propagation in the nearby liquid can be pictured as the beam's partial emergence from the solid into the liquid, its travel for some finite distance along the surface, and then its reentrance into the solid. The distance of propagation along the surface is measurable for a finite-width beam and is called the Goos-Hanchen shift). The field in the liquid, the "evanescent wave," is capable of exciting fluorescent molecules that might be present near the surface.

For an infinitely wide beam (i.e., a beam width many times the wavelength of the light, which is a very good approximation for our purposes), the intensity of the evanescent wave (measured in units of energy/area/s) exponentially decays with perpendicular distance z from the interface:

$$I(z) = I(0) \, e^{-z/d} \tag{5}$$

where

$$d = \frac{\lambda_o}{4\pi} [n_3^2 \sin^2 \theta - n_1^2]^{-1/2} \tag{6}$$

where λ_o is the wavelength of the incident light in vacuum. Depth d is independent of the polarization of the incident light and decreases with increasing θ. Except at $\theta = \theta_c$ (where $d \, l \, \infty$), d is in the order of λ_o or smaller. A physical picture of refraction at an interface shows TIR to be part of a continuum, rather than a sudden new phenomenon appearing at $\theta = \theta_c$. For small θ, the light waves in the liquid are sinusoidal, with a certain characteristic period. Consider the wave pattern seen as one moves directly away from the surface along a normal. As θ approaches θ_c, that period becomes longer as the refracted rays propagate increasingly parallel to the surface. At $\theta = \theta_c$ exactly, that period is infinite, as the wavefronts of the refracted light are normal to the surface. This situation corresponds to $d = \infty$. As θ increases beyond θ_c, the period becomes mathematically imaginary; physically, this corresponds to the exponential decay in Equation 5.

The factor $I(0)$ in Equation 4 is a function of θ and the polarization of the incident light. $I(0)$ is different for each polarization component of the evanescent wave. To write the $I(0)$ intensities that are observed polarized along three orthogonal directions, we first set up a coordinate system such that the plane of incidence is the x-z plane, with x parallel to the surface and z perpendicular to it. The y-direction is then normal to the plane of incidence. There are two independent incident polarizations possible: p- and s-, for the electric field vectors parallel or perpendicular, respectively, to the plane of incidence defined by the paths of the incident and reflected rays. For s-polarized incident light, the evanescent wave is also s-polarized, which in our coordinate system, is entirely along the y-direction. Its intensity is

$$I_y(0) = I_s \frac{4 \cos^2 \theta}{1 - n^2} \tag{7}$$

For p-polarized incident light, the evanescent wave is entirely p-polarized. But the situation is a little more complex, because the p-polarization can contain both an x- and z-component. The x- and z-directed polarized intensities for p-polarized evanescent waves are

$$I_x(0) = I_p \frac{4 \cos^2 \theta \, (\sin^2 \theta - n^2)}{n^4 \cos^2 \theta + \sin^2 \theta - n^2} \tag{8}$$

$$I_z(0) = I_p \frac{4 \cos^2 \theta \sin^2 \theta}{n^4 \cos^2 \theta + \sin^2 \theta - n^2} \tag{9}$$

Factors $I_{p,s}$ in Equations 4–6 are the polarized intensities of the incident electric field in the glass, and n is the ratio n_1/n_3, assumed less than unity. For s-polarized incident light, the total evanescent intensity $I_s(0) = I_y(0)$. For p-polarized incident light, the total evanescent intensity $I_p(0) = I_x(0) + I_z(0)$. The I_y intensity is linearly polarized, but the I_p intensity is elliptically polarized because the x- and z-components of the electric fields are 90° out of phase with each other.

Intensities $I_{p,s}(0)$ are plotted vs. θ in Figure 11, assuming the incident intensities $I_{p,s}$ are set equal to unity. The plots can be extended without breaks to the subcritical angle range (based on calculations with Fresnel coefficients), again illustrating the continuity of the transition to TIR. The evanescent intensity approaches zero as θ 1 90°. On the other hand, for super-critical angles within ten degrees of θ_c, the evanescent intensity is as great or greater than the incident light intensity.

B. TIR Through an Intermediate Dielectric Layer

In actual experiments, the interface may not be a simple one between two media but rather more stratified with at least one intermediate layer. One example is the case of a biological membrane [Reichert and Truskey, 1990] or lipid bilayer interposed between glass and aqueous media. Another example is a thin metal film coating, which has useful features discussed in the next subsection.

The rather complicated expressions for the evanescent intensities at stratified interfaces can be found in Hellen and Axelrod [1987]. Certain qualitative features can be noted for a three-layer system in which incident light travels

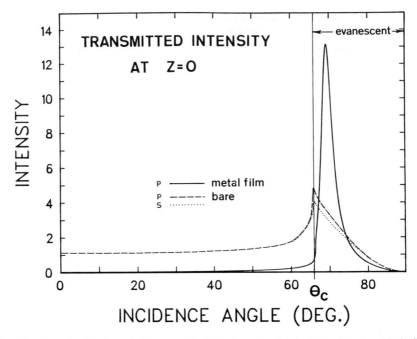

Fig. 11.   Intensity $I_p(O)$ vs. incidence angle θ at a bare fused silica/water interface (dashed) and a 20 nm aluminum film coated fused silica/water interface (solid). Intensity $I_s(O)$ is also shown for bare glass (dotted); it is essentially zero for the metal film surface. Refractive indices $n_3 = 1.46$ and $n_1 = 1.33$ are assumed, corresponding to $\theta_c = 65.7°$.

from medium 3 (refractive index n_3) through the intermediate layer (n_2) toward medium 1 (n_1):

1. Insertion of an intermediate layer never thwarts TIR, regardless of the intermediate layer's refractive index n_2. The only question is whether TIR takes place at the n_3:n_2 interface or the n_2:n_1 interface. Since the intermediate layer is likely to be very thin (no deeper than several tens of nanometers) in many applications, precisely which interface supports TIR is physically not well-defined and is not important for qualitative studies.

2. Regardless of n_2 and the thickness of the intermediate layer, the evanescent wave's characteristic depth in medium 1 will be given by Equations 5 and 6. However, the overall distance of penetration of the field measured from the surface of medium 3 is affected by the intermediate layer.

3. Irregularities in the intermediate layer can cause scattering of incident light which then propagates in all directions in medium 1. This subject has been treated theoretically by Chew et al. [1979]. Experimentally, scattering appears not to be a problem on samples even as inhomogeneous as biological cells.

Direct viewing of incident light scattered by a cell surface lying between the glass substrate and an aqueous medium confirms that the scattering intensity is many orders of magnitude dimmer than the incident or evanescent intensity, and will thereby excite a correspondingly dim contribution to the fluorescence. Nevertheless, for incidence angles very near the critical angle, discontinuities in the intermediate layer (such as the edge of cell membranes) do appear to cast a "shadow" along the interface which is particularly noticeable on metal-coated surfaces. This effect is reduced with higher incidence angles.

C. Emission of Fluorescence Near a Glass Substrate

Regardless of how a fluorophore is excited (whether by a TIR evanescent wave or an epi-illumination propagating wave), the emitted fluorescence is not released isotropically. For a fluorophore far from any interfaces, the intensity follows a simple $\sin^2\beta$ pattern, where β is the polar angle from the transiton dipole moment direction. But for fluorophores near an interface, such as those excited by TIR, the pattern is much more complicated and affects in a nontrivial manner exactly how much fluorescence is detected.

Physically, the complication arises because a portion of the electromagnetic field emitted by a fluorophore (modeled as an oscillating electric dipole) does not propagate away from the dipole but instead decays within the first wavelength or so from the dipole. In fact, this "near field" consists of a whole set of exponentially decaying fields with a continuous range of decay constants. When a fluorophore (say, in water) is near a higher refractive index surface (say, a distance z_0 from a planar glass surface), each of these exponentially decaying near field components can interact with the surface and ultimately be converted into a propagating wave in the glass at its own unique angle θ (with respect to the normal). Angle θ is always greater than the critical angle θ_c for total internal reflection at that interface. The conversion of exponentially decaying waves from the near field of the dipole in water into super-critical angle propagating waves in the glass can be significant for fluorophores within about one wavelength of the surface. The result is that fluorophores near the surface tend to deposit much of their energy into the glass in a hollow cone distribution with a half-angle greater than the critical angle, with all the remaining directions receiving a reduced portion of energy.

The light intensity distribution into the water is also affected by the interface, in this case by interference between light directly propagating from the fluorophore and light reflected from the interface. This leads to an interference pattern which alters, in a z_0-dependent manner, the total light power gathered by a finite aperture objective.

Mathematical descriptions of fluorophore emission at surfaces are given by Carniglia et al. [1972], Lee et al. [1979], Hellen and Axelrod [1987], and Burghardt [1989], and references therein.

D. TIR at Metal-Coated Glass

Glass which has been vacuum coated by a thin (~20 nm) layer of metal (e.g., aluminum) produces some potentially useful optical effects. As concerns TIR excitation, $I(0)$ is dramatically affected (Fig. 11). The s-polarized evanescent intensity $I_y(0)$ becomes negligibly small. But the p-polarized intensity is actually enchanced for a very narrow band of incidence angles, becoming an order of magnitude brighter than the incident light at its peak. This resonance-like effect is due to excitation of a surface plasmon mode at the metal/water interface. The peak is at the "surface plasmon angle," which for an aluminum film at a glass/water interface, is greater than the critical angle for TIR. This enchancement feature is rather remarkable since a 20 nm thick metal film is almost opaque to the eye.

The metal film leads to a highly polarized evanescent wave (provided $I_p \neq 0$), regardless of the purity of the incident polarization.

An interesting consequence of the effect shown in Figure 11 is that a light beam incident upon a 20 nm Al film from the glass side at a glass/aluminum film/water interface evidently does not have to be collimated to produce TIR. Those rays that are incident at the surface plasmon angle will create a strong evanescent wave; those rays that are too low or high in incidence angle will create a negligible field in the water. This phenomenon may ease the practical requirement for a collimated incident beam in TIRF.

E. Emission of Fluorescence Near Metal-Coated Glass

One main reason for using metal-coated glass is that the metal film quenches fluorescence emitted from within 10 nm of the surface. The emitted intensity into all angles (not just super-critical ones) is virtually zero for distances less than 5 nm. The excitation energy is almost entirely converted into heat in the metal film.

Nevertheless, fluorophores excited (by TIR or otherwise) at larger distances will produce observable emission, but the pattern is extremely anisotropic [Hellen and Axelrod, 1987], even more so than at a bare glass interface. A peak of intensity directed into the glass is again present, but here centered in an extremely narrow band at some $\theta = \theta_p > \theta_c$. Angle θ_p is called the "surface plasmon" angle and arises from near field waves from the dipole which are exactly matched to resonant electronic vibrations. These electronic vibrations then propagate on the metal surface [Weber and Eagan, 1979] and re-emit light into a hollow cone of vertex half angle θ_p. Dipoles perpendicular to the metal surface can furnish energy into the surface plasmons quite effectively, leading to an apparent transmission of emitted fluorescence through a virtually opaque metal film. However, dipoles parallel to the surface are very unsuccessful at coupling with surface plasmons and almost all the radiated emission appears in the water.

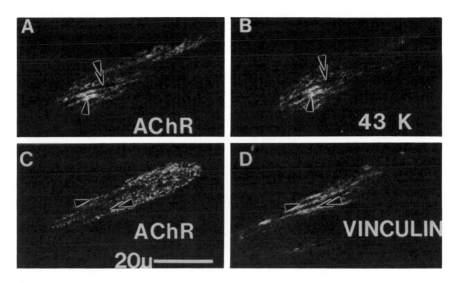

Fig. 12.   AChR clusters on a rat myotube in culture, visualized by **A,C**) fluorescein α-bungaro-toxin; **B**) rhodamine anti-43K protein; and **D**) rhodamine anti-vinculin (antibody gifts from R. Bloch). Note that AChR and 43K tend to aggregate in the same regions, whereas AChR and vinculin appear to exclude each other.

At even larger distances from the surface, the near field is too weak to inter-act with the surface, and the intensity observed through the glass substrate at supercritical angles drops toward zero.

The ability of a metal film to quench almost completely fluorescence of those fluorophores within about 10 nm allows suppression of the signal from labeled protein nonspecifically adsorbed to the metal film substrate while still permitting fluorescence from the slightly more distant labeled protein adsorbed to an adherent membrane to be detected. It is this effect which allows the erythrocyte membranes shown in Figure 10 to appear brighter than the under-lying metal film substrate despite the large amount of adsorbate on the substrate.

IV. APPLICATIONS OF TIRF

A. Qualitative Observation of Labeled Cells

The most straightforward application of TIRF is to observe the location, lateral structure, and (with time lapse video) the motion of cell/substrate con-tacts. Cells may be labeled by a membrane lipid fluorescent analog such as 3-3′-dioctadecylindocarbocyanine (diI) [Axelrod, 1981] or specific ligands such as rhodamine α-bungarotoxin (targeted to acetylcholine receptors, AChR) or fluorescent antibodies against cell surface and cytoplasmic components [Bloch et al., 1989].

Figure 12 shows double-labeled TIRF views of the relative distribution of AChR (labeled by fluorescein-α-bungarotoxin) clustered on the surface of cultured rat myotubes and of certain specific non-AChR proteins (labeled by rhodamine-antibodies) in or immediately under the membrane. Cytoplasmic filaments in particular are difficult to image in thick cells such as these by epi-illumination because of the large amount of out-of-focus fluorescence which obscures details. On the other hand, the "optical sectioning" effect of TIR creates clear, high-contrast images. Somewhat fortuitously for the use of TIR, the receptor clusters in this biological system happen to be found predominately in the general regions where the myotube plasma membrane is near the glass substrate, which allows TIR effectively to excite their fluorescence. The biological effect revealed in Figure 12 is that AChR codistributes with 43K protein but interdigitates with vinculin.

The interaction between immune system cells and their targets often involves a specific and as yet incompletely understood surface reaction. This interaction can be made optically accessible by modeling the target as a lipid monolayer or bilayer supported on glass [see Thompson and Palmer, 1988; Thompson et al., 1988, for reviews on the preparation of such surfaces]. For example, Weis et al. [1982] could visualize the contact region between basophils (which expose Fc receptors) and hapten-containing target model membranes by illuminating with TIR in the presence of fluoresceinated IgE antibodies. The contact region, where the Fc receptors are indirectly connected to the haptens through the IgE, appeared rather variegated and punctate, perhaps due to filopodia-like structures in the contact zone. This pattern could not be observed with conventional epifluorescence.

TIRF seems particularly advantageous for long term viewing of cells, since the evanescent wave minimizes exposure of the cells' organelles to excitation light. Also, in many cases, some probe molecules become internalized in living cells. Under epi-illumination, this tends to obscure the fluorescent pattern of the membrane or submembrane labeling. But since TIRF optically sections the sample, one can observe distinct surface patterns even in the presence of a large amount of internalized label.

B. Negative TIRF for Viewing Cell/Substrate Contacts

A variation of TIRF to observe cellular morphology, introduced by Gingell et al. [1985], produces essentially a negative of the standard fluorescence view of labeled cells. The solution surrounding the cells is doped with a nonadsorbing and nonpermeable fluorescent volume marker, fluorescein-labeled dextran. Focal contacts then appear as dark areas and other areas appear brighter, depending on the depth of solution illuminated by the evanescent wave in the cell/substrate gap. A quantitative theory for converting fluorescence intensities into cell/ substrate contact distances has been developed [Gingell et al., 1987]. By using a high refractive index glass (n = 1.83) as the TIR and cell substrate surface, a very shallow evanescent wave (1/e decay distance ~ 37 nm) can be produced [Todd et al., 1988], which minimizes the contribution from the cytoplasm and probes small undulations in the cell/substrate contact region.

C. Adsorption Equilibria of Proteins

TIRF has been used to study equilibrium adsorption of blood proteins to artificial surfaces [e.g., Lowe et al., 1986], mainly to explore the surface properties of various biomaterials that have medical applications (e.g., nonthrombogenic medical prostheses). Much of this work has been done in nonmicroscopic configurations, but the approach can usually be easily adapted to microscopy for proteins extrinsically labeled with visible or near-UV fluorophores such as fluorescein, rhodamine, NBD, or dansyl groups.

It is possible that such extrinsic groups might affect the physical adsorption process under investigation. To avoid this possible disruption, some nonmicroscopic studies have monitored the intrinsic fluorescence of tryptophan residues on unlabeled proteins, excited at $\lambda_o = 280$ nm [Hlady et al., 1985, 1988a; Horsley et al., 1987] and on labeled proteins [Hlady and Andrade, 1988]. In standard epi-illumination microscopy, working with intrinsic protein fluorescence requires expensive fused silica objectives to transmit the excitation light. TIR intrinsic fluorescence microscopy would require only regular objectives (which should transmit the emission at $\lambda = 330$ nm) along with fused silica excitation focusing lenses and prism, and a very bright (usually pulsed) laser source coupled with a dye laser and frequency doubler to achieve $\lambda_o = 280$ nm.

Calibration of a TIRF intensity to derive an absolute concentration of adsorbate is a nontrivial problem, mainly because fluorescence quantum efficiencies are apt to change upon adsorption to a surface. One route around this problem is to measure the depletion of bulk solute (in epi-illumination mode or in a standard spectrofluorimeter) when it is allowed to adsorb onto a known surface area [e.g., Burghardt and Axelrod, 1981]. Another route involves the use of a radioactively labeled derivative of the protein adsorbate [Hlady et al., 1986].

The emission spectra contains information on the hydrophobicity of the fluorophore environment. An investigation of TIR-excited emission shifts has successfully detected hydrophobicity gradients on silica surfaces which appear correlated with the ability of each local region to adsorb blood albumin protein [Hlady et al., 1988b].

Recently, specific (rather than nonspecific as above) reversible binding constants have been measured by TIRF for the interactions of antibodies with either haptens [Kalb et al., 1990] or Fc receptors [Poglitsch and Thompson,

1990] reconstituted into planar lipid bilayers supported on fused silica or glass. In the near future, such measurements undoubtedly will be expanded to the study of specific binding parameters between proteins and their cell surface receptors in natural biological membranes. In particular, one can measure the equilibria and reversible kinetics of fluorescein epidermal growth factor binding to the membranes of A431 cells adhered to glass surfaces [Hellen and Axelrod, 1990].

D. TIRF Combined With Fluorescence Photobleaching Recovery (FPR)

Consider a labeled molecule in equilibrium between a surface-bound state and a free solute state:

$$\text{(free solute)} + \text{(vacant surface site)} \underset{k_2}{\overset{k_1}{\rightleftarrows}} \text{(surface-bound solute)}$$

If the evanescent wave intensity is briefly flashed brightly, then some of the fluorophores associated with the surface will be photobleached. Subsequent exchange with unbleached dissolved fluorophores in equilibrium with the surface will lead to a recovery of fluorescence, excited by a continuous but much attenuated evanescent wave. The time course of this recovery is a measure of the desorption kinetic rate k_2 [Thompson et al., 1981].

If the evanescent wave intensity is variegated over a short characteristic distance on the surface, then surface diffusion coefficients can also be measured (at least if surface diffusion is fast enough to carry the molecule through the characteristic distance before desorption). Two studies have utilized TIR/FPR in this manner. The adsorption/desorption kinetics and surface diffusion of rhodamine bovine serum albumin at a glass surface have been examined by a TIR illumination area focused into a thin line [Burghardt and Axelrod, 1981]. Similar parameters have been examined for fluorescein-insulin reversibly adsorbed to erythrocyte membrane ghosts with a TIR illumination area of interference fringes created by intersecting beams such as those used in Figure 10.

E. TIRF Combined With Fluorescence Correlation Spectroscopy (FCS)

The volume defined by the depth of the evanescent wave in an area delimited by the image plane diaphragm of a microscope can be extremely small, down to about 0.01 μm^3. Within this volume, the entrance or exit of a single fluorophore can cause a significant change in the fluorescence intensity. In fact, these TIRF fluctuations are clearly visible to the "naked eye" through the microscope. By autocorrelating (on-line) the random noise arising from such statistical fluctuations (a technique called fluorescence correlation spectroscopy or FCS), one can obtain information about three parameters: the mean time of surface binding ($= 1/k_2$), the surface diffusion coefficient, and the absolute mean number of fluorescent molecules bound per surface area (even without one knowing anything about quantum efficiencies or light collection efficiencies).

Two investigations have combined TIR with FCS. The first [Hirschfeld et al., 1977] adapted TIR/FCS to measure the absolute concentration of virions in solution. The other [Thompson and Axelrod, 1983] measured the adsorption/ desorption kinetics of immunoglobulin on a protein-coated surface on the millisecond time scale.

F. Concentration of Molecules Near Surfaces

The concentration of a solute or adsorbate may be a nontrivial function of the distance to the surface, a function which contains information about the thermodynamics of the surface interaction. To explore the fluorophore concentration $C(z)$ as a function of distance z from the surface, one can record the observed fluorescent intensity F as the characteristic depth d of the evanescent wave is varied. Mathematically, the relationship is

$$F(d) = \int_0^\infty I(0)\, e^{-z/d}\, C(z)\, g(z)\, dz \qquad (10)$$

where the exponential arises from the evanescent wave intensity. The term $g(z)$ is, in general, a rather complicated function arising from the nonisotropic and z-dependent emission pattern from fluorophores near an interface. However, if we approximate $g(z)$ as a constant, then $C(z)$ can be computed simply by taking an inverse Laplace transform of an experimentally measured $F(d)$.

To vary d, the angle of incidence θ can be varied as indicated by Equation 6. Experimentally, this is not trivial, because d is a very strong function of θ within only a few degrees greater than θ_c, and therefore θ must be measured to fractions of a degree. In addition, the presence of a solute (or the cytoplasm of a biological cell) alters the refractive index n_1 from its pure water value, and this must also be known accurately. Rondelez et al. [1987] measured $F(d)$ vs. θ to obtain information on the z-dependent concentration profile of artificial polymers adsorbed to glass or silica. The ability of this general approach to correctly report concentration profiles was checked on planes of fluorophores deposited in steps between layers of Langmuir-Blodgett films [Suci and Reichert, 1988].

Another method of obtaining $C(z)$ involves varying the observation angle of emission (rather than incidence) and by utilizing the special properties of fluorophore emission near surfaces as discussed earlier. This method, which has been used in combination with TIR excitation but does not require it, is most easily compatible with nonmicroscope setups [Ausserre et al., 1985].

Deduction of the absolute distance from the surface to a labeled cell membrane at a cell/substrate contact region can be based on the variation of $F(d)$ with θ [Lanni et al., 1985]. This effort is challenging because corrections have to be made for θ-dependent reflection and transmission through four stratified layers (glass, culture medium, membrane, and cytoplasm), all with different refractive indices. For 3T3 cells, the authors derive a plasma membrane/substrate spacing of 49 nm for focal contacts and 69 nm for ''close'' contacts elsewhere. They were also able to calculate an approximate refractive index for the cytoplasm of 1.358 to 1.374.

G. Orientation and Rotation of Molecules Near Surfaces

The polarization properties of the evanescent wave can be used to excite selected orientations of fluorophores; e.g., fluorescent-labeled phosphatidylethanolamine embedded in lecithin monolayers on hydrophobic glass [Thompson et al., 1984]. When interpreted according to an approximate theory, the total fluorescence gathered by a high-aperture objective for different evanescent polarizations gives a measure of the probe's orientational order.

Both the physics and the chemistry of proximity to a surface can alter the excited state lifetime and rotational motion of a fluorophore. Itaya et al. [1987] have described a TIR system for obtaining time-resolved fluorescence decay curves induced by laser flash illumination of polymer films. Since their optical system was a standard microscope, the approach can easily be extended to studies on biological cells. On nonmicroscopic setups, some first experimental steps in combining TIRF with time-resolved polarized anisotropy decay to measure molecular rotation rates have been tested on sapphire prisms coated with fluorescence-doped polystyrene films [Masuhara et al., 1986] and on pyrene-labeled serum albumin adsorbed to artificial polymer films [Fukumura and Hayashi, 1990].

An unusual application of TIRF for measuring dye concentrations on a woven fabric of silk vs. distance into the silk led to the conclusion that dye concentrates into the interior of the silk fibers rather than on the surface [Kurahashi et al,. 1986].

Many substances preferentially concentrate at interfaces, including liquid/liquid ones. Although TIRF is most easily adaptable to solid/liquid interfaces, Morrison and Weber [1987] succeeded in observing the preferential adsorption of certain amphiphilic dyes at the interface between two immiscible and optically dissimilar liquids. Steady-state TIR fluorescence polarization in that system showed that the rotational diffusion of the interfacially adsorbed dye was restricted.

H. Fluorescence Energy Transfer and TIRF

TIRF can be combined with fluorescence energy transfer to measure distances between fluorophores on a surface in the presence of a large background of bulk-dissolved fluorophores.

Burghardt and Axelrod [1983] detected TIRF/energy transfer evidence of a change in donor/acceptor-labeled bovine serum albumin conformation upon the protein's adsorption to glass.

In a TIRF/energy transfer study of more relevance to cell biology, Watts et al. [1986] explored whether helper T-cells could force two nonidentical antigens in a target membrane into closer proximity with each other. These two antigens, one labeled with a fluorescence energy transfer donor and the other with an acceptor, were incorporated into a planar lipid bilayer on a TIR hydrophilic glass surface. Significant amounts of one of them remained in solution, so microscopic TIRF was needed to limit excitation to the region near the glass and overlaying lipid bilayer. TIRF also served to reduce the autofluorescence normally observed from the T-cells that were allowed to settle on the lipid bilayer. It was found that fluorescence energy transfer occurred only in those microscopic lipid bilayer regions where the T-cell surface came into close apposition with the bilayer. The conclusion was that the T-cell surface forces the two membrane antigens to which it binds to within a distance of 4 nm of each other.

I. Coating the TIR Surface

TIRF experiments often involve specially coated substrates. A glass surface can be chemically derivatized to yield special physi- or chemi-absorptive properties. Covalent attachment of certain specific chemicals are particularly useful in cell biology and biophysics, including: poly-l-lysine for enhanced adherence of cells; hydrocarbon chains for hydrophobicizing the surface in preparation for lipid monolayer adsorption; and antibodies, antigens, or lectins for producing specific reactivities.

Derivatization generally involves pretreatment by an organosilane (see the catalog of Petrarch Systems). The protocol for poly-l-lysine attachment to planar glass slides is similar to that described for the treatment of spherical glass beads [Jacobson et al. 1978]. The protocol for preparing lipid monolayers on hydrophobic glass is given by VonTscharner and McConnell [1981]. Methods for preparing model membranes on planar surfaces suitable for TIR are reviewed by McConnell et al. [1986] and Thompson and Palmer [1988].

Aluminum coating can be accomplished in a standard vacuum evaporator; the amount of deposition can be made reproducible by completely evaporating a premeasured constant amount of aluminum.

After deposition, the upper surface of the aluminum film spontaneously oxidizes in air very rapidly. This aluminum oxide layer appears to have some similar chemical properties to the silicon dioxide of a glass surface; it can be derivatized by organosilanes in much the same manner.

J. TIR on Flattened Biological Membranes

To examine reversible binding/unbinding of fluorescent-labeled molecules on biological membranes using TIRF, the membranes must be flattened against the TIR surface. This has been successfully accomplished with erythrocyte ghosts [Axelrod et al. 1986]. After the glass substrate is covalently coated with poly-l-lysine, erythrocytes are allowed to adhere, followed by hyposmotic shock. Rather than floating away or crumpling up on the surface, the membrane ghosts flatten into circular disks on the glass with a characteristic tear that exposes the outer surface and the cytoplasmic surface to the solution in their own distinct regions (Fig. 10). This technique or a modification of it may also work for other cell types.

V. CONCLUSIONS

TIRF is an experimentally simple technique for selective excitation of fluorophores on or near a surface. It can be set up on a standard upright or inverted microscope. It is compatible and rapidly interchangeable with bright-field, darkfield, phase contrast, and epi-illumination and accommodates a wide variety of common microscope objectives without alteration.

Confocal microscopy (CM) is another technique for apparent optical sectioning, achieved by exclusion of out-of-focus emitted light with a set of image plane pinholes. CM has the clear advantage in versatility; its method of optical sectioning works at any plane of the sample, not just at an interface between dissimilar refractive indices. However, other differences exist which, in some special applications, can favor the use of TIRF:

a. The depth of the optical section in TIRF is $\sim$0.1 μm, whereas in CM it is a relatively thick $\sim$0.6 μm.

b. In some applications (e.g., FRAP, FCS, or on cells whose viability is damaged by light), illumination and not just detected emission is best restricted to a thin section; this is possible only with TIRF.

c. Since TIRF can be adapted to and made interchangeable with existing standard microscope optics, even with ''home-made'' components, it is much less expensive than CM.

Cell-substrate contacts can be located by a nonfluoresence technique completely distinct from TIRF, known as ''internal reflection microscopy'' (IRM) [Gingell and Todd, 1979]. Using conventional illumination sources, IRM visualizes cell-substrate contacts as dark regions. IRM has the advantage that it doesn't require the cells to be labeled, but the disadvantages that it contains no information of biochemical specificities in the contact regions and that it is less sensitive to changes in contact distance (relative to TIRF) within the critical first 100 nm of the surface.

Applications of TIRF in cell biology and surface chemistry include

1. Localization of cell-substrate contact regions in cell culture.

2. High-contrast visualization of submembrane cytoskeletal structure on thick cells.

3. Measurement of the kinetic rates and surface diffusion of reversibly bound biomolecules at flattened biological membrane surfaces.

4. Measurement of the concentration and orientational distributions of fluorescent molecules as a function of distance from the surface.

5. Measurement of intermolecular distances between fluorescent surface-bound molecules in the presence of a large excess of fluorophore or background fluorescence in the bulk.

6. Reduction of cell autofluorescence relative to fluorescence excited at cell-substrate contacts.

ACKNOWLEDGMENTS

The author is grateful to Robert M. Fulbright, Andrea Stout, and Edward H. Hellen for their photography and sample preparation work, to Ariane McKiernan for her very helpful comments on the manuscript, and to Robert Bloch of the University of Maryland Medical School for his gift of antibodies. This work was supported by a USPHS NIH grant NS 14565 and an NSF grant DMB 8805296.

VI. REFERENCES

Ausserre D, Hervet H, Rondelez F (1985) Concentration profile of polymer solutions near a solid wall. Phys Rev Lett 54:1948–1951.

Axelrod D (1981) Cell-substrate contacts illuminated by total internal reflection fluorescence. J Cell Biol 89:141–145.

Axelrod D, Fulbright RM, Hellen EH (1986) Adsorption kinetics on biological membranes: measurement by total internal reflection fluorescence. In Taylor DL, Waggoner AS, Murphy RF, Lanni F, Birge RR (eds): Applications of Fluorescence in the Biomedical Sciences. New York: Alan R. Liss, Inc., pp 461–467.

Bloch RJ, Velez M, Krikorian J, Axelrod D (1989) Microfilaments and actin-associated proteins at sites of membrane-substrate attachment within acetylcholine receptor clusters. Exp Cell Res 182:583–596.

Burghardt TP (1989) Polarized fluorescent emission from probes near dielectric interfaces. Chem Phys Lipids 50:271–287.

Burghardt TP, Axelrod D (1981) Total internal reflection/fluorescence photobleaching recovery study of serum albumin adsorption dynamics. Biophys J 33:455–468.

Burghardt TP, Axelrod D (1983) Total internal reflection fluorescence study of energy transfer in surface-adsorbed and dissolved bovine serum albumin. Biochemistry 22:979–985.

Carniglia CK, Mandel L, Drexhage KH (1972) Adsorption and emission of evanescent photons. J Opt Soc Am 62:479–486.

Chew H, Wang D, Kerker M (1979) Elastic scattering of evanescent electromagnetic waves. Appl Opt 18:2679–2687.

Fukumura H, Hayashi K (1990) Time-resolved fluorescence anisotropy of labeled plasma proteins adsorbed on polymer surfaces. J Coll Interfac Sci 135:435–442.

Gingell D, Todd I (1979) Interference reflection microscopy. A quantitative theory for image interpretation and its application to cell-substratum separation measurement. Biophys J 26:507–526.

Gingell D, Todd I, Bailey J (1985) Topography of cell-glass apposition revealed by total internal reflection fluorescence of volume markers. J Cell Biol 100:1334–1338.

Gingell D, Heavens OS, Mellor JS (1987) General electromagnetic theory of internal reflection fluorescence: The quantitative basis for mapping cell-substratum topography. J Cell Sci 87:677–693.

Glass TR, Lackie S, Hirschfeld T (1987) Effect of numerical aperture on signal level in cylindrical waveguide evanescent fluorosensors. Appl Opt 26:2181–2187.

Harrick NJ, Loeb GI (1973) Multiple internal reflection spectrometry. Anal Chem 45:687–691.

Hellen EH, Axelrod D (1987) Fluorescence emission at dielectric and metal-film interfaces. J Opt Soc Am B 4:337–350.

Hellen EH, Axelrod D (1990) Dissociation rate constant of epidermal growth factor specifically bound to its receptor measured with prismless TIR/FPR. Biophys J 57:293a.

Hirschfeld T (1965) Total reflection fluorescence (TRF). Can Spect 10:128.

Hirschfeld T, Block MJ, Mueller W (1977) Virometer: An optical instrument for visual observation, measurement and classification of free viruses. J Histochem Cytochem 25:719–723.

Hlady V, Andrade JD (1988) Fluorescence emission from adsorbed bovine albumin and albumin-bound 1-anilinonaphthalene-8-sulfonate studied by TIRF. Coll Surf 32:359–368.

Hlady V, Golander C, Andrade JD (1988b): Hydrophobicity gradient on silica surfaces: A study using total internal reflection fluorescence spectroscopy. Coll Surf 33:185–190.

Hlady V, Reinecke DR, Andrade JD (1986) Fluorescence of adsorbed protein layers: Quantitation of total internal reflection fluorescence. J Coll Interfac Sci 111:555–569.

Hlady V, Rickel J, Andrade JD (1988a) Fluorescence of adsorbed protein layers. II. Adsorption of human lipoproteins studied by total internal reflection intrinsic fluorescence. Coll Surf 34:171–183.

Hlady V, Van Wagenen RA, Andrade JD (1985) Total internal reflection intrinsic fluorescence (TIRIF) spectroscopy applied to protein adsorption. In Andrade J (ed): Surface and Interfacial Properties of Biomedical Polymers, Vol. 2: Protein Adsorption. New York: Plenum Press, Inc., pp 81–119.

Horsley D, Herron J, Hlady V, Andrade JD (1987) Human and hen lysozyme adsorption: A comparative study using total internal reflection fluorescence spectroscopy and molecular graphics. In Brash JL, Horbett TA (eds): "Proteins at Interfaces: Physicochemical and Biochemical Studies. ACS Symp. Series #343, Washington DC: American Chemical Society, pp 290–305.

Itaya A, Kurahashi A, Masuhara H, Tamai N, Yamazaki I (1987) Dynamic fluorescence microprobe method utilizing total internal reflection phenomena. Chem Lett (Jpn) 1987:1079–1082.

Jacobson BS, Cronin J, Branton D (1978) Coupling polylysine to glass beads for plasma membrane isolation. Biochim Biophys Acta 506:81–96.

Kalb E, Engel J, Tamm LK (1990) Binding proteins to specific target sites in membranes measured by total internal reflection fluorescence microscopy. Biochemistry 29:1607–1613.

Kurahashi A, Itaya A, Masuhara H, Sato M, Yamada T, Koto C (1986) Depth distribution of fluorescent species in silk fabrics as revealed by total internal reflection fluorescence spectroscopy. Chem Lett (Jpn) 1986:1413–1416.

Lanni F, Waggoner AS, Taylor DL (1985) Structural organization of interphase 3T3 fibroblasts studied by total internal reflection fluorescence microscopy. J Cell Biol 100:1091–1102.

Lee E-H, Benner RE, Fen JB, Chang RK (1979) Angular distribution of fluorescence from liquids and monodispersed spheres by evanescent wave excitation. Appl Opt 18:862–868.

Lowe R, Hlady V, Andrade JD, Van Wagenen RA (1986) Human haptoglobin adsorption by a total internal reflection fluorescence method. Biomaterials 7:41–44.

Masuhara H, Tazuke S, Tamai N, Yamazaki I (1986) Time-resolved total internal reflection fluorescence spectroscopy for surface photophysics studies. J Phys Chem 90:5830–5835.

McConnell HM, Watts TH, Weis RM, Brian AA (1986) Supported planar membranes in studies of cell-cell recognition in the immune system. Biochim Biophys Acta 864:95–106.

Morrison LE, Weber G (1987) Biological membrane modeling with a liquid/liquid interface. Probing mobility and environment with total internal reflection excited fluorescence. Biophys J 52:367–379.

Poglitsch CL, Thompson N (1990) Interaction of antibodies with Fc receptors in substrate-supported planar membranes measured by total internal reflection fluorescence microscopy. Biochemistry 29:248–254.

Reichert WM, Truskey GA (1990) Total internal reflection fluorescence (TIRF) microscopy. I. Modeling of cell contact region fluorescence. J Cell Sci (in press).

Rondelez F, Ausserre D, Hervet H (1987) Experimental studies of polymer concentration profiles at solid-liquid and liquid-gas interfaces by optical and x-ray evanescent wave techniques. Annu Rev Phys Chem 38:317–347.

Stout A, Axelrod D (1989) Evanescent field excitation of fluorescence by epi-illumination microscopy. Appl Opt 28:5237–5242.

Suci PA, Reichert WM (1988) Determination of fluorescence density profiles of Langmuir-Blodgett deposited films using standing light waves. Langmuir 4:1131–1141.

Thompson NL, Axelrod D (1983) Immunoglobulin surface-binding kinetics studied by total internal reflection with fluorescence correlation spectroscopy. Biophys J 43:103–114.

Thompson NL, Burghardt TP, Axelrod D (1982) Measuring surface dynamics of biomolecules by total internal reflection with photobleaching recovery or correlation spectroscopy. Biophys J 33:435–454.

Thompson NL, McConnell HM, Burghardt TP (1984) Order in supported phospholipid monolayers detected by the dichroism of fluorescence excited with polarized evanescent illumination. Biophys J 46:739–747.

Thompson NL, Palmer AG (1988) Model cell membranes on planar substrates. Comm Mol Cell Biophys 5:39–56.

Thompson NL, Palmer AG, Wright LL, Scarborough PE (1988) Fluorescence techniques for supported planar model membranes. Comm Mol Cell Biophys 5:109–113.

Todd I, Mellor JS, Gingell D (1988) Mapping cell-glass contacts of Dicyostelium amoebae by total internal reflection aqueous fluorescence overcomes a basic ambiguity of interference reflection microscopy. J Cell Sci 89:107–114.

VonTscharner V, McConnell HM (1981) Physical properties of lipid monolayers on alkylated planar glass surfaces. Biophys J 36:421–427.

Watts TH, Gaub HE, McConnell HM (1986) T-cell-mediated association of peptide antigen and major histocompatibility complex protein detected by energy transfer in an evanescent wave-field. Nature 320:176–179.

Weber WH, Eagan CF (1979) Energy transfer from an excited dye molecule to the surface plasmons of an adjacent metal. Opt Lett 4:236–238.

Weis RM, Balakrishnan K, Smith BA, McConnell HM (1982) Stimulation of fluorescence in a small contact region between rat basophil leukemia cells and planar lipid membrane targets by coherent evanescent radiation. J Biol Chem 257:6440–6445.

Noninvasive Techniques in Cell Biology: 129–152
© 1990 Wiley-Liss, Inc.

6. Fluorescence Photobleaching Recovery: A Probe of Membrane Dynamics

James Thomas and Watt W. Webb

Departments of Physics (J.T.) and Applied Physics (W.W.W.), Cornell University, Ithaca, New York 14853

I. INTRODUCTION

The fluidity of the bilayer lipid membrane permits lateral molecular mobility of its protein constituents. This mobility is known to be essential for many cellular signal transduction systems in which cross-linking of cell surface receptors is required for signal initiation; such cross-linking cannot occur without molecular motion. Molecular mobility is required for cellular motility, in which protein reorganization is essential in the dissolution and reformation of adhesion plaques. Molecular mobility is essential to the internalization of ligands and their receptors at coated pits, since many receptors appear remote from the pits to which they eventually bind.

But in addition to these many cellular processes that require some degree of lateral diffusibility and whose dynamics may be limited by that diffusibility, there are other processes for which lateral mobility must be inhibited. Acetylcholine receptors must be localized to the neuromuscular junction. Polarized cell lines, such as intestinal epithelia, must maintain polarity in protein distributions in order that unidirectional transport of nutrients can be ensured. The capability to maintain permanent structures while permitting randomizing dynamics is a salient feature of living systems.

The experimental technique that has so far provided the bulk of information on the lateral diffusibility of cell surface constituents is fluorescence photobleaching recovery, or FPR. The broad range of biological systems to which FPR has been applied can be found in any of the numerous reviews on cellular diffusion [Cherry, 1979; Jacobson et al., 1982; Axelrod, 1983; McCloskey and Poo, 1984; Kapitza and Jacobson, 1986; Edidin, 1987; Jovin and Vaz, 1989]; here we will discuss technical issues in FPR instrumentation and analysis.

Fluorescence photobleaching recovery provides a measurement of tracer diffusion in biological systems. The cellular component of interest is first labeled with a fluorescent tag. Often this tag is created by conjugating the fluorescent moiety fluorescein or rhodamine to an antibody that binds to a surface protein. Certain proteins are known to have specific, strong affinities for various ligands; in these cases, the ligand may be fluorescently labeled and allowed to bind to its receptor. Once the cell has been labeled, a bright light pulse of a wavelength absorbed by the fluorophore (the *bleach beam*) selectively destroys that fluorescence in a small region of the cell. The diffusibility of the labeled species permits that bleached region to recover its fluorescence through the exchange of proteins having bleached fluorescent moieties with proteins initially remote from the region of bleach. This fluorescence recovery is observed with a greatly attenuated laser beam, the *monitor beam*. A gradient of fluorescence is established; the decay of this gradient characterizes the transport processes of the labeled species.

The mechanism by which fluorescence is restored to the bleached region need not be purely diffusive, but may involve bulk translation of the fluorescent marker, or flow. Cell membranes are complex, organized structures whose lipids are constantly recycled to the cells' internal lipid pool. These internalization events must establish flow fields in the membrane; the extent of such fields will depend on the detailed organization of internalization and reinsertion processes. In certain cases, such as motile or spreading cells, reinsertion may occur predominantly at the active cell edge, generating a retrograde or centripetal flow [Bretscher, 1984; Bergman et al., 1983]. Alternatively, concerted protein motion could result from multiple attachment to an internal cytoskeletal structure that itself undergoes a driven, energy consuming motion. While evidence of concerted motion has been seen in single particle tracking experiments [Ghosh and Webb, 1989], it has not been reported in conventional fluorescence recovery experiments on cells. As will be discussed later, the presence of flow in the shape of the recovery of a bleached spot is difficult to detect, especially if diffusion is occurring simultaneously. The absence of such observations in the literature should not be taken as a proof that such flows do not occur, but merely as a not too stringent limit of the extent and velocity of possible flows. Certain fluorescence photobleaching experimental strategies have been developed to improve sensitivity to lateral flow; some of these strategies will be discussed in the conclusion to this chapter.

Other techniques have been effectively utilized to characterize cell surface dynamics. Single particle labeling, either with colloidal gold for brightfield tracking, or with bright fluorescent aggregates for fluorescence tracking, has been useful in determining trajectories of individual protein molecules across the cell surface. Statistical analysis of many such trajectories can provide information about protein diffusibility, and can reveal non-diffusive, correlated modes of behavior. However, not all surface proteins are amenable to these labeling techniques. While the attachment of a large fluorescent aggregate or a 200 nm colloidal gold marker is not expected to substantially alter the diffusion coefficient of an ideal protein in an ideal membrane [Saffman and Delbrück, 1975], cell surfaces are distinctly non-ideal. Proteins do interact with other proteins with varying degrees of specificity and binding energy. These interactions may be further hindered by the intrusion of a large label. Alternatively, the label may nonspecifically interact with components of the extracellular matrix, glycocalyx, or even the cell substrate more strongly than the native protein. In addition, gold markers may be multivalent, cross-linking several receptors to the same marker. While any label whatsoever has the potential to alter the behavior of the protein under scrutiny (even the small fluorescent probes used in FPR), the larger labels are more likely to have a significant effect.

The behavior of lipids in the cell membrane is an important aspect of dynamic membrane structure. It appears likely that large probes attached to lipids would provide more information about the probe behavior than the lipid behavior.

Another fluorescent technique used to probe cell surface dynamics is post-electrophoresis relaxation, or PER. Cell surface proteins can often be induced to migrate cathodally in a DC electric field of some 10 V/cm. The removal of this field permits the reestablishment of equilibrium through back-diffusion. This is PER. It should be noted, however, that this diffusion process is subtly different from that measured by single particle tracking or FPR. In post-electrophoretic relaxation, the diffusion of protein is driven by a protein concentration gradient; the cell surface protein distribution is non-equilibrium and non-uniform. Local diffusion may depend on local protein concentration, or on the strength or duration of the electric field required to produce these global cellular redistributions [Poo, 1981]. In FPR, the cell is not perturbed from its steady state. It is the bleached, non-equilibrium distribution of *label* that permits characterization of the transport coefficients.

Even at equilibrium, the number of fluorescently labeled proteins in a small region of a cell's membrane will statistically fluctuate for purely thermodynamic reasons. The temporal decay of a small fluctuation is governed by the same rate constants that describe "macroscopic" relaxation. In this case, that rate is proportional to the diffusion coefficient of the protein divided by the area of the region. If the protein is labeled with a fluorescent marker, the technique is fluorescence correlation spectroscopy. Fluorescence correlation spectroscopy has proven to be a useful tool for the study of reaction-diffusion coupling in solutions [Elson, 1985]; difficulties in applying the technique to cells seem to be due to the dynamic behavior of living cells, which prevents lengthy observation of a single, stationary region.

II. INSTRUMENTATION

The photobleaching workhorse used for many years in our laboratory is outlined in Figure 1. In overview, it consisted of

• a laser light source for the bleaching and subsequent monitoring excitation of fluorescence,

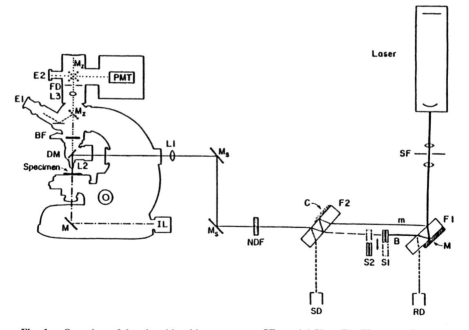

Fig. 1. Overview of the photobleaching apparatus. SF, spatial filter; F1, F2, quartz flats; M, aluminized surface; m,B, monitor and bleach beams; S1, S2, shutters; RD, SD, reference and sample diodes; NDF, neutral density filters (if needed); M_s, beam steering mirrors; L1, L2, L3, lenses; DM, dichroic mirror; BF, barrier filter; E1, E2, eyepieces; FD, field diaphragm; M_z, movable mirrors. The flats are aligned by ensuring that the beams to SD and RD are parallel.

• an attenuation system for rapid, reproducible switching from bleach intensities to monitor intensities,
• a Zeiss Universal research microscope equipped with epifluorescence capability, and
• a cooled photomultiplier tube and associated counting electronics to quantiatively measure the fluorescence intensity.

The importance of each of these system components will now be discussed.

A. Bleaching Light Source

The enormous luminous density required for the rapid destruction of fluorophores may be estimated from the absorption cross section and photo-bleaching yield for these molecules. The extinction coefficient of rhodamine isothiocyanate is approximately 55,000 $M^{-1}cm^{-1}$, which corresponds to an absorption cross section of 0.9Å^2. The quantum yield (QY) is about 0.2, and one can expect roughly 10^5 emitted photons per bleaching event. The bleaching rate is therefore

$$R = \frac{N_\sigma}{\Delta t} \frac{QY}{10^5}$$

where N_σ is the number of photons passing through the absorption cross section in time Δt. The intensity is the energy per unit time, per unit area, in this photon flux:

$$I = \frac{N_\sigma \epsilon_\gamma}{\Delta t \sigma}$$

where ϵ_γ is the energy per photon and σ is the cross section. Then

$$R = \frac{I\sigma}{\epsilon_\gamma} \cdot \frac{QY}{10^5}$$

$\epsilon_\gamma = 2.5 \, eV = 4 \times 10^{-19} J$. Consequently,

$$R = I(W/cm^2) \cdot 4.5 \times 10^{-4}$$

In order that the fluorescence destruction occur in a fraction of a second, which is essential for rapidly diffusing species, power densities of many thousands of watts per square centimeter must be achieved. While a mercury arc lamp may radiate several watts of visible power, this power is radiated into the full 4π solid angle from an arc area of nearly 1 cm^2. No set of lenses or collimators can increase the luminous density above that at the source. As a result, bleaching from an arc lamp generally requires hundreds or thousands of seconds, preventing their use as FPR sources. A laser, however, radiates in a single electromagnetic mode, unlike incoherent sources. The maximum luminous density obtainable with a laser source is limited only by the size of the spot to which it is focused, which is in turn limited by the wave nature of light through diffraction to ~ 1 μm^2, with good microscope optics. Yet 10 mW of power so focused yields a power density of 1 MW/cm^2. Consequently, the true advantage of the laser source is not that it provides more power, but that it provides greater power density, i.e., greater local intensity.

In addition to high power density, a single mode TEM$_{00}$ laser source facilitates the analysis of FPR data. The TEM$_{00}$ mode has a well-defined Gaussian beam profile in the focal plane, and above and below, as illustrated in Figure 3. This illumination profile gives recovery curves that are well fitted by a simple reciprocal function, as described by Yguerabide et al. [1982]; other illumination profiles yield more complex recovery functions and more computationally intensive fitting procedures. In addition, knowledge of the full three-dimensional beam profile permits determination of the effective volume of acceptance for fluorescence, as indicated by the cross-hatched region in Figure 3. This acceptance volume is critical in three-dimensional applications, e.g., diffusion measurements of a cytosolic constituent. It is also important in membrane studies, since it dictates whether dorsal or ventral cell surfaces, *or both,* contribute to the observed diffusive recovery.

The claim that a laser provides a single transverse electromagnetic mode is not strictly true in all circumstances. While a CW argon ion laser will lase in the TEM$_{00}$ Gaussian mode at moderate tube current, at high current densities, we have observed mode degradation in our Coherent Innova-90 laser. The TEM$_{00}$ mode is supplanted by a superposition of TEM$_{10}$ modes, giving rise to an annular radiation pattern of slightly greater diameter than the TEM$_{00}$ beam. This mode degradation is irreversible, in that no combination of optics, spatial filters, etc., can restore the orthogonal TEM$_{00}$ state. However, the onset of mode degradation occurs only at greater than 2 W single line (514 nm) light output; consequently, we have not been limited by this factor.

Lasers are subject to pointing fluctuations, as well. Small changes in the beam direction need not pose a problem for FPR applications, however. FPR optics generally places a beam waist in the field plane. With this arrangement, changes in the beam direction do not affect the position of the focused spot.

Regardless of the specific choice of laser for an FPR application, a flexible apparatus requires some degree of wavelength tuneability, since different fluorophores have different excitation spectra. Argon ion lasers can be tuned to any one of several visible lines by means of an intracavity prism; with an alternate choice of end mirrors several UV lines are obtainable. Krypton ion lasers have a different set of resonant lines that reach farther into the long wavelength region of the visible spectrum (Fig. 2). If complete wavelength tuneability is required, a dye laser may be used in conjunction with a CW ion laser or a frequency doubled or tripled Nd:YAG laser.

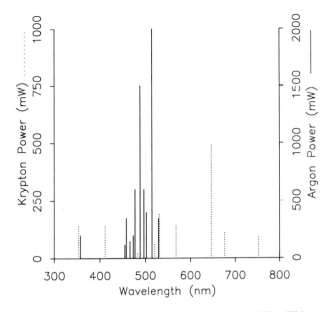

Fig. 2. Wavelengths available with a 5 W argon ion laser and a 800 mW krypton ion laser (Coherent, Inc.) Special optics and testing are required for all krypton lines except 676 nm and 647 nm, and for the argon lines at 528 nm, 454 nm, and broadband UV.

B. Attenuation System

The primary requirement of an attenuation system for FPR experiments is that the monitor beam must suffer no displacement from the bleaching beam. While an adequate shutter can be as simple as a solenoid to remove and replace a neutral density filter from the beam path, this procedure requires a high-quality, optically flat filter with parallel surfaces. If this method is used, the filter should be placed in an auxilliary field plane. Small deviations from planarity will then not disturb the position of the spot.

A more robust approach is to use quartz optical flats to separate the laser into two parallel beams, one bleach and one monitor, recombining them after independent shutters. The alignment of the flats can be performed by ensuring that stray reflected beams, as shown in Figure 1 (to SD and RD), are parallel. The intensity ratio between monitor and bleach beams is 1:1,000 in our apparatus. It is important that the monitor beam be reduced to an intensity at which very little fluorophore bleaching occurs. Such bleaching will result in erroneous estimates of the diffusion coefficient and the diffusible fraction. Our 1:1,000 ratio is attained by coating the back of the first flat with aluminum. The monitor beam is reflected off the front surface, while the bleach beam reflects off the aluminized back surface. The monitor beam undergoes additional attenuation when it reflects off the internal surfaces of the second flat during recombination. The monitor and bleach beams are independently shuttered in the region between the two flats.

Alternatively, an acousto-optic modulator (AOM) may be used to switch rapidly between monitor and bleach intensities. An AOM consists of a crystal driven by an acoustic (piezoelectric) transducer at MHz frequencies. An appropriately directed laser beam will undergo Bragg diffraction off the density variations in the crystal. The intensity of the diffracted beam is varied by changing the amplitude of the acoustic wave. The intensity of the first order diffracted beam may be continuously varied from 0 to 80 or 90% of the incident intensity. The response time, usually quoted as a bandwidth, is the transit time of the acoustic wave across the beam diameter, typically 175 ns for a beam diameter of 1 mm. Acousto-optic modulators and the necessary drive electronics are available from many commercial vendors at prices ranging from several hundred to several thousand dollars, depending on the aperture size, wavelength tolerance, power tolerance, and other factors. In an optical system employing acousto-optic modulation, the bleach and monitor beams are always coincident. The principal disadvantage is that the alignment of the AOM is wavelength dependent, due to the Bragg condition. Note that the zero order, undiffracted beam cannot be used for FPR, since this beam can only be attenuated by about a factor of ten with the highest acoustic wave amplitudes. This ratio is too small for FPR, since excessive bleaching would result during monitor exposure. Consequently, the first order diffracted beam should be used.

C. Photomultiplier

The fluorescence of the sample is measured before and after bleaching with an RCA C31034A photomultiplier tube. To prevent tube damage, the first dynode is shorted to the photocathode during the bleach. The photomultiplier (PMT) is the detector of choice for this application, in spite of the fact that other detectors, notably photodiodes, have slightly higher quantum efficiency. The reason is that only a photomultiplier provides high quantum efficiency signal detection *and* amplification. A photodiode or charge-coupled device has a gain of unity (1 e$^-$/detected photon), whie a photomultiplier has a typical gain of 10^6. In addition, no additional noise is added to the measurement when a photomultiplier is used in a photon counting configuration, above the statistical shot noise intrinsic in the signal. Regardless of the efficiency of detection, if the signal amplification process introduces substantial noise, the S/N ratio will suffer. Additional improvement in the noise is obtained by cooling the photomultiplier with dry ice, which reduces the dark current to less than 2 counts/s. Cooling the PMT requires that the photocathode be kept dry and isolated from atmospheric moisture, which would condense on the face and scatter incoming radiation. The commercial housing we use (RF-TSA, Products for Research, Danvers, MA) includes a heating element on the window to the housing to prevent such condensation. The photocathode is sealed within the housing and kept cold, of course.

The photomultiplier signal must be discriminated, to remove low amplitude pulses caused by thermionic emission in the lower dynodes, and counted. In addition, it may be necessary to correct for fluctuations in the laser intensity, unless the laser is equipped with light intensity feedback control [Engstrom, 1980].

In certain applications, with sufficiently bright fluorescence labeling, the photomultiplier may be replaced with more economical (photodiode) or more versatile (video) detectors.

D. Optical Alignment

Fluorescence photobleaching experiments are optically confocal. The laser must be made to achieve a focus, or waist, on the sample when the sample is in focus. Additionally, an aperture (FD in the figure) is positioned in an auxiliary field plane and is closed down to permit only light from the region illuminated by the laser to reach the PMT. This combination of focused illumination and apertured detection provides uncompromised rejection of out-of-focus fluorescence. This is a result of two factors: one, the excitation intensity drops off rapidly as one moves away from the laser waist; and two, the acceptance for fluorescence emitted by out-of-focus fluorophores drops off as well, since much of this light is blocked by the aperture. The resultant ellipsoid of acceptance is illustrated in Figure 3. The rejection of out-of-focus fluorescence allows the experimenter to select which membrane to study in sufficiently thick specimens, reduces background fluorescence and scatter, and allows the generalization of FPR to three dimensions.

The use of epifluorescence geometry bears some discussion. A brightly labeled cell will have a density of several thousand fluorophores per square micron of membrane area. With a tyical absorption cross section of ~ 1 Å^2, the fraction of excitation light absorbed by the sample is $\sim 10^{-5}$. The fluorescence quantum yield can easily reduce the emitted energy by another factor of ten, and the fraction of this light that will be admitted to the objective is less than 1/3. Consequently, the fluorescence signal is inevitably attenuated by more than 10^6 from the excitation; much more attenuation will occur for dimly stained cells. If the excitation beam is directed into the objective, this light must be removed from the signal. The process of removing the scattered excitation is greatly simplified by the use of epifluorescence geometry, in which the excitation light is brought through the objective by means of a dichroic mirror in the microscope nosepiece. In this arrangement, only light back-scattered off refraction index mismatches must be removed, and this is easily achieved by the dichroic mirror with an additional colloidal colored glass barrier filter (Schott Glass Co.) in the detector path. Standard, 3 mm thick long-pass filters will transmit $<10^{-5}$ of the incident light more than 20 nm below the 50% pass wavelength. Colloidal glass filters do autofluoresce; this background must be substracted from the signal.

The alignment of any focused illumination system begins with the determi-

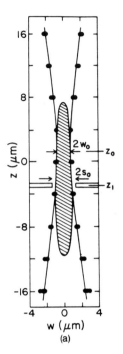

Fig. 3. Profile of focused laser illumination. The shaded ellipsoid is the region of $1/e^2$ acceptance or better, with a confocal aperture. Smaller spot sizes generate better resolution in z. (Reprinted from Schneider and Webb, [1981] with permission from the Optical Society of America.)

nation of the auxiliary field and conjugate planes in the illumination path. This is most easily achieved by back projection. A graticule is placed on the microscope stage and focused to the eyepiece(s). The objective lens used is unimportant; a low-power lens will suffice. With a dichroic mirror in position, some fraction of the broadband condenser (tungsten filament) illumination will be diverted back through the epi-illuminator. When a long focal length lens is placed in this beam, it will be brought to a focus, where a projected image of the graticule can be seen. The location of the conjugate (back) focal planes can be found from the image of the phase plate in a phase objective, or from the image of the condenser aperture when the condenser has been properly aligned for Köhler illumination. The laser must be brought to a waist in an auxiliary image plane. The final size of the focused spot is determined by the extent to which the back aperture is filled. If the beam fills very little of the back aperture, the convergence will be gentle and the focused spot relatively large. If the beam nearly fills the back aperture, convergence will be abrupt, and the spot very small. When the laser fills the back aperture, diffraction will occur and the resulting spot will lose its pure Gaussian character, tending to an Airy function in the limit of a uniformly filled back aperture.

The filled back aperture is the standard optical configuration in confocal scanning microscopy. Turn-key confocal scanning systems can be used for spot FPR experiments if provision is made to disable the scan and to shield the photomultiplier during the bleaching pulse. Although existing analysis algorithms require either Gaussian or "top hat" (disc) illumination profiles, the central maximum of the Airy pattern, which contains 84% of the radiated power, is well approximated by a Gaussian. Therefore, the analysis algorithms discussed here (section 3) can be applied to these FPR recoveries, with only a small systematic error at short times due to the high spatial frequency Airy rings at the periphery of the bleached spot.

Once coarse alignment has been achieved, the beam should be centered through the epi-illuminator. This is easily aaccomplished with a beam steering device, which provides the four degrees of freedom necessary for the arbitrary positioning (2) and pointing (2) of a beam. With the microscope nosepiece removed, the centration of the beam through variable apertures at the entrance and exit of the epi-illuminator can be seen in the projection of the beam on a wall or screen. As with all optical alignment, the farthest "downstream" element, here the top beam steering mirror, is used to align through the farthest "downstream" aperture. The upstream element is then used to center through the upstream aperture. This will de-align the downstream aperture. Repeated iteration of this procedure will converge, however.

At this point, the nosepiece is replaced. The objective lens and the stage and condenser are removed, in order that the dichroic mirror may be adjusted to project the beam along the optic axis, through the center of the substage field aperture. After replacing the objective lens, the diverging cone of light emanating from it may be centered with an adjustable lens at the epi-illuminator exit port.

A fluorescent film, such as a thin sheet of Formvar plastic with embedded fluorophore [Schneider and Webb, 1981], can be used for the final alignment. It should be positioned on the microscope stage, and the fluorescence from *attenuated* laser illumination can be viewed through the eyepiece(s). *We recommend the safer procedure of using a video camera to perform any laser alignment.* The top beam steering mirror may be used to center the spot in the field of view, while the epi-port lens will orient the diverging cone along the optic axis. (Removal of the condensor facilitates the alignment of this cone.) With any laser light source, utmost caution is advised in all stages of alignment and use. If possible, an interlock should be devised to prevent high-intensity laser illumination from entering the microscope when a light path to the eyepieces is open.

The photomultiplier photocathode should be positioned in a conjugate plane, and should overfill that plane, in order that all photons captured by the optical system strike the photocathode. In a conjugate plane, the sample is fully defocused, so that the photon flux is spread over the entire photocathode and not localized. This promotes longer tube life; in addition, the response of the system will not depend on the position of the spot in the field plane, as it would for an improperly placed photocathode.

E. Labels

While this is not intended to be a complete treatise on protein modification chemistry, some discussion of fluorescent labels is appropriate. Specifically, the reader should be aware of what properties a good fluorescence label for FPR should have. Clearly, the brightness of the label is a primary concern. Brighter labels will produce larger signals in the same biological context. Brightness consists of three factors:

• the molar absorption of each fluorophore;
• the quantum yield; and
• the number of useful fluorophores that can be attached to the protein of interest.

Molar absorption should exceed $50,000 \, M^{-1}cm^{-1}$, as it does for the popular fluorescein and rhodamine derivatives. Quantum yield should exceed 0.1. This parameter is environment sensitive, and the presence of quenching agents, such as oxygen, can dramatically reduce the quantum yield. Unfortunately, it is not possible to remove molecular oxygen from living biological specimens without compromising viability. However, in certain cases oxygen has been removed from specimens with a marked increase in quantum yield [Bloom and Webb, 1984]. The third factor, the label multiplicity, will depend on the specifics of the fluorophore and the protein carrier. With the isothiocyanates, the multiplicity will depend on the availability of primary amines, principally lysine, to undergo the coupling reaction shown in Figure 4. However, self-quenching can occur at high conjugation ratios or concentrations [Goding, 1983; Hirschfeld, 1976], so that it is sometimes beneficial to avoid pushing the reaction to saturation.

Paradoxically, a good label for photobleaching is one that resists being photobleached. During the recovery of a bleached region, an accurate estimate of the diffusion coefficient and mobile fraction can only be obtained if the illuminating source does not destroy the fluorescence signal it is to excite. At the present time, the detailed dynamics of photobleaching chemistry in biological and histological specimens is poorly understood, but under normal illumination conditions from 10^4 to 10^5 fluorescent photons may be expected before a typical fluorophore undergoes an irreversible bleaching reaction [Mathies and Stryer, 1986]. Fluorescein seems to be more susceptible to bleaching than rhodamine, which would contraindicate its use as an FPR label. However, the ability to obtain much higher conjugation ratios with fluorescein circumvents this difficulty.

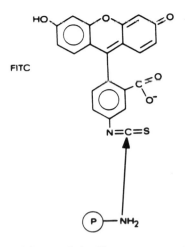

FITC

Fig. 4. Mechanism of coupling of fluorescein isothiocyanate to ϵ-amino groups of proteins. The isothiocyanate carbon undergoes nucleophilic attack by the unbonded electron pair on the amino nitrogen. A hydrogen atom is transferred from the amino nitrogen to the isothiocyanate nitrogen.

The preceding discussion of isothiocyanate-activated fluorophore applies to their conjugation to a protein, usually a ligand, antibody, or antibody fragment, which is then allowed to bind with its receptor or antigen on the cell surface. For the study of the behavior of the cell membrane itself, an excellent probe is the synthetic lipid analog 3,3,3',3' tetramethyl dioctodecylindo-carbocyanine, or diI(3)-C_{18}. This probe, like the membrane lipids themselves, has two long chain hydrocarbon tails whose hydrophobicity facilitates the incorporation of molecule into the bilayer. Studies have demonstrated a lipid like orientation for the incorporated dye [Wolf, 1985; Axelrod, 1979]. Its molar absorption is $133,000 \text{ M}^{-1}\text{cm}^{-1}$, and it is highly resistant to bleaching.

III. ANALYSIS

The time-dependent fluorescence signal during recovery is obtained from the integral of the Gaussian monitor intensity, $I(r)$, times the fluorescence concentration, $C(r,t)$:

$$F(t) = q \int I(r) \cdot C(r,t)d^2r$$

The factor q accounts for the detection efficiency of the system. If the bleach were very shallow, the depletion profile would be gaussian, and the integrand would be a simple product of gaussians. In practice, deeper bleaches are required to generate sufficient signal. When the bleach is deeper, the fluorescence pro-

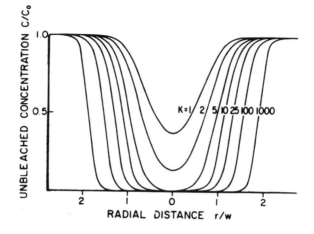

Fig. 5. Normalized post-bleach fluorophore concentration profiles for various values of the parameter K, which is proportional to the bleach pulse energy. (Reprinted from Axelrod et al. [1976] by copyright permission of The Biophysical Society.)

file becomes substantially non-Gaussian, due to local depletion of bleachable fluorophore and consequent decrease in the overall bleaching rate near the center of the gaussian illumination. The net effect is a broadening of the profile, as shown in Figure 5. In this case, a series solution to the above integral has been presented by Axelrod et al. [1976].

$$F(t) = F_0 \sum_{n=0}^{\infty} [(-K)^n/n!] [1 + n(1 + 2t/\tau D)]^{-1}$$

Here F_0 is the fluorescence intensity before bleaching, and τ_D is the characteristic diffusion time given by

$$\tau_D = w^2/4D$$

D is the lateral diffusion coefficient and w is the laser beam half-width at e^{-2} intensity. K is a measure of the amount of bleaching that has occured, and is proportional to the total energy deposited by the bleaching pulse. Yguerabide et al. [1982] have shown that the reciprocal function

$$R(t) = \frac{F_\infty}{F_\infty - F(t)}$$

where F_∞ is the fluorescence intensity as $t \to \infty$, is very nearly linear for all but the most severe bleaches. The intercept divided by the slope of $R(t)$ yields $t_{1/2}$, the half-time for fluorescence recovery. The correction for bleach depth appears in the relation between $t_{1/2}$ and τ_D:

$$t_{1/2} = \beta \tau_D$$

where β is a numerically determined factor that corrects for the broadening of the spot with increasing bleach depth. (A table of β values is compiled in the reference by Yguerabide et al. [1982].) This simplifies the fitting of FPR data.

A. Spot Size

In any diffusion process, the most important parameter is the length scale over which diffusion is occurring. Diffusion times vary as the square of the length scale, so that small errors in length determination are amplified. An accurate estimate of the size of the beam waist is therefore essential. This can be obtained by measuring the divergence of the laser from the objective, to the $1/e^2$ points. A diode that can be positioned along a line at the base of the microscope can be used after removing the stage and condenser. The diode must be rotated to remain normal to the propagating wavefronts. The divergence half angle of a propagating TEM_{00} mode is given by

$$tan\frac{\Theta}{2} = \frac{w}{z}$$

The diffraction limited spot must then have a radius of

$$w_0 = \frac{2\lambda}{\pi\Theta}$$

[Verdeyen, 1981]. It is important for this method that the laser reach a waist in the sample plane; otherwise, the effective radius will be larger than that calculated here. A more general method for measuring the diameter of the FPR spot has been developed by Schneider and Webb [1981], which is valid even in the case where the laser does not focus in the sample.

It is informative to consider the effect of flow on diffusion. Figure 6 illustrates this effect. It is clear that, without an independent estimate of the diffusion coefficient, recovery by diffusion and flow is difficult to distinguish from recovery by pure diffusion. In Figure 6, a simulation of recovery by diffusion and flow is well fitted by a diffusional recovery alone, when 3% shot noise is present (which corresponds to over 1,000 counts per time bin, a relatively high signal level). Here the diffusion and flow recoveries have the same time

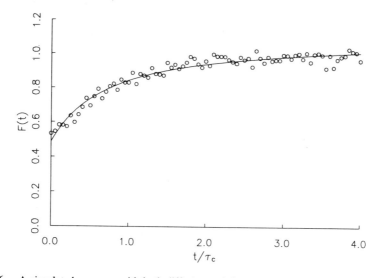

Fig. 6.   A simulated recovery with both diffusion and flow, $\tau_D = \tau_F$ in the low K limit. The line is a fit curve using diffusion alone; the χ^2 for this curve is no larger than for a similar curve with no flow.

constants. If the flow recovery is slower, even lower noise levels are required for flow detectability. For this reason, it is difficult to rigorously exclude the presence of any flow recovery in an FPR experiment. Alternative experimental geometries designed to improve flow detectability will be discussed in a later section.

It is an empirical fact that fluorescence recoveries of proteins on cell surfaces are not complete. The fluorescence never attains its prebleach level, even at very long times after the bleach. This reflects the presence of an immobile fraction of the labeled protein that is not free to diffuse on the time scale of the experiment. At the present time, the exact biochemical causes of this immobility are not understood. Experiments in our laboratory have shown dramatic increases in both D and mobile fraction on cell membrane blebs [Tank et al., 1982], suggesting that the cellular cytoskeleton, which is absent in blebs, may play a role in membrane dynamics. Recent experiments, however, have shown that association with the cytoskeleton, as observed by detergent insolubility, need not correlate with FPR immobile fraction. Certainly, there might be noncovalent protein anchorage that is susceptible to detergent lysis, and conversely, cytoskeletal elements flexible enough or motile enough to permit some lateral diffusibility of any attached membrane proteins.

Even if the causes of immobilization are not understood, at least one important consequence is clear. Reactions involving immobilized species will have reduced rates, especially in cases of cross-linking. Cells may manipulate the mobility of their membrane receptor systems to regulate their sensitivity.

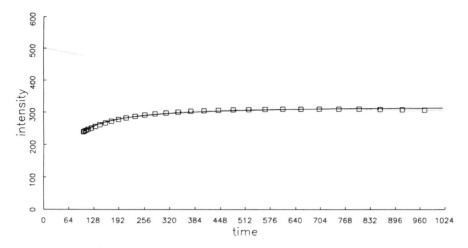

Fig. 7.   Effect of monitor beam bleaching. Simulated recovery with 900 fluorophores/μm, 10^4 fluorescent photons per bleaching event, and a 2% detection efficiency. The boxes are averages of nearby points; this local averaging is performed to reduce fitting computation. The line is the best fit recovery. It yields a value of D that is twice the actual value, and an immobile fraction of 66% rather than 50%.

B. Monitor Beam Effects

In spite of the fact that the monitor intensity is several orders of magnitude dimmer than the bleach, it is still capable of fluorescence destruction. This process will occur continuously throughout the duration of the recovery. For weakly fluorescent samples, the problem becomes limiting: the monitoring laser intensity cannot be increased without so severely altering the recovery shape that proper fitting is no longer possible. Although an analytical solution to the bleaching problem is not possible, numerical simulations serve to illustrate its characteristics. In Figure 7, a simulated recovery is shown for a bleach-sensitive fluorophore, such as fluorescein, taking into account the overall detection efficiency, estimated at about 2%. This curve, when fitted with a standard recovery, generates radically erroneous values of D and immobile fraction. The fitted value of D is nearly twice the true value, while the fitted immobile fraction is 66% rather than the true value of 50%. Increasing the ratio of bleach to monitor intensities will not help, since decreasing the monitor will reduce the signal size and increase the shot noise. (Increasing the bleach intensity can have no effect during the recovery.) A higher collection efficiency will help somewhat, and can be maximized by using high NA, non-phase objectives. Limiting exposure to the monitor beam, for example, by taking fewer data points at long times and shuttering the monitor beam, will

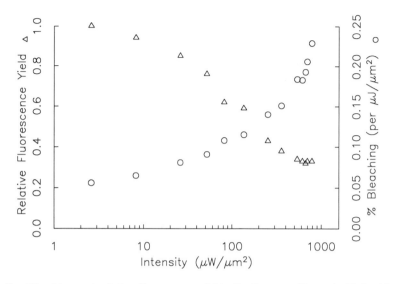

Fig. 8. Bleaching and relative fluorescence yield of a Formvar film embedded with the fluorophore diI-C_{16}, using an expanded laser beam.

also help. An exponential acquisition schedule, in addition to reducing sample exposure at long times, can also provide greater precision in the measurement of a larger range of diffusion coefficients [Donaldson, 1989]. Finally, any procedure that reduces the fluorophore bleachability should be employed, if it is compatible with the biological requirements of the system under study.

C. Photochemistry of Photobleaching

In fact, FPR scans are often taken at signal levels comparable with those used to generate Figure 8. These scans often show little qualitative evidence of monitor bleaching, such as the sharp decrease in the pre-bleach monitor level and the overturn at long times. They are well fit by zero monitor bleach theory.

Implicit in Figure 7 is the assumption that bleaching rates are proportional to intensity, but there is no a priori reason to assume that this is always the case. We have found, for example, that both the quantum efficiency for bleaching and the quantum yield from a fluorescent film vary with the incident intensity (Fig. 8). This figure shows that photodestruction is favored at high intensities, while photoemission is enhanced in dim light. Photodestruction and other excited state reactions may be autocatalytic; e.g., photoproducts, such as free radicals, may catalyze these reactions. These reactions will compete with singlet fluorescence emission, so that as their rates increase, a decrease in fluorescence emission is expected. Although these non-linearities will slightly

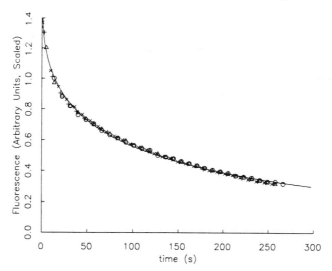

Fig. 9. Fluorescence as a function of time for rat basophilic leukemia cells labeled with rhodamine-immunoglobulin E, illuminated with an expanded 10 mW laser beam. The traces from four different cells were scaled in intensity and superimposed. The line is a stretched exponential curve, with a time constant τ of 95 s, and stretch factor β of 0.43.

diminish the size of a bleached hole (depending on bleaching intensities, bleach times, fluor sensitivities, and perhaps other variables as well), the overall chemistry is especially favorable for fluorescence recovery experiments, since bleaching by the dim monitor beam is suppressed.

Bleachability will depend on environment, as does quantum yield. This is suggested by the non-exponential time course of the bleaching of fluorecently labeled cells under continuous illumination, Figure 9. The fluorescence intensity decays not as a single exponential, but is well fit by a stretched exponential,

$$I = I_0 e^{-(t/\tau)^\beta}$$

The same *qualitative* fading kinetics (i.e., a faster initial fluorescence decay and a slow late decay) have been observed with other dyes in other systems [Johnson et al., 1982; Enerback and Johansson, 1973]. Quantitatively, a stretched exponential time dependence can arise when a broad spectrum of parallel exponential processes contribute to the overall reaction, or when relaxation (bleaching) depends on the diffusion of randomly placed "defects" (perhaps triplet oxygen) to the fluorophore [Klafter and Shlesinger, 1986]. It is likely that fluorophores on a cell surface have a range of susceptibilities to bleaching, perhaps depending on their accessibility to the solvent and solvent-borne bleaching catalysts. However, there is no evidence at this time to suggest that the bleachability of a fluorophore is correlated in any way with the dynamics of the protein to which it is bound.

IV. ALTERNATIVE GEOMETRIES

The complexity of cell surface behavior, and the difficulty in identifying and quantifying non-diffusive recovery with single spot FPR, has lead many researchers to develop alternative strategies. In general, these geometries abandon the confocality of spot FPR in order to sample a larger area of the cell surface. The most direct extension of conventional FPR is to replace the aperture and PMT with a low-light-level video camera, using a mercury arc lamp to give uniform field illumination during the recovery phase. This procedure has been implemented by Kapitza et al. [1985]. The analysis of the resultant sequence of video images is complex. The basic strategy is to assume specific values for the transport parameters: D_x, D_y, v_x, v_y. The first post bleach image is then time evolved by fourier transform techniques. Subsequent test images are correlated with data images, and the parameter values are adjusted in an iterative cycle. While this is computationally intensive, it is a fully general procedure, so that any combination of recovery mechanisms may be elucidated.

In our laboratory, we have chosen to use the interference pattern of two intersecting laser beams to create a spatially periodic bleaching pattern for video analysis. Fringe pattern systems have been used by many researchers, including Smith and McConnell [1978] and Davoust et al. [1982]. The analysis of a pattern recovery is mathematically simple. Starting from the diffusion equation,

$$\frac{\partial C}{\partial t} = D \frac{\partial^2 C}{\partial x^2}$$

We write the Fourier transform of C:

$$C(x,t) = \frac{1}{2\pi} \int e^{-ikx} \tilde{C}(k,t) dk$$

Taking derivatives and substituting

$$\int e^{-ikx} \frac{\partial \tilde{C}(k,t)}{\partial t} dk = -D \int k^2 e^{-ikx} \tilde{C}(k,t) dk$$

Consequently

$$\frac{\partial \tilde{C}(k,t)}{\partial t} = -Dk^2\tilde{C}(k,t)$$

$$\tilde{C}(k,t) = \tilde{C}_0(k,t)e^{-Dk^2 t}$$

The bleaching of a spatially periodic pattern creates a large Fourier peak at one spatial frequency. The analysis simply involves monitoring the amplitude and phase of that single Fourier component. Changes in the phase during recovery are the signature of a flow process. Note, however, that only the projection of the flow velocity perpendicular to the pattern lines is observable by this technique.

Our beam splitting optics for generating interference fringes is illustrated in Figure 10. A movable silvered cube, M1, allows the beam separation and hence the pattern spacing to be easily adjusted, with minor corrections required to compensate for the spherical aberration in lens L1.

Any video technique is rate limited by the video signal standard—30 frames/s. Davoust et al. [1982] use a photomultiplier tube to increase accessible rates. A two-beam fringe pattern is used to bleach the cell surface fluorescence. The pattern is attenuated and spatially shifted back and forth during the recovery, by means of a vibrating mirror in the path of one beam. The fluorescence

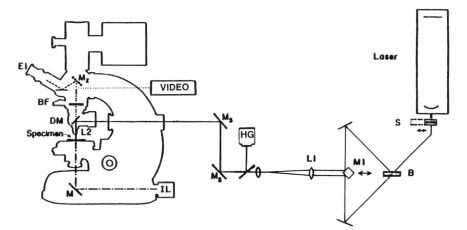

Fig. 10. Two beam interference pattern optics. Beam splitter B creates two beams, recombined by movable mirrored cube M1. Lens L1 crosses these beams in a a field plane. Moving M1 adjusts the beam separation, and therefore the interference pattern spacing. Mercury arc illumination (HG) is brought into the optical path by a mirror in a back focal plane with a small slot cut to permit the laser beams to pass. Video observation of recoveries is effected through a custom port in the microscope. Shutter (S) controls bleaching. Other components are described in Figure 1.

signal is detected with a photomultiplier; the signal varies in time as the illumination stripes alternately align with the bleached and then the unbleached regions of the cell. The amplitude of this signal is found with a lock-in amplifier, gated to the vibrating mirror. The amplitude decays as the bleached pattern diffuses away. Furthermore, the presence of flow would shift the phase of the bleached regions relative to the monitor, generating a second harmonic which can also be detected with a lock-in. In this scheme, the inability of the photomultiplier to image different regions of the cell is overcome by illuminating different regions at different times, much in the same way that a scanning confocal microscope acquires spatial information. In FPR, however, only two illumination states must be sampled to detect a diffusibility and a flow, so the system can be pushed to much higher frequencies.

All three techniques discussed offer improved detectability for non-diffusive recoveries. Nonetheless, such recovery mechanisms must be large scale, on a cellular level, to be observable. In addition, these schemes sacrifice confocality, and consequently z resolution, in order to gather the global information necessary to resolve flow. Finally, video-based systems are rate limited by the video format to 30 frames/s. As a simple biophysical diagnostic, confocal photobleaching should not be overlooked.

V. CONCLUSIONS

In the last decade, FPR has proven to be an invaluable biophysical tool for the study of cell surface dynamics. Many questions remain to be answered, however. The slow diffusion of typical surface proteins remains refractory. In specific cases, diffusional effects have been seen from cytoskeletal interactions; in another, removal of a protein's cytoplasmic tail seems to have no effect. Some enhancement of diffusion has been reported on reduction of glycosylation, but not enough to account fully for deviations from ideality. Nonspecific interactions have been shown to retard diffusion in crowded systems; it remains to be seen whether such effects can be observed on the cell surface. It is likely that many effects contribute to the observed non-ideal behavior of proteins on cell surfaces. Perhaps the most fruitful approach will be to elucidate fully the role of diffusion in specific cell surface protein systems, to gain an understanding of both the effects of slow diffusion and cellular control of it.

A prevalent observation in FPR on biological specimens is the presence of excess noise, occasionally in individual FPR scans, but ubiquitously from scan to scan. Diffusion coefficients and immobile fractions vary enormously from cell to cell in a homogeneous population, and even from region to region on a single cell surface. Uncertainty in diffusion coefficients is typically reported as a standard error of the mean, which, although appropriate as a statistical measure of the certainty of the mean, tends to obscure the population variability. This variability can be an order of magnitude in D and 30% in immobile fraction. We have as yet no understanding for these variations, which are present on otherwise homogeneous cell surfaces, such as cultured fibroblasts. It is reasonable to expect that the cell membrane has as yet uncharacterized and unappreciated diversity, and that this diversity has functional correlates. It is also reasonable to expect that, as FPR is integrated into the arsenal of tools available to the cell biologist, we will understand this diversity.

ACKNOWLEDGMENTS

This work was supported by the Office of Naval Research, grants ONR-N00014-84-K-0390 and ONR-N00014-89-J-1656, and by the National Science Foundation, grants DIR 8800278 and DMB 86-09084.

VI. REFERENCES

Axelrod DD, Koppel DE, Schlessinger J, Elson E, Webb WW (1976) Motility measurement by analysis of fluorescence photobleaching recovery kinetics. Biophys J 16:1055–1068.

Axelrod D (1979) Carbocyanine dye orientation in red cell membrane studied by microscopic fluorescence polarization. Biophys J 26:557–574.

Axelrod D (1983) Lateral motion of membrane proteins and biological function. J Membr Biol 75:1–10.

Bergman JE, Kupfer A, Singer SJ (1983) Membrane insertion at the leading edge of motile fibroblasts. Proc Natl Acad Sci USA 80:1367–1371.

Bloom JA, Webb WW (1984) Photodamage to intact erythrocyte membranes at high laser intensities: Methods of assay and suppression. J Histochem Cytochem 32(6):608–616.

Bretscher MS (1984) Endocytosis: Relation to capping and cell locomotion. Science 224:681–686.

Cherry RJ (1979) Rotational and lateral diffusion of membrane proteins. Biochim Biophys Acta 559:289–327.

Davoust J, Devaux PF, Leger L (1982) Fringe pattern photobleaching: A new method for the measurement of transport coefficients of biological macromolecules. EMBO Journal 1(10):1233–1238.

Donaldson P (1989) Modulation of Lateral Diffusion on Lipid Bilayer Membranes. Ph.D. Thesis, Cornell University.

Edidin M (1987) Rotational and lateral diffusion of membrane proteins and lipids: Phenomena and function. Curr Top Membr Transport 29:91–127.

Elson EL (1985) Fluorescence correlation spectroscopy and photobleaching recovery. Annu Rev Phys Chem 36:379–406.

Enerback L, Johansson K-A (1973) Fluorescence fading in quantitative microscopy: A cytofluorometer for the automatic recording of fluorescence peaks of very short duration. Histochem J 5:351–362.

Engstrom RW (1980) RCA Photomultiplier Handbook. Lancaster, PA: Burle Industries, pp 3–10.

Ghosh RN, Webb WW (1989) Automated tracking of LDL receptors on cell surfaces with nanometer precision. Biophys J 55(2,2):498a (abstract).

Goding JW (1983) Monoclonal Antibodies: Principles and Practice. Orlando, FL: Academic Press, pp 208–247.

152   Thomas and Webb

Hirschfeld T (1976) Quantum efficiency independence of the time-integrated emission from a fluorescent molecule. Appl Opt 15(12):3135–3139.

Jacobson K, Elson E, Koppel D, Webb W (1982) Fluorescence photobleaching in cell biology. Nature (Lond) 295:283.

Johnson GD, Davidson RS, McNamee KC, Russell G, Goodwin D, Holborow EL (1982) Fading of immunofluorescence during microscopy: A study of the phenomenon and its remedy. J Immunol Methods 55:231–242.

Jovin TM, Vaz WLC (1989) Rotational and translational diffusion in membranes measured by fluorescence and phosphorescence methods. Meth Enzymology 172:471–513.

Kapitza HG, McGregor G, Jacobson KA (1985) Direct measurement of lateral transport in membranes by using time-resolved spatial photometry. Proc Natl Acad Sci USA 82:4122–4126.

Kapitza HG, Jacobson K (1986) Lateral motion of membrane proteins. In Ragan CI, Cherry RJ (eds): Techniques for the Analysis of Membrane Proteins. London: Chapman and Hall.

Klafter J, Schlesinger MF (1986) On the relationship among three theories of relaxation in disordered systems. Proc Natl Acad Sci USA 83:848–851.

Mathies RA, Stryer L (1986) Single-molecule fluorescence detection: A feasibility study using phycoerythrin. In Taylor DL, Waggoner AS, Murphy RF, Lanni F, Birge RR (eds): Applications of Fluorescence in the Biomedical Sciences.'' New York: Alan R. Liss, Inc., pp 129–140.

McCloskey M, Poo M-m (1984) Protein diffusion in cell membranes: Some biological implications. Int Rev Cytol 87:19–81.

Poo M-m (1981) In-situ electrophoresis of membrane components. Annu Rev Biophys Bioeng 10:245–276.

Saffman PG, Delbrück M (1975) Brownian motion in biological membranes. Proc Natl Acad Sci USA 72:3111–3113.

Schneider MB, Webb WW (1981) Measurement of submicron laser beam radii. Appl Opt 20:1382–1388.

Smith BA, McConnell HM (1978) Determination of molecular motion in membranes using periodic pattern photobleaching. Proc Natl Acad Sci USA 75:2759–2763.

Tank DW, Wu E-S, Webb WW (1982) Enhanced molecular diffusibility in muscle blebs: Release of lateral constraints. J Cell Biol 92:207–212.

Verdeyen JT (1981) Laser Electronics. Englewood Cliffs, NJ: Prentice-Hall, Inc., pp 53–64.

Wolf DE (1985) Determination of the sidedness of carbocyanine dye labeling of membranes. Biochemistry 24:582–586.

Yguerabide J, Schmidt JA, Yguerabide EE (1982) Lateral mobility in membranes as detected by fluorescence recovery after photobleaching. Biophys J 39:69–75.

Noninvasive Techniques in Cell Biology: 153–176
© 1990 Wiley-Liss, Inc.

7. Use of Fluorescence Microscopy in the Study of Receptor-Mediated Endocytosis

Kenneth W. Dunn and Frederick R. Maxfield

Departments of Pathology and Physiology, Columbia University, New York, New York 10032

I. INTRODUCTION

Fluorescence microscopy has proven to be a powerful tool in the study of cell biology. By taking advantage of the specific fluorescent properties of fluorophores and the sensitivity and specificity of various probes, including antibodies and analogs of proteins or lipids, researchers have been able to characterize the distributions of particular molecules and structures in tissues and cells [see review, Osborne and Weber, 1982]. Additionally, the sensitivity of some fluorophores to their ionic environment has allowed subcellular characterizations of physiologically important parameters such as pH and calcium concentration [see review, Maxfield, 1989a]. Once limited mainly to staining fixed preparations, fluorescent probes are increasingly used to characterize dynamic processes in living cells.

Fluorescence microscopy is particularly valuable for the study of endocytosis. Since endocytosis involves internalization of extracellular macromolecules and fluid, live cells may be labeled utilizing the cell's own endocytic machinery. The closed topology of endocytic compartments ensures very specific labeling of the structures of the endocytic pathway with membrane-impermeant probes. Since different ligands follow different endocytic pathways, specific branches of the pathway can be labeled with fluorophore-ligand conjugates. Fluorescence microscopy can be used to provide an intensive, detailed view of individual cells labeled in this way. Alternatively, large numbers of fluorescently labeled cells may be analyzed by means of fluorescence-activated cell sorting [e.g., Murphy et al., 1984; Sipe and Murphy, 1987] or through spectrofluorometry [Ohkuma and Poole, 1978; Salzman and Maxfield, 1989].

From its beginnings as a descriptive tool in support of biochemical data, fluorescence microscopy has developed into a quantitative technique which has enhanced our understanding of endocytosis. In studies of endocytosis conducted in our laboratory, we have used fluorescence microscopy to characterize the kinetics of endocytic routing in several cells lines, to measure the pH of particular endocytic compartments in normal and mutant cell lines, to quantify fusion between endocytic vesicles, and to characterize the process of endocytic sorting. In this review we will discuss these and other examples of applications of fluorescence microscopy to the study of receptor-mediated endocytosis. Discussions of general aspects of fluorescence microscopy can be found in other chapters of this volume and in two recent volumes [Wang and Taylor, 1989; Taylor and Wang, 1989].

II. SUMMARY OF THE ENDOCYTIC PATHWAY

Cells internalize a variety of nutrients, growth factors, toxins, and viruses through the process of receptor-mediated endocytosis [see reviews, Steinman et al., 1983; Goldstein et al., 1985; Pastan and Willingham, 1985; Steer and Hanover, 1990], depicted schematically in Figure 1. Receptors specific for particular ligands migrate freely over the surface of the cell until encountering a clathrin-coated pit where they may become trapped. Some, for example, low-density lipoprotein (LDL), are partially concentrated in coated pits prior to binding ligand [Anderson et al., 1982]. For others, such as epidermal growth factor (EGF), occupancy triggers entry into coated pits [Schlessinger, 1980]. Various receptors and receptor-ligand complexes accumulate in common coated pits which rapidly invaginate and pinch off at the plasma membrane to form vesicles. These early endocytic vesicles fuse with one another to form an endosome involved in the process of sorting ligands and receptors. This sorting endosome contains both ligands destined for lysosomes, such as LDL and the serum protease inhibitor α-2-macroglobulin (α_2M), and ligands to be recycled

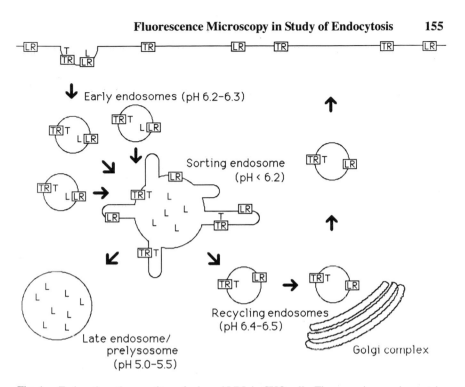

Fig. 1. Endocytic pathways of transferrin and LDL in CHO cells. The serum iron carrier protein transferrin (T) and the cholesterol carrier protein, LDL (L) each binds to a specific receptor on the cell surface. These receptors cluster in clathrin-coated pits which invaginate to form endocytic vesicles. Endocytic vesicles fuse to form sorting endosomes. Endocytic compartments become progressively more acidic with time, which causes the dissociation of certain receptors from their ligands and the release of iron from transferrin. Unlike LDL, transferrin remains bound to its receptor in endosomes. Both lysosomally directed ligands (LDL) and recycling ligands (transferrin) and receptors (LR, TR) are found in the same sorting endosomes, but at this point part company. Lysosomally directed ligands are next found in prelysosomes and later in lysosomes. Recycling ligands and receptors are next found in recycling endosomes from which they return to the cell surface. There is no standard nomenclature for the different organelles involved in endocytosis. The terminology used in this figure is consistent with previous work from this laboratory [e.g., Salzman and Maxfield, 1989].

to the cell surface, such as transferrin. In the sorting endosome lysosomally directed ligands are segregated from those to be recycled back to the cell surface. Transferrin and recycling receptors are directed from the sorting endosome to recycling endosomes whereas LDL and $\alpha_2 M$ are next found in late endosomes and ultimately in lysosomes where they are degraded. Compartments similar to the "sorting endosome" have been described in various models of endocytosis as "CURL" [Geuze et al., 1983], "early endosomes" [Schmid et al., 1988], and "receptosomes" [Pastan and Willingham, 1983]. There is presently no standard nomenclature for the different organelles involved in endocytosis. The terminology used here is consistent with previous work from this laboratory [e.g., Salzman and Maxfield, 1989].

Chinese hamster ovary (CHO) cells and TRVb-1 cells (a transfectant CHO cell line expressing large numbers of the human transferrin receptor, but none of its own [McGraw et al., 1987]) are especially suitable for fluorescent studies of endocytic routing because the recycling and lysosomal pathways are morphologically distinguishable from one another. Late endosomes and lysosomes are distributed in a punctate pattern, but recycling endosomes largely accumulate in a single region near the Golgi apparatus [Yamashiro et al., 1984].

While LDL and α_2M are routed to lysosomes, their receptors are recycled by the same pathway taken by transferrin. The mechanism by which receptors and ligands are separated from one another is not fully understood, but involves a pH-dependent dissociation of receptor and ligand mediated by the acidification of endosomes [Tycko et al., 1983; Brown et al., 1983; Harford et al., 1983a,b]. The binding of iron to transferrin is likewise pH sensitive, and the release of iron from transferrin occurs in an acidic endosome [Dautry-Varsat et al., 1983; Klausner et al., 1983].

III. TYPES OF FLUORESCENT PROBES USED TO STUDY THE ENDOCYTIC PATHWAY

A. Fluorescent Ligand Conjugates

The process of endocytosis is particularly amenable to fluorescent microscopic analysis because, rather than depending upon delicate fixations and permeabilizations, particular endocytic pathways of live cells can be labeled with ligand-fluorophore conjugates using the cell's own endocytic apparatus.

In order to represent accurately the normal endocytic pathway taken by a ligand, ligand conjugates must be carefully prepared. A complete discussion of fluorescent conjugate preparation is given by Maxfield [1989b] and by Haugland and Minta (this volume). Inadequate labeling may result from an underderivatized conjugate, but overderivatization may reduce ligand activity and/or specificity. New conjugate preparations must be assayed for specificity and to ensure that the behavior of the fluorescent analog closely mimics the native molecule. The most effective method for testing specificity is by incubating cells in solutions containing both the new conjugate and excess unlabeled ligand. If the ligand conjugate is internalized by cells as a result of binding to specific, saturable receptors, the presence of excess unlabeled ligand will reduce the uptake of the fluorescent conjugate. If the new conjugate is internalized through a receptor-independent mechanism, the presence of competitor will have no effect on its incorporation by cells. Ideally, the competition should be checked quantitatively and should be similar to competition of specific radiolabeled probes.

Rhodamine and fluorescein are the fluorophores most commonly conjugated to ligands in the study of endocytosis. When illuminated with a mercury arc lamp, rhodamine will generally produce a stronger signal than fluorescein because of the match between rhodamine's absorption spectrum and the mercury lamp's strong 546 nm transmission line. In contrast, rhodamine is relatively poorly stimulated by the emission of the argon laser frequently used in confocal microscopy, while fluorescein's absorption is well matched to the 488 nm emission line.

Although greater care must be taken during the preparation of rhodamine conjugates to remove unconjugated rhodamine, perhaps due to its greater hydrophobicity, rhodamine has several advantages over fluorescein. First, rhodamine fluorescence is much less susceptible to photobleaching, the attenuation of fluorescence due to photo-oxidation. Photo-oxidation not only reduces the strength of fluorescence, but also produces toxic compounds that may compromise studies of living cells. Second, rhodamine fluorescence is less likely to be obscured by autofluorescence, the intrinsic cellular fluorescence induced most strongly by wavelengths of light under 500 nm [see review, Aubin, 1979]. In most cells, the 530-560 nm illumination used to excite rhodamine fluorescence stimulates far less autofluorescence than does the 450-490 nm light typically used to stimulate fluorescein. Finally, in studies of live cells the low pH of endosomes substantially attenuates the fluorescence of fluorescein but not rhodamine. This effect can be obviated by treating labelled cells with a weak base, such as methylamine, which dissipates proton gradients across biological membranes and thus raises the pH of endosomes to near neutrality [e.g., Yamashiro and Maxfield, 1987a,b; see Maxfield, 1985, for discussion]. However, dissipating the pH gradients of every intracellular organelle will have far-ranging effects on cell physiology, which must be acknowledged when interpreting such studies. The utility of the pH sensitivity of fluorescein in measuring endosomal pH will be discussed later.

The excitation and emission spectra of rhodamine and fluorescein are reasonably well spaced so that with the appropriate filter sets the two can be used together to characterize the distribution of two ligands in the same cells. Rhodamine is slightly fluorescent when illuminated and observed with conventional fluorescein optics lacking a band-pass filter on the emission light path. This signal contamination can be largely avoided through the use of narrower band-pass emission filters, or by exciting fluorescein with light of shorter wavelengths. Each solution imposes an appreciable attenuation in signal strength. Fluorescein fluorescence is only slightly stimulated by the longer wavelengths of light used to excite rhodamine.

Some alternative dyes offer better resolution in multiple color applications. Both Texas Red [Swanson, 1989] and rhodamine X [Roederer et al., 1987] have longer wavelength excitation and emission than rhodamine, permitting better distinction from fluorescein fluorescence. Because of this better separation from fluorescein, Texas Red is recommended for use with fluorescein in dual-label confocal microscopy, despite being poorly excited by 514 nm argon laser emission. Bodipy, a new fluorophore (Molecular Probes, Eugene, OR), has fluorescence properties similar to fluorescein but its narrow emission spectrum provides less overlap with rhodamine. In addition, Bodipy fluorescence is pH insensitive and less susceptible to photobleaching. There is still strong need for additional fluorescent probes. These include photostable fluorophores with spectroscopic properties well matched to light sources and detectors, but well resolved from existing probes. Additional requirements include ease of conjugation and relatively minor changes in the overall properties of the protein (e.g., charge, hydrophobicity, etc.). In addition, probes with selective sensitivity to endosome contents (e.g., pH, [Ca], enzymes) are needed. The reader is referred to the chapter by Haugland and Minta (this volume) for a more complete survey of available fluorophores.

Rhodamine, fluorescein, and Texas red conjugates are coupled to proteins through a covalent bond formed through a reactive residue on the fluorophore. The large lipid core of LDL provides another means of fluorescent labeling; the lipid core can be extensively loaded with fluorescent lipids to produce a very brightly fluorescent LDL probe. The intense fluorescence of two such probes, diI-LDL and diO-LDL (LDL conjugated to 3,3′-dioctadecylindocarbocyanine and 3,3′-dioctadecyloxacarbocyanine, respectively), has permitted the characterization of early events in endocytosis [Dunn et al., 1990, described later] and even the tracking of single molecules of LDL on the surface of cells [Barak and Webb, 1981]. Pitas et al. [1981] describe an effective, gentle method for the preparation of diI-LDL which has also been used to prepare diO-LDL [Tabas et al., submitted]. The fluorescence spectra of diI and diO are sufficiently similar to rhodamine and fluorescein, respectively, that no special optics are required for their use. For single labeling studies, diI is preferable to diO as it is more resistant to photobleaching, more easily distinguished from autofluorescence, and its absorption better matched to the output of a mercury lamp.

In practice, cells are labeled with fluorescent conjugates of particular ligands by incubating them in medium containing a concentration of conjugate sufficient to saturate partially the receptor population. Excessively high ligand concentrations should be avoided in order to ensure that uptake is only, or at least primarily, receptor-mediated. If serum contains homologues of the ligand being studied it should be omitted from labeling media. Nonspecific binding of ligand to cell surfaces can be reduced by incubating in 1 to 2 mg/ml of carrier protein, such as albumin (which, depending upon its origin and purity may also contain interfering ligand homologues). At the end of the labeling interval, cells are extensively rinsed in unlabeled medium and then either fixed or observed while alive.

When cells are incubated with ligand-fluorophore conjugates, surface-bound label may obscure internalized label. The pH dependence of receptor binding found for some ligands can be exploited by the experimenter to avoid this problem. By treating cells with acidic solutions of non-permeant buffers (pH 5.5 or below), certain surface-bound ligand-fluorophore conjugates can be removed to permit better resolution of the distribution of internalized conjugates. For example, transferrin bound to the surface of cells may be removed by exposing labeled cells to a 50 mM solution of 2-(n-morpholino) ethanesulfonic acid (pH 5.0), followed by a neutral rinse [Salzman and Maxfield, 1988]. Surface-bound fluorophore fluorescence can also be reduced by the addition of certain membrane-impermeant quenchers [Chalpin and Kleinfeld, 1983; Wolf, 1985].

The sequential passage of ligands through the various endocytic organelles can be exploited to label selectively certain compartments by providing label as a pulse and then following the progress of the pulse through the various endocytic compartments. Because the passage between certain compartments is temperature dependent, label can also be made to accumulate at particular points in the endocytic pathway by manipulating temperature.

Low-temperature binding is commonly used to increase the amount of label internalized in a pulse-chase experiment. At 4°C cells are capable of binding extracellular ligands (although perhaps with different affinity from that at 37°C), but internalization is prevented. In this way the population of surface receptors may be saturated with ligand conjugate during protracted 4°C incubations. Upon warming to 37°C, the goal is to have endocytosis proceed with probe entering the pathway as a more concentrated pulse than would be possible with a short term exposure at physiological temperature. Endocytosis appears to proceed normally once cells are warmed to 37°C, but the side effects of this temperature treatment have not been well characterized. Rather than synchronizing the delivery of probe through the sequence of compartments, low-temperature treatment may actually blur the compartment boundaries since the endocytic machinery slowly recuperates from the low-temperature treatment [Weigel and Oka, 1983; Willingham and Pastan, 1985].

Incubating cells at 18°C can be used to prevent delivery of ligands to lysosomes. Dunn et al. [1980] showed that rat liver cells incubated at 18°C were capable of internalizing asialoglycoprotein, but that delivery of ligand to lysosomes was blocked. Wolkoff et al. [1984] and Mueller and Hubbard [1986] showed that in rat liver cells incubated at 16°–18°C not only is delivery of asialoglycoprotein to lysosomes prevented, but segregation of receptor from ligand is also blocked. In one example of how this low-temperature manipulation can be used, Salzman and Maxfield [1989] incubated cells at 18°C to

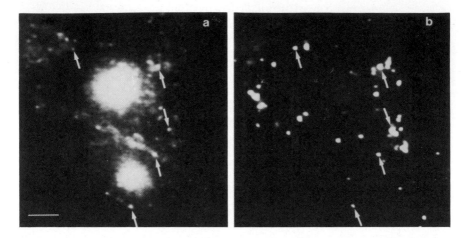

Fig. 2. Labeling of the endocytic pathway through long-term (10 min) incubations with transferrin and LDL. Long-term incubations result in the labeling of many endocytic compartments. By comparing the rhodamine-transferrin image of **panel a** with the diO-LDL image of **panel b**, one can discern several different types of compartments. Endosomes containing both labels (indicated by arrows) can be identified as either the early endosomes or sorting endosomes of Figure 1. Late endosomes and lysosomes are distinguished by the presence of LDL, but not transferrin. Recycling endosomes, containing transferrin and little, if any, detectible LDL densely aggregate in the para-Golgi region. TRVb-1 cells were incubated in 20 μg/ml rhodamine-transferrin and 5 μg/ml diO-LDL in medium 1 (150 mM NaCl, 20 mM HEPES, pH 7.4, 1 mM $CaCl_2$, 5 mM KCl, 1 mM $MgCl_2$) with 2 mg/ml ovalbumin for 10 min, rinsed, and fixed. DiO-LDL was visualized using fluorescein optics. Bar is 5 μm in length.

demonstrate the existence of functionally distinct late endosomes which are no longer competent to fuse with early endosomes but have not yet acquired lysosomal hydrolase activity.

Incubating cells with fluorescent ligands for prolonged periods results in labeling of an entire endocytic pathway. Figure 2 shows examples of TRVb-1 cells incubated with rhodamine-transferrin and diO-LDL to label the recycling and lysosomal endocytic pathways, respectively. Sorting endosomes can be identified by the presence of both recycling rhodamine-transferrin and lysosomally destined diO-LDL. Recycling endosomes, aggregated in the para-Golgi region of CHO cells, are prominently labeled with rhodamine-transferrin, but are not detectibly labeled with diO-LDL. Other punctate, widely distributed organelles are labeled with diO-LDL, but not rhodamine-transferrin. These are late endosomes and lysosomes.

By incubating cells with label for a short period of time and then continuing the incubation for various periods in the absence of label (pulse-chase), one can fairly selectively label particular compartments. TRVb-1 cells shown in Figure 3 were labeled with both rhodamine-transferrin and diO-LDL for 2 min at 37°C and chased for various periods of time. In cells fixed immedi-

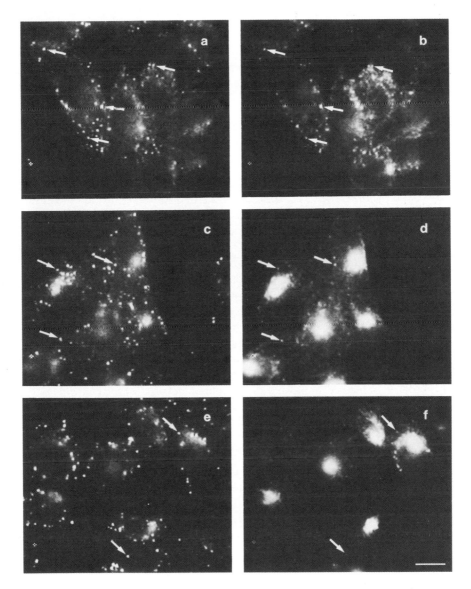

Fig. 3. Labeling of particular endocytic compartments through pulse-chase incubations. TRVb-1 cells were incubated in 20 μg/ml rhodamine-transferrin and 5 μg/ml diO-LDL in medium 1 with ovalbumin for 2 min, rinsed, and either fixed immediately (**panels a** and **b**), or incubated in unlabeled medium for another 2 min (**panels c** and **d**) or for another 4 min (**panels e** and **f**) prior to fixation. It can be seen that rhodamine-transferrin (shown in panels b, d, and f) moves from the sorting endosomes it initially shares with diO-LDL (shown in panels a, c, and e) into the para-Golgi aggregation of recycling endosomes. Initially nearly all rhodamine-transferrin and diO-LDL is localized in common early endosomes and sorting endosomes, but by 4 min of chase almost all of the cell-associated transferrin is localized in the para-Golgi region. Bar is 10 μm in length. (From Dunn et al. [1990]; reprinted with permission.)

ately after the labeling period, rhodamine-transferrin and diO-LDL appear in the same small, punctate, widely distributed early and sorting endosomes. With a 2 min chase, transferrin is less apparent in these endosomes and has begun to appear in the post-sorting recycling endosomes. With a 4 min chase, nearly all internalized transferrin has passed through the punctate sorting endosomes and into the recycling endosomes. Throughout this period diO-LDL remains in the punctate endosomes characteristic of the lysosomal pathway.

As an example of a more complex variation of this strategy we used a pulse-chase-pulse protocol to demonstrate that the LDL-containing compartment is accessible to subsequently endocytosed transferrin [Dunn et al., 1990]. Cells were incubated with diO-LDL, chased for 2 min in unlabeled medium, and then incubated for another 2 min in rhodamine-transferrin. We found that nearly all of the diO-LDL containing endosomes also became labeled with rhodamine-transferrin.

It should be noted that the studies discussed above in which the sorting endosome has been identified and the flux of ligands through the sorting endosome described have only recently become technically feasible. Compartments that could only be poorly labeled with transferrin in normal CHO cells have been satisfactorily labeled using TRVb-1 cells, a transfectant CHO cell line in which the endogenous hamster transferrin receptor has been replaced with the human transferrin receptor [McGraw et al., 1987]. Labeling is enhanced both because of the larger number of receptors expressed in TRVb-1 cells and also because of the higher affinity of the human receptor for the commercially available human form of transferrin used. The intense fluorescence of diI and diO has facilitated the detection of LDL in very early endosomes. Finally, improvements in the optics and detectors used in fluorescence microscopy have allowed satisfactory imaging of even weakly fluorescent structures. This example illustrates how advances in genetic engineering of cell lines, probe chemistry and electronic imaging have dramatically improved the power of this approach over the past few years. Further improvements can be expected in the near future.

B. Fluorescent Fluid Phase Labels

The lysosomal pathway can be labeled with lysosomally destined ligand conjugates such as diO-LDL (Figs. 3,4) or fluorescein-α_2M. In some cell types, this pathway can also be effectively labeled with fluid phase probes which are internalized in a non-selective bulk fashion. Highly polar tracers such as lucifer yellow or high molecular weight tracers such as fluorescent conjugates of dextran are most commonly used [see review, Swanson, 1989]. Since ligands become concentrated during receptor-mediated endocytosis, ligand conjugates will yield more intense labelling of early endocytic structures. However, fluid phase probes can be used to provide specific and very intense labelling of lysosomes where, due to their resistance to lysosomal hydrolases, they may

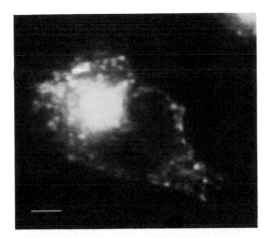

Fig. 4. TRVb-1 cells labeled with NBD-SM. In TRVb-1 cells, NBD-SM appears to follow the same recycling pathway taken by transferrin (compare to Fig. 2a). TRVb-1 cells were incubated in a 10 μM NBD-SM liposome solution in 20 mM HEPES buffered F-12 (HF-12, pH 7.4) for 30 min on ice. At the end of the incubation, cells were rinsed four times, and incubated in unlabeled HF-12 for 10 min at 37°C to internalize label. Cells were next rinsed and then incubated in six changes of HF-12 containing 10% fetal calf serum on ice for 5 min each, to back-exchange uninternalized NBD-SM from the cell surface to the medium. Finally cells were rinsed and fixed for observation. Fixed cells were treated with phenylenediamine to stabilize the rapidly bleaching fluorescence, and visualized using fluorescein optics. Bar is 5 μm in length.

accumulate for hours or even days. At the end of these labeling periods the entire endocytic pathway leading to lysosomes will be labeled. Probe may be chased into lysosomes by continued incubation for approximately 1 h in the absence of label.

It should be recognized that specific lysosomally destined probes, such as diO-LDL, and nonspecific probes, such as lucifer yellow, may not label the same early endocytic compartments. Fluid uptake occurs not only as a by-product of receptor-mediated endocytosis, but also via pinocytosis [see review, Steinman et al., 1983], so that fluid phase probes may label vesicles that do not contain receptor-mediated ligands.

Although all of these probes are internalized via bulk uptake, differences between them should be considered when choosing a probe for a particular application. Care must be taken to demonstrate the specificity of labeling when using lucifer yellow, which has been shown to cross intracellular membranes in some cell types [see discussion in Swanson, 1989]. Additionally, Wang and Goren [1987] found that sulforhodamine and fluorescein-dextran, internalized into the same early compartments, are delivered to lysosomes at different rates. Consequently, it is possible that fluid phase probes may not always act strictly as bulk labels and that properties such as molecular weight or binding to membranes within endosomes may influence their trafficking. Unless one is interested in measuring the pH of lysosomes, rhodamine-dextran is preferred over fluorescein-dextran for the reasons discussed before. The very low pH of lysosomes (≤ 5.0) makes the pH-dependent quenching of fluorescein fluorescence a particular problem in studies of live cells. As discussed previously, the fluorescence of fluorescein in acidic compartments can be increased by the addition of weak bases which disturb intracellular pH gradients.

C. Fluorescent Lipid Membrane Probes

Pagano has pioneered the use of fluorescent lipid analogs to label cellular structures [see review, Pagano and Sleight, 1985]. Various lipids (such as phosphatidylcholine and sphingomyelin) can be conjugated to 7-nitrobenz-2-oxa-1, 3-diazole (NBD), a fluorophore with excitation and emission characteristics similar to fluorescein. Cells incubated in solutions containing C_6-NBD-lipid liposomes acquire the fluorescent label in the plasma membrane as the fluorescent lipid exchanges with plasma membrane lipids. The probe is then endocytosed as part of the plasma membrane. Sleight and Pagano [1984] found that endocytosed C_6-NBD-phosphatidylcholine is transported to a region near the Golgi apparatus of Chinese hamster V79 fibroblasts. To study recycling of plasma membrane lipids, Koval and Pagano [1989] developed a C_6-NBD-lipid that is more resistant to hydrolysis (C_6-NBD-sphingomyelin, C_6-NBD-SM) and thus can be followed during prolonged incubations. They found that the intracellular distribution of C_6-NBD-SM is similar to that of fluorescent transferrin conjugates, but that the lipid recycles to the cell surface with slower kinetics. When using a lipid probe it is important to ascertain that the probe's movements reflect bulk membrane transport and not lipid movements outside the membrane bilayer (either through diffusion or via carrier molecules). Koval and Pagano [1989] found that NBD-SM is slowly hydrolyzed to NBD-ceramide, which is capable of transbilayer movement and diffusion to the Golgi apparatus, where it is converted back to NBD-SM. NBD-ceramide labeling of the Golgi apparatus is known to occur independent of endocytosis since it occurs even in fixed cells [Pagano, 1989]. In order to label endocytic structures with C_6-NBD-SM, cells are incubated in solutions containing C_6-NBD-SM liposomes at 4°C, rinsed, and then warmed for various periods of time to allow endocytosis to proceed. Whatever fluorescent lipid remains at the cell surface at the end of the warming period is removed by back-exchange of the surface fluorescent lipid with unlabeled liposomes provided in the incubation medium during a 4°C incubation. Figure 4 shows an example of cells prepared by this method.

D. Immunofluorescence

Proteins involved in endocytosis may also be labeled via indirect immunofluorescence or with fluorescent lectin conjugates. While indirect immunofluorescence is capable of exquisite specificity, fluorescent lectins offer the ability to characterize entire classes of glycoproteins.

Immunofluorescence methods have been used to localize many receptors such as those for asialoglycoproteins and LDL which, unlike the transferrin receptor, separate from their ligands after internalization [Breitfeld et al., 1985; Brown et al., 1983]. While their ligands are directed to lysosomes, these receptors recycle back to the cell surface, following the same pathway as that taken by both transferrin and its receptor. The distribution of transferrin receptors in TRVb-1 cells localized through immunofluorescence is identical to the steady-state distribution of transferrin localized with fluorescent ligand conjugates.

Live cells may also be treated with antibodies, but the endocytic pathway labeled in this way may not reflect the normal pathway of the antigen of interest. In several cases [Anderson et al., 1982; Mellman and Plutner, 1984; Schwartz et al., 1986], antibodies have been shown to cause misrouting of normally recycling receptors to lysosomes. Antibody-induced misrouting of receptors is apparently related to antibody valency, since cells treated with monovalent antibody fragments show normal patterns of receptor routing [Anderson et al., 1982; Mellman et al., 1984; Schwartz et al., 1986].

IV. QUANTITATIVE FLUORESCENCE CHARACTERIZATIONS OF ENDOCYTIC PROCESSES
A. Measurement of pH of Endocytic Compartments

Endocytic compartments are characterized by an acidic pH which appears to be critical to a variety of functions. Low pH facilitates dissociation of many ligands from receptors and hence sorting of recycling receptors from lysosomally destined ligands [Tycko et al., 1983; Brown et al., 1983; Harford et al., 1983a,b]. The acid pH of endosomes causes iron to dissociate from its carrier protein transferrin [Dautry-Varsat et al., 1983; Klausner et al., 1983]. The acidity of lysosomes optimizes the activity of various lysosomal hydrolases [Barrett, 1972]. Certain viruses and toxins penetrate from endosomes to the cytosol only after the endosome has acidified [White et al., 1983; Marsh, 1984; Olsnes and Sandvig, 1985]. Characterizing the pH of various endocytic compartments has thus been a central focus of endocytosis research.

One of the oldest methods used to morphologically identify lysosomes depends upon the acidic nature of these compartments. Cells incubated with the fluorescent weak base, acridine orange, concentrate the label in their lysosomes [Robbins and Marcus, 1963; Hart and Young, 1975]. At neutral pH these bases exist in an unprotonated state, and freely cross cell membranes. Upon encountering the acidic interior of an intracellular compartment, they acquire a positive charge which diminishes their ability to cross membranes, causing them to accumulate in the acidic interiors of endosomes and lysosomes. Acridine orange forms multimers at high concentrations,causing a shift in its fluorescence emission spectrum from green to red allowing easy distinction of acidic compartments.

Although the pH of a compartment can be estimated according to its accumulation of a weak base [see review, Roos and Boron, 1981], a more commonly used method involves fluorescence ratio measurements. The use of fluorescence ratios to measure pH is reviewed by Bright et al. [1989] and Maxfield [1985, 1989a]. Our discussion will be limited to its applications to the study of endosomal pH.

The intensity of fluorescein fluorescence is strongly dependent upon the pH of its environment such that when excited with 490 nm light the intensity is much higher at pH 7 than at pH 5 (Fig. 5). Fluorescein fluorescence is much less pH sensitive when excited with 450 nm light. By measuring the fluorescence intensity for each excitation wavelength, and calculating the ratio of fluorescence generated with 490 nm excitation to that with 450 nm excitation, one

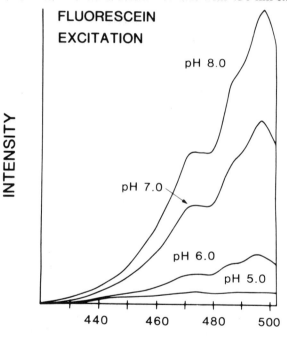

Fig. 5.   Fluorescein fluorescence emission spectra for fluorescein solutions of different pH values. Fluorescein fluorescence emission is plotted as a function of excitation wavelength for aqueous fluorescein-α-2-macroglobulin solutions of pH 5, 6, 7, and 8. Fluorescein fluorescence is not only attenuated in a low pH environment, but its intensity is also much less dependent upon excitation wavelength at low pH.

obtains a quantity which is sensitive to pH but nearly independent of all the other sources of fluorescence intensity variation (e.g., concentration of dye, degree of focus, photobleaching). The method can be calibrated in vitro using fluorescein solutions buffered to a range of pH values. While fluorescein fluorescence is relatively independent of dye concentration as well as to the concentration of ions other than protons [Heiple and Taylor, 1980], in situ calibrations are preferable as they correct for any potential artifacts induced by the endosomal environment. In situ calibrations can be made by labeling cells with fluorescein conjugates and then using the weak base methylamine to equilibrate the labeled compartments to the pH of a range of buffers. The fluorescein fluorescence ratio is most accurate over a range of pH from around 5.0 to 6.5. Outside of this range the fluorescence ratio becomes nearly independent of pH. Additionally, fluorescein fluorescence is very weak at low pH.

Fluorescein can be used to produce conjugate probes that are specific (e.g., ligand conjugates, fluorescein-transferrin, fluorescein-α_2M) or nonspecific fluid phase markers (e.g., fluorescein-dextran). Fluorescein-dextran is delivered to lysosomes, where it accumulates since it is refractory to degradation by lysosomal hydrolases. Cells may be incubated in relatively high concentrations of fluorescein-dextran (10–20 mg/ml) for hours to obtain strongly labeled lysosomes. In the first application of fluorescein to the measurement of pH of intracellular compartments, Ohkuma and Poole [1978] measured the fluorescence of fluorescein-dextran labeled cells using a fluorometer. They estimated the pH of lysosomes of mouse peritoneal macrophages to be between 4.7 and 4.8.

Fluorescence ratio measurements of whole, single cells can be made using either a microscope spectrofluorometer or digital image processing techniques. Whole cell measurements report an average endosomal pH, but specific populations of endosomes may be labeled by exploiting the kinetics of transfer between the various compartments through careful pulse-chase experiments. Yamashiro and Maxfield [1987a] labeled CHO cells with either fluorescein-α_2M or fluorescein-dextran, then incubated them for various periods of time (1 to 10 min) in unlabeled medium. By this method they were able to demonstrate the progressive acidification of endosomes on the lysosomal pathway. Yamashiro et al. [1984] measured a pH of 6.4 for the para-Golgi recycling endosomes by labelling cells with fluorescein-transferrin for 18 min and then briefly chasing the cells in unlabeled medium, a procedure that results in nearly all cellular transferrins being found in the para-Golgi recycling endosomes. By manipulating the length of the labeling period, Yamashiro and Maxfield [1987a] were also able to show that recycling components are taken into acidic endosomes with an average pH of 6.2 and then encounter compartments of somewhat higher pH prior to return to the cell surface. The average pH of endosomes in cells labeled with fluorescein-transferrin for 5 min was found to be slightly lower than that of cells labeled for 18 min. The kinetics of endosome acidification can also be studied using flow cytometry, which permits measurement of the average endosome pH of large numbers of individual cells. Using flow cytometry, Murphy and his colleagues found patterns of endosome acidification along the lysosomal and recycling pathways of BALB/c 3T3 cells [Murphy et al., 1984; Sipe and Murphy, 1987] similar to those found in the microscopic studies of CHO cells discussed above.

A difficulty with whole cell spectrophotometric measurements is that some fraction of measured fluorescence derives from the endogenous fluorescence of the cell (autofluorescence). Measurements may be corrected by subtracting the average fluorescence measured in unlabeled cells, but this is an approximate correction that can only be applied when cell to cell variation in autofluorescence is a relatively small fraction of the total measured fluorescence.

A "null point" technique of pH measurement that circumvents this problem has been developed [Yamashiro and Maxfield, 1987a]. Cells are incubated with a fluorescent conjugate, and then placed in medium buffered to a "test pH" with a membrane impermeant buffer. Fluorescence is measured before and then again after cellular pH gradients have been collapsed by the addition of methylamine and ammonium acetate. The addition of the weak acid and base rapidly equilibrates the endocytic compartments to the test pH. Since this treatment has no effect on surface-bound label fluorescence, and since autofluorescence is largely pH independent, any change in measured fluorescence following the addition of the weak acid and base solution is due to a change in the pH of an internal, fluorescein-containing endocytic compartment. An increase in fluorescence occurs upon collapse of the pH gradients of endocytic compartments whose internal pH is more acidic than the test pH. If the compartments being studied are more alkaline than the test pH, a decrease in fluorescence will be measured as the pH gradient is dissipated. By careful bracketing with a series of test buffers, the pH of particular compartments can thus be measured. By labeling cells with fluorescein-dextran for as little as 3 min Yamashiro and Maxfield [1987a] were able to use the null point method to measure the pH of faintly labeled early endosomes whose fluorescence was not much brighter than the level of cellular autofluorescence.

Tycko et al. [1983] developed digital image processing techniques for eliminating autofluorescence from digitized images of fluorescently labeled cells. Since cellular autofluorescence is frequently distributed diffusely throughout a cell, it may be treated as the "background" level of fluorescence in labeled cells. Since endosomes occupy a small fraction of an image of a cell, the background fluorescence can be estimated as the median fluorescence in a local region, typically 5 μm square. Using digital image analysis, the background level of fluorescence can be calculated for each point, or pixel, in a digital image and the contributions of autofluorescence to the total fluorescence in the image may be eliminated point by point by subtracting the background level from the total measured intensity for each pixel [Tycko et al.,

1983]. Using this technique, Tycko et al. [1983] were able to measure an average pH of 5.4 in fluorescein-asialo-orosomucoid labelled endosomes in highly autofluorescent HepG2 cells.

Digital image processing also allows the measurement of pH of endosomes located in particular regions of a cell, or even of particular single endosomes. Using digital image processing techniques, Yamashiro and Maxfield [1987b] measured the pH of the para-Golgi recycling structure of CHO cells labeled with fluorescein-transferrin. By combining the pixel-by-pixel flexibility offered by digital image processing with the sensitivity of the null point technique, they were also able to measure separately the pH of fluorescein-dextran-labeled endosomes appearing either in a punctate pattern or as a diffuse fluorescence. In this way, they were able to measure pH in early endosomes so small and faintly labeled that they were not individually resolvable. In studies of mutant cell lines with endocytosis defects, heterogeneity in punctate endosomal pH was found even though the mean pH of these endosomes was only slightly higher than that of normal CHO cells. This heterogeneity could not be detected using whole cell methods.

Estimates of average endosomal pH derived from fluorescein fluorescence ratio measurements of whole cells can be seriously biased by heterogeneity in the pH of endocytic compartments. The fluorescence intensity of fluorescein increases non-linearly with increasing pH. Consequently, in measurements of whole cells, high pH compartments will be disproportionately represented, leading to overestimates of average endosomal pH [see Maxfield, 1985]. This effect is easily detected and avoided when pH is measured in individual endosomes by digital image analysis.

B. Fluorescence Assays of Endosome Fusion

Using enhanced video digital image processing techniques, Pastan and Willingham [1985] described the movement of individual fluorescently labeled endosomes and observed fusion between early endocytic vesicles. By viewing cells every 18 s shortly after exposure to rhodamine-α_2M, they were able to reconstruct the paths taken by individual endocytic vesicles through the cell. After a brief period of stasis (perhaps corresponding to the time required for an endocytic vesicle to pinch off from the plasma membrane), endocytic vesicles were observed to move rapidly and sometimes fuse with one another. Since then, several in vitro studies have indicated that endocytic vesicles are highly fusogenic shortly after endocytosis but quickly lose the capacity for fusion [Gruenberg and Howell, 1986; Braell, 1987; Diaz et al., 1988].

In our laboratory, we have used fluorescence techniques to characterize the kinetics and magnitude of endocytic fusion in vivo. We have employed two different approaches, both based on changes in the fluorescence of fluorescently labeled endosomes upon fusion with subsequently formed endocytic vesicles.

Salzman and Maxfield [1988] studied fusion of endocytic vesicles along the recycling pathway using an in vivo assay of endocytic fusion based on the reduction of fluorescein fluorescence upon binding to anti-fluorescein antibody. CHO cells were incubated in a solution of fluorescein-transferrin, rinsed, and chased for varying periods of time and then incubated with an anti-fluorescein antibody. Fusion between endocytic vesicles containing fluorescein-transferrin and the subsequently formed vesicles containing pino-cytosed anti-fluorescein antibody was indicated by the reduction of fluorescein fluorescence caused by the binding of anti-fluorescein antibody to fluorescein-transferrin. The reduction in fluorescence can only result from close molecular contact between the two, indicating the fusion of their two compartments. By comparing the endosomal fluorescence of labeled cells incubated with anti-fluorescein to that of comparably labeled cells incubated without antibody it was possible to measure the degree of fusion between endocytic compartments formed at different times. Prior to measurements methylamine was added to cells to dissipate pH gradients, thus eliminating the effects of endosomal pH on fluorescein fluorescence. These experiments demonstrated that fluorescein-transferrin-containing compartments were accessible to subsequently endocy-tosed anti-fluorescein antibodies, but that fusion accessibility rapidly decreased with time. Measurements were taken both using a microscope spectrofluoro-meter, measuring the overall fluorescence levels of whole fields, and also using digital image processing of videotaped images of cells in an effort to identify the intracellular compartment(s) in which fusion occurs. From these studies it could not be determined whether fusion occurred prior to or upon delivery of transferrin to the para-Golgi recycling endosomes of the CHO cells.

Antibody binding not only reduces the intensity, but also nearly eliminates the pH dependence of fluorescein fluorescence. Salzman and Maxfield [1989] exploited this characteristic to quantify endosome fusion by the decrease in the ratio of fluorescence at high pH to that at low pH as measured in a spectrofluorometer. Experiments were similar to those described above, but cellular fluorescence was measured after equilibration of intracellular com-partments to pH 6, 7 and 8 by the addition of methylamine and ammonium acetate. Using this technique, they were able to characterize the kinetics of endocytic fusion along both the recycling and lysosomal pathways.

Dunn et al. [1990] quantified endosome fusion by measuring the accumu-lation of ligand in single endosomes. Using highly fluorescent diI-LDL, we measured the fluorescence intensity of individual endosomes after various peri-ods of continuous uptake. We found that during a continuous uptake, individ-ual endosomes accumulate diI-LDL for approximately 10 min. After ten minutes of internalization, the amount of LDL per endosome is approximately 40 times that after 1 min, indicating that during this time endosomes experience approx-imately 40 fusions. Even this large value is probably an underestimate of the actual number of endocytic fusions. First, the kinetics of accumulation suggest that an appreciable amount of fusion might occur during the first minute of internalization, prior to our initial measurements. Second, it is known that some fraction of internalized LDL is recycled back to the cell surface, so endosomes might not retain all of the LDL delivered into them.

In parallel experiments, cells were incubated with rhodamine-transferrin rather than diI-LDL. In contrast to the results with diI-LDL, the average brightness of rhodamine-transferrin labeled endosomes increased only 3–4-fold over time, reaching steady state within 2 min of uptake. Taken together with the results discussed in section III.A, which demonstrated that transferrin rapidly passes through LDL-containing compartments, these results were used to develop a model of endocytic sorting. Since transferrin and LDL are internalized via the same endocytic vesicles (appearing in the same early endocytic compartments, see Figs. 3a,b), the relatively small accumulation of transferrin suggests that transferrin passes through the compartment that accumulates LDL (note in fig. 2 that transferrin can be found in compartments showing every level of LDL accumulation). According to this model, early endocytic vesicles continuously fuse with a sorting endosome that accumulates lysosomally directed ligands as it repeatedly sorts and exports recycling components.

Measuring the fluorescence intensities of individual endosomes required overcoming some difficulties not present in ratio methods. The measured intensity of an endosome varies not only with fluorophore concentration, but also with its position in the field, due to inhomogeneities in the field of illumination or image collection, with the plane of focus from which the endosome is imaged and with the degree to which the fluorophore is photobleached. Reducing these sources of intensity variation required developing the techniques discussed below.

Spatial inhomogeneity in a fluorescence microscopic image may arise from an improperly focussed lamp, aberrations in the objective lens or other optics in the excitation or emission path, and spatial nonuniformity in camera sensitivity [for more detailed discussion see Inoue, 1986]. In order to reduce spatial inhomogeneities in the field of measured endosomes, the fluorescence illuminator was carefully centered and focused. Homogeneity was assessed by measuring the spatial brightness distribution in digital images of a fluorescent solution. Measurements were made on endosomes located in the central portion of the field which showed less than 10% spatial variation in intensity from the center to the edges.

The effects of mis-focus on endosome fluorescence intensity measurements were reduced to less than 10% through the use of a technique devised to identify the in-focus image from a set of closely spaced serial optical sections combined with the use of a quantification parameter that is relatively insensitive to changes in focal plane. In order to analyze large numbers of endosomes a computer algorithm was devised that identifies the in-focus image of a particular endosome from a set of ten serial focal plane images spaced 1.2 μM apart in the vertical axis. Rather than being directed at serial reconstruction of three-dimensional information on the distribution of endosomes, our technique was devised strictly to eliminate all out-of-focus images of endosomes before quantification. The procedure has the advantage of being computationally simple and having modest image processing equipment needs.

Fluorescently labeled, fixed cells were examined microscopically using brightfield optics (Leitz Diavert fluorescent microscope equipped with a 63 × oil immersion objective, NA 1.4). Well spread cells were then centered in the microscope field. The optics were changed for rhodamine epifluorescence (530–560 nm band-pass excitation filter, 580 nm dichroic mirror, 580 nm long-pass emission filter), and the focus was adjusted to slightly above the plane of the tops of the cells. The field was then illuminated for 2 s, and its image recorded on videotape (JVC CR665OU video cassette recorder, using a Videoscope VS2000N camera and a Videoscope KS1380 image intensifier). Videotaping permits collection of much more image data than is possible when images are stored digitally (a single 512 by 512 pixel image, with 256 levels of gray level resolution, requires approximately one-quarter megabyte of digital storage space). We have found that with high-quality videotape equipment, the loss of image resolution is negligible. This procedure was repeated for nine more focal planes, each 1.2 μM lower than the previous, finally extending the focus to just below the focal plane of the bottoms of the cells. Focal plane was adjusted by means of a microstepping motor Z-axis controller (Kinetek, Yonkers, NY).

These recorded images were digitized, and background fluorescence was digitally removed as previously described. The central portion of each field (see previous discussion) was isolated for measurement. The ten serial focal plane images were then analyzed to identify endosomes for which focussed images were collected. If an endosome appears in three successive focal planes and its image is brightest in the middle plane, the image from that plane is identified as the focused image for that endosome. Spots that do not satisfy these criteria, as well as spots that are outside the size limits observed for endosomes, are eliminated from analysis. Subjective comparisons of the serial focal plane images showed this procedure to be highly successful at eliminating out-of-focus material from the final composite image of a field of cells.

Finally, the brightness of each endosome was quantified as the sum of all pixel intensity values greater than or equal to 60% of that of the brightest pixel in each spot. This parameter was found to be relatively insensitive to changes in focus [Maxfield and Dunn, 1990].

The endosomes in this study showed an approximately 1,000-fold range in fluorescence intensity, far greater than the useable linear range of our Videoscope VS2000N camera. A set of neutral density excitation filters, transmitting 2.5%

to 50% of the lamp output, was used to keep the fluorescence intensity of brighter specimens within the linear range of the camera. In this way we were able to effectively extend the linear range of the camera 40-fold.

Limiting the level of illumination with neutral density filters helped to minimize label photobleaching. Photobleaching was further limited by minimizing the duration of illumination and by employing relatively photostable fluorophores, diI and rhodamine.

While out-of-focus contributions to images may be eliminated via scanning confocal microscopy (see chapter by Steltzer, this volume) or through the use of mathematical deconvolution algorithms (see chapter by Carrington et al., this volume), this relatively simple and economical technique may be sufficient for many users with more modest facilities. Although we have used this technique to measure the accumulation of fluorescently labeled ligands in endosomes, it could as easily be used to measure the intensity of any small punctate, fluorescently labeled microscopic bodies.

V. PROSPECTS FOR FUTURE DIRECTIONS IN FLUORESCENCE STUDIES OF ENDOCYTOSIS

Fluorescence microscopy has developed from its beginnings as a strictly descriptive tool into a critical component of many quantitative methodologies. In our laboratory we have used fluorescence microscopy to measure the pH of endocytic compartments of normal and mutant cultured cells, to quantify the kinetics and magnitude of fusion between endocytic compartments, and to characterize the process of endocytic sorting. In other laboratories fluorescence microscopy has been applied to these topics and a variety of others, e.g., measurement of the mobility of receptors and receptor-ligand complexes on the surfaces of cells (see chapter by Thomas and Webb, this volume), characterization of the movements of early endosomes [Pastan and Willingham, 1985], and quantification of the kinetics of plasma membrane lipid recycling [Koval and Pagano, 1989].

In most cases, fluorescence studies in endocytosis have been limited to characterizing events that occur long after internalization, in part because of the speed of the process and in part because endosomes have not been detectible until they have sufficiently concentrated fluorophore labelled probes. However, as discussed above, many critical events in vesicle fusion, ligand sorting, and vesicle acidification occur very quickly after internalization in endosomes which have accumulated little ligand. Consequently, methods capable of rapidly resolving and characterizing what are likely to be faintly labeled, transient structures need to be developed. Detectors are near the theoretical limits of sensitivity, with noise levels approaching the Poisson limits of an ideal detector. It is not likely that fluorescence optics will improve more than incrementally in the future. Much of the advances, then, must come in the form of new, brighter, and more stable fluorophores and in the development of appropriate cell lines.

ACKNOWLEDGMENTS

We would like to thank Dr. Laurence Borden and Dr. Bill Hendey for their helpful suggestions on the manuscript. This work was supported by NIH grant DK 27083 (F.M.) and NIH fellowship F32 GM12148 (K.D.).

VI. REFERENCES

Anderson RGW, Brown MS, Beisiegel U, Goldstein JL (1982) Surface distribution and recycling of the LDL receptor as visualized by anti-receptor antibodies. J Cell Biol 93:523–531.

Aubin JE (1979) Autofluorescence of viable cultured mammalian cells. J Histochem Cytochem 27:36–43.

Barak LS, Webb WW (1981) Fluorescent low density lipoprotein for observation of dynamics of individual receptor complexes on cultured human fibroblasts. J Cell Biol 90:595–604.

Barrett AJ (1972) Lysosomal enzymes. In Dingle JT (ed): Lysosomes: A Laboratory Handbook. New York: North-Holland/American Elsevier, pp 46–135.

Braell WA (1987) Fusion between endocytic vesicles in a cell free system. Proc Natl Acad Sci USA 84:1137–1141.

Breitfeld PP, Simmons CF, Strous GJ, Geuze HJ, Schwartz AL (1985) Cell biology of the asialoglycoprotein system: A model of receptor-mediated endocytosis. Int Rev Cytol 97:47–95.

Bright GR, Fisher GW, Rogowska J, Taylor DL (1989) Fluorescence ratio imaging microscopy. Methods Cell Biol 30:157–192.

Brown MS, Anderson RGW, Goldstein JL (1983) Recycling receptors: The round trip itinerary of migrant membrane proteins. Cell 32:663–667.

Chalpin DB, Kleinfeld AM (1983) Interaction of fluorescence quenchers with the n(9-anthroyloxy): Fatty acid membrane probes. Biochim Biophys Acta 731:465–474.

Dautry-Varsat A, Ciechanover A, Lodish HF (1983) pH and the recycling of transferrin during receptor-mediated endocytosis. Proc Natl Acad Sci USA 80:2258–2262.

Diaz R, Mayorga L, Stahl P (1988) *In vitro* fusion of endosomes following receptor-mediated endocytosis. J Biol Chem 263:6093–6100.

Dunn KW, McGraw TE, Maxfield FR (1990) Iterative fractionation of recycling receptors from lysosomally destined ligands in an early sorting endosome. J Cell Biol 109:3303–3314.

Dunn WA, Hubbard AL, Aronson NN (1980) Low temperature selectively inhibits fusion between pinocytic vesicles and lysosomes during heterophagy of ^{125}I asialofetuin by the perfused rat liver. J Biol Chem 255:5971–5978.

Geuze HJ, Slot JW, Strous GJAM, Lodish HF, Schwartz AL (1983) Intracellular site of asialoglycoprotein receptor-ligand uncoupling: Double label immunoelectron microscopy during receptor mediated endocytosis. Cell 32:277–287.

Goldstein JL, Brown MS, Anderson RGW, Russell DW, Schneider WJ (1985) Receptor mediated endocytosis: Concepts emerging from the LDL receptor system. Annu Rev Cell Biol 1:1–39.

Gruenberg JE, Howell KE (1986) Reconstitution of vesicle fusions occurring in endocytosis with a cell free system. EMBO J 5:3091–3101.

Harford J, Bridges K, Ashwell G, Klausner RD (1983a) Intracellular dissociation of receptor-bound asialoglycoproteins in cultured hepatocytes: A pH-mediated nonlysosomal event. J Biol Chem 258:3191–3197.

Harford J, Wolkoff AW, Ashwell G, Klausner RD (1983b) Monensin inhibits intracellular dissociation of asialoglycoproteins from their receptor. J Cell Biol 96:1824–1828.

Hart PD, Young MR (1975) Interference with normal phagosome-lysosome fusion in macrophages, using ingested yeast cells and suramin. Nature 256:47–49.

Heiple JM, Taylor DL (1980) Intracellular pH in single motile cells. J Cell Biol 86:885–890.

Inoue S (1986) Video Microscopy. New York: Plenum Press.

Klausner RD, Ashwell G, Van Renswoude J, Harford JB, Bridges KR (1983) Binding of apotransferrin to K562 cells: Explanation of the transferrin cycle. Proc Natl Acad Sci USA 80:2263–2266.

Koval M, Pagano RE (1989) Lipid recycling between the plasma membrane and intracellular compartments: Transport and metabolism of fluorescent sphingomyelin analogues in cultured fibroblasts. J Cell Biol 108:2169–2181.

Marsh M (1984) The entry of enveloped viruses into cells by endocytosis. Biochem J 218:1–10.

Maxfield FR (1985) Acidification of endocytic vesicles and lysosomes. In Pastan I, Willingham M (eds): Endocytosis. New York: Plenum Press, pp 235–257.

Maxfield FR (1989a) Measurement of vacuolar pH and cytoplasmic calcium in living cells using fluorescence microscopy. Methods Enzymol 173:745–771.

Maxfield FR (1989b) Fluorescent analogs of peptides and hormones. Methods Cell Biol 29:13–28.

Maxfield FR, Dunn KW (1990) Studies of endocytosis using image intensification fluorescence microscopy and digital image analysis. In Herman B, Jacobson K (eds): Optical Microscopy for Biology. New York: Wiley-Liss, pp 357–371.

McGraw TE, Greenfield L, Maxfield FR (1987) Functional expression of the human transferrin receptor cDNA in Chinese hamster ovary cells deficient in endogenous transferrin receptor. J Cell Biol 105:207–214.

Mellman I, Plutner H (1984) Internalization and fate of macrophage Fc receptors bound to polyvalent immune complexes. J Cell Biol 98:1170–1177.

Mellman I, Plutner H, Ukkonen P (1984) Internalization and rapid recycling of macrophage Fc receptors tagged with monovalent antireceptor antibody: Possible role of a prelysosomal compartment. J Cell Biol 98:1163–1169.

Mueller SC, Hubbard AL (1986) Receptor-mediated endocytosis of asialoglycoproteins by rat hepatocytes: Receptor-positive and receptor-negative endosomes. J Cell Biol 102:932–942.

Murphy RF, Powers S, Cantor CR (1984) Endosome pH measured in single cells by dual fluorescence flow cytometry: Rapid acidification of insulin to pH 6. J Cell Biol 98:1757–1762.

Ohkuma S, Poole B (1978) Fluorescence probe measurements of the intralysosomal pH in living cells and the perturbation of pH by various agents. Proc Natl Acad Sci USA 75:3327–3331.

Olsnes S, Sandvig K (1985) Entry of polypeptide toxins into animal cells. In Pastan I, Willingham M (eds): Endocytosis. New York: Plenum Press, pp 195–234.

Osborne M, Weber K (1982) Immunofluorescence and immunocytochemical procedures with affinity purified antibodies: Tubulin-containing structures. Methods Cell Biol 24:97–132.

Pagano R (1989) A fluorescent derivative of ceramide: Physical properties and use in studying the Golgi apparatus of animal cells. Methods Cell Biol 29:75–85.

Pagano R, Sleight RG (1985) Defining lipid transport pathways in animal cells. Science 229:1051–1057.

Pastan I, Willingham MC (1983) Receptor mediated endocytosis: Coated pits, receptosomes and the Golgi. Trends Biochem Sci 8:250–254.

Pastan I, Willingham MC (1985) The pathway of endocytosis. In Pastan I, Willingham M (eds): Endocytosis. New York: Plenum Press, pp 1–44.

Pitas RE, Innerarity TL, Weinstein JN, Mahley RW (1981) Acetoacetylated lipoproteins used to distinguish fibroblasts from macrophages in vitro by fluorescence microscopy. Arteriosclerosis 1:177–185.

Robbins E, Marcus PI (1963) Dynamics of acridine orange-cell interaction I. Interrelationships of acridine orange particles and cytoplasmic reddening. J Cell Biol 18:237–250.

Roederer M, Bowser R, Murphy RM (1987) Kinetics and temperature dependence of exposure of endocytosed material to proteolytic enzymes and low pH: Evidence for a maturation model for the formation of lysosomes. J Cell Physiol 131:200–209.

Roos A, Boron WF (1981) Intracellular pH. Physiol Rev 61:296–434.

Salzman NH, Maxfield FR (1988) Intracellular fusion of sequentially formed endocytic compartments. J Cell Biol 106:1083–1091.

Salzman NH, Maxfield FR (1989) Fusion-accessibility of endocytic compartments along the recycling and lysosomal pathways in intact cells. J Cell Biol 109:2097–2104.

Schlessinger J (1980) The mechanisms and role of hormone-induced clustering of membrane receptors. Trends Biochem Sci 5:210–214.

Schmid SL, Fuchs R, Male P, Mellman I (1988) Two distinct subpopulations of endosomes involved in membrane recycling and transport to lysosomes. Cell 52:73–83.

Schwartz AL, Ciechanover A, Merritt S, Turkewitz A (1986) Antibody-induced receptor loss: Different fates for asialoglycoproteins and the asialoglycoprotein receptor in HepG2 cells. J Biol Chem 261:15225–15232.

Sipe DM, Murphy RF (1987) High resolution kinetics of transferrin acidification in Balb/c 3T3 cells: Exposure to pH 6 followed by temperature sensitive alkalinization during recycling. Proc Natl Acad Sci USA 84:7119–7123.

Sleight RG, Pagano RE (1984) Transport of a fluorescent phosphatidylcholine analog from the plasma membrane to the Golgi apparatus. J Cell Biol 99:742–751.

Steer C, Hanover J (1990) Intracellular Trafficking of Proteins. New York: Cambridge Univ Press, in press.

Steinman RM, Mellman IS, Muller WA, Cohn ZA (1983) Endocytosis and recycling of plasma membrane. J Cell Biol 96:1–27.

Swanson J (1989) Fluorescent labeling of endocytic compartments. Methods Cell Biol 29:137–151.

Tabas I, Maxfield FR, Lim S. Endocytosed β-VLDL and LDL are delivered to different intracellular vesicles in mouse peritoneal macrophages. J Cell Biol (submitted).

Taylor DL, Wang Y (eds) (1989) "Fluorescence Microscopy of Living Cells in Culture, Part B. Methods in Cell Biology, Volume 30." New York: Academic Press.

Tycko B, Keith CH, Maxfield FR (1983) Rapid acidification of endocytic vesicles containing asialoglycoproteins in cells of a human hepatoma line. J Cell Biol 97:1762–1776.

Wang Y, Goren MB (1987) Differential and sequential delivery of fluorescent lysosomal probes into phagosomes in mouse peritoneal macrophages. J Cell Biol 104:1749–1754.

Wang Y, Taylor DL (eds) (1989) "Fluorescence Microscopy of Living Cells in Culture, Part A. Methods in Cell Biology, Volume 29." New York: Academic Press.

Weigel PH, Oka JA (1983) The surface content of asialoglycoprotein receptors on isolated hepatocytes is reversibly modulated by changes in temperature. J Biol Chem 258:5089–5094.

White J, Kielian M, Helenius A (1983) Membrane fusion proteins of enveloped animal viruses. Q Rev Biophys 16:151–195.

Willingham M, Pastan I (1985) Morphologic methods in the study of endocytosis in cultured cells. In Pastan I, Willingham M (eds): "Endocytosis." New York: Plenum Press, pp 281–321.

Wolf DE (1985) Determination of the sidedness of carbocyanine labeling of membranes. Biochemistry 24:582–586.

Wolkoff AW, Klausner RD, Ashwell G, Harford J (1984) Intracellular segregation of asialoglycoproteins and their receptor: A prelysosomal event subsequent to dissociation of the ligand-receptor complex. J Cell Biol 98:375–381.

Yamashiro DJ, Maxfield FR (1987a) Kinetics of endosome acidification in mutant and wild type chinese hamster ovary cells. J Cell Biol 105:2713–2721.

Yamashiro DJ, Maxfield FR (1987b) Acidification of morphologically distinct endosomes in mutant and wild type Chinese hamster ovary cells. J Cell Biol 105:2723–2733.

Yamashiro DJ, Tycko B, Fluss SR, Maxfield FR (1984) Segregation of transferrin to a mildly acidic (pH 6.5) para-Golgi compartment in the recycling pathway. Cell 37:789–800.

Noninvasive Techniques in Cell Biology: 177–212
© 1990 Wiley-Liss, Inc.

8. Analysis of Cytoskeletal Structures by the Microinjection of Fluorescent Probes

Yu-Li Wang and Mitchell C. Sanders

Cell Biology Group, Worcester Foundation for Experimental Biology,
Shrewsbury, Massachusetts 01545

I. INTRODUCTION

The cytoskeleton is composed of three kinds of filamentous structures: actin filaments, microtubules, and intermediate filaments. The former two are known to be involved in a number of crucial functions such as cell locomotion, organelle transport, and mitosis. During the past two decades, a large amount of information has become available on the biochemical and morphological properties of these filamentous structures. It is known that all of them are formed by non-covalent associations of protein subunits and accessory proteins. In addition, actin filaments and microtubules have well-defined polarities, with the two opposite ends showing different rates of assembly. The association of filaments with other structures, such as kinetochores for microtubules and plasma membrane for actin filaments, often involves a uniform filament polarity, indicating that polarity is important for structural organization. Each of the filament systems also has its characteristic distribution in the cell. For example, actin filaments in cultured non-muscle cells often form large bundles (stress fibers), which are associated with the plasma membrane at the ends, at sites where cells make focal contacts with the substrate on the exterior surface (adhesion plaques). On the other hand, microtubules and intermediate filaments appear to emanate from discrete "organizing centers" near the nucleus.

The most interesting aspect of cytoskeletal structures in living cells is their high degree of dynamics. Many cytoskeletal structures in non-muscle cells are known to be transient: they assemble in response to stimuli at specific sites, and disassemble when specific functions are completed. Many factors may contribute to the dynamic behavior. For example, both actin and microtubule subunits are known to undergo constant assembly-disassembly reactions at steady state in vitro. The existence of a defined polarity of the filament, coupled to the involvement of nucleotide hydrolysis in structural assembly, confers peculiar patterns to such reactions, such as a preferential assembly or disassembly at the opposite ends [treadmilling; Wegner, 1976], and the manifestation of discrete phases of assembly and disassembly [dynamic instability; Mitchison and Kirschner, 1984]. In addition, the assembly can be modulated by post-translational modifications such as phosphorylation [Suzuki et al., 1978]. The potential of reorganization is further increased by the wide variety of accessory proteins which are capable of severing the filaments into fragments, capping the ends of filaments, inhibiting the polymerization of monomers, nucleating the polymerization reaction, cross-linking filaments into networks or bundles, or inducing movements of filaments [see, e.g., Pollard and Cooper, 1986, for a review of actin-binding proteins]. On the one hand, it is clear that such interactions are critical to account for the versatility of the structures. On the other hand, they also represent a tremendous challenge to cell biologists. Structures are often impossible to isolate in a functional form to allow biochemical analyses, like those performed with ribosomes or mito-chondria. Ultrastructural and immunocytochemical studies are also limited by the tendency of structures to reorganize during fixation, and by the diffi-culty in determining dynamic reorganizations based on stationary images.

The application of fluorescent probes that bind or incorporate into specific structures represents a unique approach for the study of dynamic structures. Fluorescence techniques have the important advantages of being highly spe-cific and sensitive, yielding excellent signal/noise ratio, and causing minimal disruptions to living cells. These advantages are complemented by the improve-ment of microinjection techniques, which allows delivery of fluorescent probes into many types of cultured living cells, and by the development of low-light-level detectors and image processing techniques, which allows the detection of extremely low signals from single cells on a real time basis. Furthermore, fluorescence signals can also be analyzed with spectroscopic and photobleaching techniques, and provide information far beyond the resolution of a light microscope.

The purpose of this review is to examine how the microinjection of fluores-cent probes has been used to study the dynamics of cytoskeletal proteins in living cells, and what new information has been provided as a result. Closely related experiments, such as microinjection of biotinylated proteins, will also be examined. However, technical aspects will be discussed solely fo the pur-pose of evaluating existing data and future potentials. The readers are referred to several earlier reviews on various aspects of this technique and on related methods such as the microinjection of antibodies [Taylor and Wang, 1980; Wang et al., 1982a; Kreis and Birchmeier, 1982; Taylor et al., 1984; Jockusch et al., 1985; Simon and Taylor, 1986; Kreis, 1986; Wadsworth and Salmon, 1986a; Wang, 1989].

II. EXPERIMENTAL APPROACHES

In a typical experiment, fluorescent probes are microinjected into living cells. Following the association of the probes with cellular structures, fluo-rescence images are recorded and analyzed. In the following sections, we will discuss both different fluorescent probes that have been used for microinjec-tion, and different methods for the collection and analysis of data.

A. Fluorescent Probes

1. Fluorescent analogs of cellular components. To date, most probes used for cytoskeletal structures are fluorescently labeled structural components, such as actin, myosin, tubulin, and their accessory proteins (a list of microinjected analogs is shown in Table I). This approach has been referred to as fluorescent analog cytochemistry [Taylor and Wang, 1978; Wang et al., 1982a; Taylor et

TABLE I. Microinjected Fluorescent Analogs of the Cytoskeleton

Protein Derivatives	Reference
Actin	
Carboxytetramethylrhodamine succinimidyl ester	Kellogg et al., 1988.
Fluorescein isothiocyanate	Dome et al., 1988; Hamaguchi and Mabuchi, 1988; Kukulies et al., 1984; McKenna et al., 1985a.
5-Iodoacetamidofluorescein	Amato et al., 1986, 1983; DeBiasio et al., 1987, 1988; Gawlitta et al., 1980, 1981; Hamaguchi and Mabuchi, 1988; Kukulies et al., 1984; Simon et al., 1988; Stockem et al., 1983; Taylor and Wang, 1978; Taylor et al., 1980; Wang and Taylor, 1979, 1980; Wang et al., 1982b; Wehland and Weber, 1980; Wehland et al., 1980.
Lissamine rhodamine B sulphonyl chloride	Dome et al., 1988; Kreis, 1986; Kreis et al., 1982; McKenna and Wang, 1986; McKenna et al., 1985a.
N-(7-dimethylamino-4-methylcoumarinyl) maleimide	Hamaguchi and Mabuchi, 1988.
Tetramethylrhodamine iodoacetamide	DeBiasio et al., 1988; Dome et al., 1988; Glacy, 1983a, 1983b; Simon et al., 1988; Wang, 1984, 1985.
Tetramethylrhodamine isothiocyanate	Kreis et al., 1979; Kreis, 1986; Kukulies and Stockem, 1985; Kukulies et al., 1984, 1985; Sanger et al., 1980.
Myosin	
5-Iodoacetamidofluorescein	DeBiasio et al., 1988; Johnson et al., 1988; McKenna et al., 1989a.
Lissamine rhodamine B sulfonyl chloride	Glascott et al., 1987.
Tetramethylrhodamine iodoacetamide	DeBiasio et al., 1988; McKenna et al., 1989a, 1989b.
Myosin light chains	
Fluorescein isothiocyanate	McKenna et al., 1989a.
Tetramethylrhodamine iodoacetamide	Mittal et al., 1987a.
Alpha-actinin	
5-(4,6-Dichlorotriazinyl)aminofluorescein	Kreis, 1986.
Flourescein isothiocyanate	Mittal et al., 1987a; Sanger et al., 1986a,b, 1987.
5-Iodoacetamidofluorescein	Simon and Taylor, 1986; Simon et al., 1988; Stickel and Wang, 1988.
Lissaminerhodamine B sulfonyl chloride	Glascott et al., 1987; Kreis, 1986; Mittal et al., 1987a; Sanger et al., 1984a; Sanger et al., 1986a,b, 1987.
Lucifer yellow VS	Sanger et al., 1986a.

(continued)

TABLE I. Microinjected Fluorescent Analogs of the Cytoskeleton *(continued)*

Protein Derivatives	Reference
Tetramethylrhodamine iodoacetamide	Johnson et al., 1988; McKenna and Wang, 1986; MeKenna et al., 1985b, 1986; Meigs and Wang, 1986; Stickel and Wang, 1987, 1988; Wang, 1986.
Tetramethylrhodamine isothiocyanate	Feramisco, 1979; Feramisco and Blose, 1980; Hamaguchi and Mabuchi, 1986; Kreis and Birchmeier, 1980; Kreis et al., 1979; Mabuchi et al., 1985.
Filamin	
5-Iodoacteamidofluorescein	Mittal et al., 1987b.
Tetramethylrhodamine iodoacetamide	Mittal et al., 1987b; Sanger et al., 1986b.
Gelsolin	
Lissamine rhodamine B sulfonyl chloride	Cooper et al., 1988.
Metavinculin	
Tetramethylrhodamine isothiocyanate	Saga et al., 1985.
Tropomyosin	
Fluorescein isothiocyanate	Dome et al., 1988; Warren et al., 1985.
5-Iodoacetamidofluorescein	Wehland and Weber, 1980.
Lissamine rhodamine B sulfonyl chloride	Dome et al., 1988.
Vinculin	
5-(4,6,-Dichlorotriazinyl)aminofluorescein	Kreis, 1986.
Fluorescein isothiocyanate	Burridge and Feramisco, 1980; Meigs and Wang, 1986; Saga et al., 1985.
Lissamine rhodamine B sulfonyl chloride	Glascott et al., 1987; Kreis, 1986.
Tetramethylrhodamine iodoacetamide	Meigs and Wang, 1986.
Tetramethylrhodamine isothiocyanate	Stickel and Wang, 1988; Wang, 1986.
Tubulin	
Bis-caged-carboxyfluorescein	Mitchison, 1989.
Carboxyfluorescein succinimidyl ester	Vigers et al., 1988.
Carboxytetramethylrhodamine succinimidyl ester	Kellogg et al., 1988; Schulze and Kirschner, 1988; Vigers et al., 1988.
Carboxy-X-rhodamine succinimidyl ester	Gorbsky et al., 1988; Gorbsky and Borisy 1989b; Lim et al., 1989; Sammak and Borisy, 1988; Vigers et al., 1988.
5-(4,6-Dichlorotriazinyl)aminofluorescein	Cassimeris et al., 1988a; Gorbsky et al., 1987; Hamaguchi et al., 1987; Keith, 1987, 1988; Keith et al., 1981; Kellogg et al., 1988;

(continued)

TABLE I. Microinjected Fluorescent Analogs of the Cytoskeleton *(continued)*

Protein Derivatives	Reference
	Leslie et al., 1984; Salmon and Wadsworth, 1986; Salmon et al., 1984a,b; Sammak et al., 1987; Saxton and McIntosh, 1987; Saxton et al., 1984; Soltys and Borisy, 1985; Vigers et al., 1988; Wadsworth and Sloboda, 1983; Wadsworth and Salmon, 1986a,b, 1988; Webster et al, 1987.
Flourescein isothiocyanate	Hamaguchi et al., 1985, 1987.
Lissamine rhodamine B sulfonyl chloride	Kellogg et al., 1988.
MAPs	
Iodoacetamidofluorescein	Olmsted et al., 1989; Sherson et al., 1984; Vandenbunder and Borisy, 1986.
Tetramethylrhodamine iodoacetamide	Vandenbunder and Borisy, 1986.
Desmin	
Tetramethylrhodamine iodoacetamide	Mittal et al., 1989.

al., 1984; Wang, 1989]. The probes are usually prepared by reacting a limited number of the amino or sulfhydryl groups on the protein molecules with fluorescent reagents, which consist of a fluorophore attached to a reactive group such as isothiocyanate, sulfonyl chloride, or iodoacetamide [Haugland, 1989]. Following microinjection, the conjugates behave as analogs of the endogenous components and often become incorporated into physiological structures. The major advantage in using fluorescent analogues is the stable, direct association of the fluorophore to the protein under study, the potential of direct participation of the probes in structural formation, and the likely low toxicity due to the close similarity of the probe to the endogenous components. However, it is critical to ascertain that the fluorescent analogs have maintained the physiological properties and that they behave in a manner similar to the endogenous components after microinjection.

While well-characterized biochemical properties, such as self-assembly and ATPase activities, can be readily assayed in vitro, there are often interesting components, such as vinculin, whose biochemical functions are not entirely certain. In addition, some proteins, such as actin and tubulin, are likely to interact with a large number of accessory proteins, and it is impossible to be totally sure that all functions have been preserved. Fluorescent labeling may also result in quantitative changes, such as the level of ATPase activities of myosin [Reisler, 1982; Johnson et al., 1988]. It is often difficult to judge how such changes might affect the outcome of the experiments, especially if the physiological roles of the interactions affected are unclear or if the analog co-assembles with a large excess of endogenous proteins. An example is the decrease in the affinity of actin for profilin following modification of actin at cys-374 [Malm, 1984]. Even if the functions in vitro appear to be intact, it is still possible that the analog may not behave in a fashion identical to its endogenous counterpart following the microinjection. Possible factors affecting the incorporation include accessibility to assembly sites, rates of on-off equilibrium, presence of unoccupied binding sites, induction of adventitious binding by the chemical modification, and involvement of complex interactions in the incorporation process. In addition, fluorescent labeling may also have a profound effect on the rate of protein degradation [Kellogg et al., 1988].

Therefore, equally important is the analysis of the interactions between the analogs and cellular structures. Isolated myofibrils or cell models have been used for characterizing many analogs of the actin filament system [Sanger et al., 1984b,c]. However, associations of structural components with extracted structures may be very different from those with native structures in living cells [Dome et al., 1988; Johnson et al., 1988], due to the requirements of energy or protein factors, disruptions during extraction, or differences in the solution environment. Thus it is usually more informative to observe microinjected cells. Several types of observations have been performed to assess the incorporation and toxic effects of the analogs. First, the distribution of the analog has been compared to that of the endogenous protein, and to that of control proteins such as fluorescent BSA or ovalbumin, within the same cell [Taylor and Wang, 1978; Wang and Taylor, 1979]. The simplest approach to observe endogenous proteins is immunofluorescence, using a fluorophore different from that microinjected. However, immunofluorescence itself may sometimes be affected by problems such as fixation artifacts [Cooper et al., 1988], epitope shielding, or antibody accessibility [midbodies in mitotic cells fail to stain with tubulin antibodies; Saxton et al., 1984]. In addition, the results of immunofluorescence may be misleading if a large amount of the analog, relative to the endogenous counterpart, is microinjected. An equally important experiment is the comparison of structural organization in injected cells with that in uninjected cells. This will also allow the detection of disruptions induced by the analog. Artifacts observed so far include the formation of paracrystals upon the overinjection of actin into some cell lines [Wehland and Weber, 1980], and the formation of abnormally large asters following the overinjection of tubulin [Hamaguchi et al., 1985].

Several other types of assays are also useful. For example, drugs can be used to disrupt the distribution of cytoskeletal structures in microinjected cells, and the response of the analogs compared with that of endogenous components [Wehland and Weber, 1980]. It is also informative to vary the amount of microinjection [Schulze and Kirschner, 1986]; lack of effects would suggest that the process under study is not driven by the microinjected analogs. A particularly novel way to assay functions is to microinject the analog into mutants lacking the corresponding functional protein [Saxton and McIntosh, 1987]. Rescue of cellular function or morphology would suggest that the analogs are functional.

2. Fluorescent affinity probes. Besides fluorescent analogs, various types of fluorescent affinity probes have been used to label components in living cells. The assay for the binding properties of fluorescent affinity probes is usually easier than that for the functional properties of fluorescent analogs. In addition, since the fluorophore in affinity probes is usually farther separated from the labeled structural components, radiation damage may be reduced [Olmsted et al., 1989]. However, these foreign agents may themselves induce cellular disruptions, especially when used at a high concentration. In addition, experiments examining the dynamic exchange of components may be complicated by the on-off reactions of the probe with the target protein [Wang, 1987].

The best example of affinity probes is fluorescent phalloidin [reviewed by Faulstich et al., 1988]. When microinjected at a low concentration, it labels actin filaments without inducing detectable disruptions [Wehland and Weber, 1981; Hamaguchi and Mabuchi, 1982; Kukulies et al., 1985; Wang, 1987]. However, it causes dramatic changes in cellular morphology and behavior at high concentrations [Wehland et al., 1977]. Fluorescently labeled MAP2, due to its absence (or extremely low concentration) in non-neuronal cells [Vallee et al., 1984], has also been miccroinjected as a probe for microtubules in fibroblasts [Scherson et al., 1984]. Other probes such as fluorescently labeled colcemid [Hiratsuka and Kato, 1987], colchicine [Moll et al., 1982], heavy meromyosin [Sanger, 1975], DNase I [Haugland, 1989], and taxol are also worth consideration for specific purposes.

A related approach is the microinjection of fluorescently labeled antibodies against specific proteins. Two such studies involve the injection of monoclonal antibodies against tubulin [Wehland et al., 1983; Warn et al., 1987]. At low concentrations, the fluorescent antibodies cause no detectable disruptions to 3T3 cells and *Drosophila* embryos, and allow clear observations of microtubules through mitotic cycles. However, disruptions do occur at high concentrations. A second concern in using antibodies is of course the possible limitation in the accessibility of the binding sites. These problems may be minimized by using monoclonal antibodies against specific, nonfunctional domains and by using Fab fragments.

Finally, a new possibility emerges with the design of "anti-peptides" [Chaussepied and Morales, 1988]. By analyzing the amino acid sequence of target proteins, it may be possible to design peptides that recognize specific sites based on their charge or hydrophobic properties, and yield small, specific probes while minimizing perturbations to living cells.

B. Detection of Fluorescence Signals

Fluorescence signals from microinjected cells are usually very weak, due to the limited number of probes that can be microinjected without disrupting the cell. This problem cannot be solved by simply increasing the intensity of the excitation light, since the emission is proportional to the excitation intensity only to a certain extent, after which the number of fluorophores at the ground state may become a limiting factor. Intense illumination also induces photobleaching of fluorophores and damage to the structures.

Photodamage is an important concern for fluorescence microscopy of living cells. Detailed studies on the photodamage of actin and tubulin in vitro have been reported recently [Vigers et al., 1988; Simon et al., 1988]. Although photodamage of cells is usually manifested as readily detectable changes in morphology [Vigers et al., 1988], there may also be subtle changes, such as cross-linking of proteins [Leslie et al., 1984], that are difficult to detect by direct observations. However, the extent of damage can be greatly reduced by the presence of reducing agents, and by mixing fluorescent probes with excess unlabeled molecules [Leslie et al., 1984; Vigers et al., 1988]; both conditions are fulfilled in living cells. It also appears that the extent of photodamage is highly dependent on the condition of illumination [Leslie et al., 1984], on the fluorophore used, and on the protein under study. Experimentally, the possibility of photodamage may be ruled out by varying the frequency and duration of illumination (which should not affect the results), by performing immunofluorescence or high-resolution optical and electron microscopy [McKenna et al., 1985a; Saxton et al., 1984], and by testing the response of the illuminated structures to agents that induce structural reorganization.

Photobleaching and photodamage represent the major limitations in earlier studies, where direct photography is used for image recording. Low-light-level video cameras are now used extensively [Spring and Lowy, 1989], allowing intermittent recording of single living cells for several days without detectable disruptive effects [McKenna et al., 1986; Sanger et al., 1986a]. In addition, video signals can be easily processed by digital image processors to improve the signal-to-noise ratio, to enhance the contrast, and to perform quantitative measurements of the fluorescence intensity in specific areas [Arndt-Jovin et al., 1985]. However, in some studies, the signal may remain too weak despite the use of these techniques. One approach to amplify the fluorescence signal is to stain the cell with antibodies against the fluorophores [Gorbsky and Borisy, 1989a]. Although the cells have to be fixed and dynamic processes cannot be followed directly, this method has proven very useful in identifying a small number of incorporated molecules against a high level of unincorporated molecules [Soltys and Borisy, 1985; Amato and Taylor, 1986].

C. Analysis of Fluorescence Signals

What types of information can be obtained from the microinjection of fluorescent probes? Successful incorporation of an analog into physiological structures would indicate that the binding sites are available and accessible. Often this is indicative of a dynamic exchange of components between the soluble and incorporated states. By studying the pattern of fluorescence distribution during incorporation, it is also possible to identify active sites of structural assembly [Soltys and Borisy, 1985]. Furthermore, comparisons can be made between different protein isoforms and between proteolytic domains regarding the ability and pattern of incorporation [McKenna et al., 1985a; Johnson et al., 1988]. Conversely, it may be possible to study the turnover of protein analogs following their incorporation into structures.

The main advantage in studying living cells, of course, is the ability to follow dynamic processes. Both fluorescent analogs and affinity probes have been used extensively for this purpose. Conspicuous structures, such as stress fibers and myofibrils, can be observed directly, and pathways for the assembly or drug-induced disruptions can be determined [e.g., Wang, 1984; Meigs and Wang, 1986]. In addition, it should be possible to study the transport of soluble protein components, by carefully microinjecting fluorescent probes within a localized area and examining the subsequent pattern and rate of dispersal [Cao and Wang, unpublished results].

Photobleaching with laser microbeams, first applied to membrane components [Jacobson et al., 1987], has become a powerful tool for analyzing dynamics at the molecular level. Typically, a laser microbeam at a wavelength absorbed by the fluorophore is used to bleach fluorescence from a small area, 1–5 μm in diameter, without disrupting the integrity of the structures. The recovery of fluorescence then reflects the movement of unbleached molecules into the bleached area. In addition, recovery along discrete structures suggests that structural components undergo an equilibrium between incorporated and soluble states. The bleached spots have been observed either directly or after staining cells with antibodies that recognize unbleached, but not bleached, fluorophores [Sammak et al., 1987]. Light-induced artifacts can be largely ruled out by varying the extent of photobleaching and by comparing the recovery of multiple photobleachings at the same site [Wolf et al., 1980; Salmon et al., 1984b]; neither should affect the outcome of the experiments. In addition, control experiments as discussed above can be performed to rule out light-induced structural damage.

While the half-time and the extent of recovery are useful for qualitative comparisons, quantitative analyses of the recovery kinetics are limited by our knowledge of the way cytoskeletal components move in the cell. The movement may not follow the rule of diffusion. However, equally informative is the location of the bleached spot and the pattern of recovery, which can yield clues on the occurrence of directional transport and the polarity of structural assembly [e.g., Wang, 1985]. In addition, different patterns, such as multiple lines or spots [Saxton and McIntosh, 1987; McKenna and Wang, 1986] have been used to analyze not only the movement within one specific area but also the relative movement among different regions.

The complementary technique to photobleaching, photoactivation [Ware et al., 1986; Kraft et al., 1988], has recently been applied to cytoskeletal structures [Mitchison, 1989]. Analogs are prepared with ''caged'' fluorophores, which remain nonfluorescent until activation by a pulse of light. The fate of the bright spot is then followed after photoactivation. The major advantage of this approach is the superb signal to noise ratio: observations are made with a bright spot against a dark background, instead of a faintly fluorescent spot against a bright background as for photobleaching. This greatly facilitates the identification of small fractions of slowly moving molecules, which are extremely difficult to detect with photobleaching techniques [Mitchison, 1989]. In addition, the generation of active fluorophores is likely to be less perturbing to the structures than does photobleaching [Ware et al., 1986; Kraft et al., 1988]. The main drawback of photoactivation is the difficulty in knowing the location or distribution of specific structures. For example, it will be difficult to position the site of activation on a specific fiber, or to measure the movement relative to a particular site. However, this problem may be overcome by microinjecting a mixture of photoactivatable probes and fluorescent probes that absorb or emit at different wavelengths.

Many powerful spectroscopic techniques still await explorations in living cells. For example, the excitation/emission spectra or quantum yield of fluorophores are often affected by the local environment and protein conformation. Good examples are pyrene- and NBD-labeled actins, which show a large increase in fluorescence intensity upon polymerization [Kouyama and Mihashi, 1981; Detmers et al., 1981]. Thus ratio imaging techniques, similar to those used currently for mapping the distributions of intracellular ions [Bright et al., 1989; Tsien, 1989], can be used to map the conformation of proteins in the cell. If fluorophores are rigidly attached to the probe, polarization measurements may also be useful [Axelrod, 1989], at least qualitatively, for studying the assembly of structures from subunits. For example, regions with an increased polarization may be identified before discrete structures become detectable. Resonance energy transfer has also been used to study protein interactions in vitro [Stryer, 1978], as well as the dynamics of lipid molecules and proteins on the membrane of living cells [Uster and Pagano, 1988; Herman, 1989]. If donors and acceptors are located at different sites of the same protein molecule, energy transfer between them (intramolecular) can be used as a highly sensitive tool for detecting conformational changes within living cells.

The application of intermolecular energy transfer will be more difficult, due to the serious dilution of probes upon microinjection. However, the approach may be useful for following the dissociation of macromolecular complexes. If preformed complexes, containing interacting donors and acceptors, are microinjected, subsequent dissociation would be manifested as a loss of resonance energy transfer.

III. MICROINJECTION OF FLUORESCENT PROBES FOR ACTIN FILAMENTS

Fluorescent analogs of various components of the actin filament system have been microinjected into living cells. Valuable insights have been gained by examining interactions between the fluorescent analogs and cellular structures, especially concerning the possible roles of protein isoforms. In addition, important information about the dynamics of lamellipodia, stress fibers, and developing myofibrils has been obtained by studying the reorganization of fluorescently labeled structures, using both direct observations and photobleaching techniques.

A. Association of Fluorescently Labeled Actin and Accessory Proteins With Cellular Structures

1. Incorporation of microinjected actin analogs. The earliest microinjection of a fluorescent analog of the cytoskeletal components was performed with actin labeled with 5-iodoacetamidofluorescein at cys-374 [Taylor and Wang, 1978]. Successful incorporation subsequently was obtained with actin labeled by a variety of fluorescent reagents (Table I). Assays performed on some of the conjugates indicated normal polymerizability and activation of the myosin Mg-ATPase [Wang and Taylor, 1980].

Following microinjection, fluorescent analogs of actin incorporate, within 30 min, into structures in amoeba [Taylor and Wang, 1978], acellular slime molds [Taylor and Wang, 1978], sea urchin eggs [Wang and Taylor, 1979], macrophages [Amato et al., 1983], muscle cells [Glacy, 1983a], and fibroblasts [Kreis et al., 1979]. The distribution of the analog is sensitive to cytochalasin B, as expected for cytoplasmic actin filaments [Kreis et al., 1979; Wehland and Weber, 1980]. At steady state, the distribution of labeled structures is very similar to that revealed by fluorescent phalloidin staining or immunofluorescence [Kreis et aal., 1982]. In addition, there is a diffusely distributed signal throughout the cytoplasm, which probably represents a combination of filament networks and soluble molecules. The latter have also been suggested based on the large mobile fraction detected by photobleaching recovery techniques [Kreis et al., 1982; Wang et al., 1982b]. Thus, it is likely that the polymerization of actin analogs is regulated by cytoplasmic factors.

In fibroblasts, microinjected actin analogs are first incorporated into lamellipodia, and then into adhesion plaques and stress fibers. A steady state is reached within 20 min [Glacy, 1983b]. Under the light microscope, the incorporation takes place simultaneously along the entire length of stress fibers [Glacy, 1983b]. However, when immunoelectron microscopy is used to detect the labeled actin, periodic sites of incorporation can be observed along the stress fiber 5 min after microinjection [Amato and Taylor, 1986]. Such sites may correlate with the periodic localization of alpha actinin or other accessory proteins along stress fibers.

The incorporation of fluorescent actin has also been examined in muscle cells. An early study yielded the unexpected results that the analog is incorporated into the I-bands (where endogenous actin filaments are located) and the M-lines (in the middle of the A-bands, where myosin filaments are located), but not into the Z-lines (where the ends of actin filaments are anchored), of myofibrils in cardiac myocytes [Glacy, 1983b]. However, more recent studies indicate that fluorescent actin incorporates into the I-bands and the Z-lines, but not the M-lines [McKenna et al., 1985a; Dome et al., 1988]. These discrepancies may be related to the metabolic state of the cell or to the handling of actin before microinjection [McKenna et al., 1985a].

The microinjection approach has been used to compare the incorporation properties of actin isoforms [McKenna et al., 1985a]. Brain actin and skeletal muscle actin, labeled with different fluorophores, were co-injected into muscle and non-muscle cells. At the resolution of a light microscope, no difference was detected between the isoforms along myofibrils in cardiac myocytes or stress fibers in fibroblasts, suggesting that the binding sites on these structures cannot differentiate between the isoforms. It was suggested that differential localization of isoforms, as observed with immunofluorescence in adult muscle tissues [Lubit and Schwartz, 1980; Pardo et al., 1983], may be achieved through differential synthesis, stabilization, or degradation. However, it is also possible that subtle differences between isoforms may exist between submicroscopic domains [Otey et al., 1988] or in some specific structures such as lamellipodia [DeNofrio et al., 1989].

2. Incorporation of microinjected myosin analogs. Myosins from skeletal or smooth muscles have been fluorescently labeled and microinjected into muscle and non-muscle cells [Johnson et al., 1988; DeBiasio et al., 1988; McKenna et al., 1989a,b]. Labeled smooth muscle myosin was used as an analog for non-muscle myosin since the two share similar biochemical properties [Scholey et al., 1982]. However, McKenna et al. [1989b] reported a partial inhibition of the ATP-induced conformational change [Suzuki et al., 1978], upon fluorescent labeling of the smooth muscle myosin.

Following microinjection into myotubes, skeletal muscle myosin disperses over a period of 1/2 to 3 h and becomes incorporated into A-bands [Johnson et al., 1988]. This is significant since skeletal muscle myosin is relatively insoluble under physiological ionic conditions. Once incorporation reaches a steady state, the exchange occurs very slowly as indicated by a very slow recovery of fluorescence after photobleaching [Johnson et al., 1988]. Thus, the association of the myosin analog with myofibrils probably represents addition to existing unoccupied binding sites, rather than an on-off reaction of the myosin molecules. Johnson et al. also microinjected different proteolytic fragments of myosin. Incorporation was observed ith the light meromyosin portion but not with heavy meromyosin, indicating that actin-myosin interactions are probably not required.

Both labeled skeletal and smooth muscle myosins have been microinjected into fibroblasts [DeBiasio et al., 1988; McKenna et al., 1989a,b]. Both analogs were incorporated into periodically arranged beads along stress fibers, indistinguishable from those shown by immunofluorescence [McKenna et al., 1989a,b]. In addition, a large number of elongated beads, either poorly organized or organized as fine linear arrays or networks, can be discerned in thin, spread regions of the cell [DeBiasio et al., 1988; McKenna et al., 1989a,b]. Judging from their length, ~0.7 μm, the beads probably represent single myosin filaments.

Each myosin molecule is known to consist of two heavy chains and four light chains. These polypeptides appear to associate stably in vitro and undergo only very slow exchange [Burke and Sivaramakrishnan, 1981]. However, when fluorescently labeled light chains of skeletal muscle myosin were microinjected, the analogs became colocalized with A-bands of myofibrils in myotubes and with stress fibers in epithelial cells within 3–5 h [Mittal et al., 1987a]. In dividing cells, the fluorescent light chains became concentrated in the equatorial region, as expected with myosin molecules. Similar observations were made when labeled regulatory (20 kD) light chain of smooth muscle myosin was microinjected into 3T3 cells [McKenna et al., 1989a]. The incorporation into stress fibers in this case reached a steady state within 1 h [McKenna et al., 1989b]. Thus, although it is difficult to demonstrate that the light chain analogs have associated correctly with heavy chains, these results strongly suggest an ability of myosin molecules, even different isoforms, to exchange their light chains in vivo. In addition, since the phosphorylation of light chains plays an important role in the regulation of smooth and non-muscle myosins, the results raise the interesting possibility that one phosphorylated light chain may be able to activate multiple myosin molecules [McKenna et al., 1989b]. Another interesting question is whether muscle and non-muscle myosin heavy chains are known to locate in different regions [Fallon and Nachmias, 1980].

3. Incorporation of microinjected actin-binding proteins. A number of actin binding proteins have been fluorescently labeled and microinjected into cultured muscle and non-muscle cells. The distribution of alpha actinin analogs in fibroblasts mimics that of endogenous alpha actinin as detected by immunofluorescence [Feramisco, 1979; Feramisco and Blose, 1980]. Since the distribution of alpha actinin closely parallels that of actin filaments but shows a lower diffuse signal, alpha actinin analogs have been used as an alternative probe for actin filaments. In addition, the clarity of the punctate alpha actinin structures along stress fibers has greatly facilitated the observation of contractile events [Kreis and Birchmeier, 1980; Wang, 1986; Sanger et al., 1986b].

When microinjected into muscle cells [Kreis and Birchmeier, 1980; Sanger et al., 1984a], analogs of alpha actinin were incorporated within 30 min into Z-lines of myofibrils. As for myosin analogs, fluorescent alpha actinin probably binds to myofibrils through the association with unoccupied binding sites rather than the on-off equilibrium. Once associated with Z-lines, the rate of exchange is low as shown by photobleaching recovery techniques [McKenna et al., 1985b].

Non-muscle alpha actinin, unlike muscle alpha actinins, cross-links actin filaments in a calcium-sensitive manner [Burridge and Feramisco, 1981]. However, alpha actinin analogs prepared from smooth muscles are able to associate with non-muscle structures [Feramisco, 1979]. In addition, when a mixture of muscle and non-muscle alpha-actinins was microinjected into myotubes, both types of alpha actinins incorporated into Z-lines with an apparently equal efficiency [Sanger et al., 1986a]. Therefore, these experiments fail to identify a physiological role of the calcium sensitivity of the non-muscle alpha actinin in resting cells. However, it is still possible that the different isoforms may behave differently in some Ca-activated events.

Filamin, another actin filament cross-linking protein, has been fluorescently labeled without affecting its cross-linking activity [Mittal et al., 1987b]. Following microinjection into epithelial cells (PtK2) and fibroblasts (3T3), the filamin analog became associated with stress fibers in a punctate pattern. In muscle cells, fluorescent filamin is associated with the Z-line and its precursors, consistent with immunofluorescence observations [Gomer and Lazarides, 1981].

Gizzard tropomyosin has been derivatized with iodoacetamidofluorescein and microinjected into epithelial cells and fibroblasts [Wehland and Weber, 1980]. The initial study indicates that the fluorescent analog is associated with stress fibers but not with lamellipodia. However, Warren et al. [1985], using a combination of immunofluorescence and fluorescent analog cytochemistry, demonstrated that tropomyosin is present in the lamellipodia during early stages of spreading, and becomes depleted from the area as the cell spreads out and assembles large stress fibers. Thus it appears that the absence of tropomyosin in the lamellipodia of well spread cells may not represent an inherent property of the lamellipodia. One possible explanation is that tropomyosin associates tightly with actin filaments and undergoes only slow exchanges among the filaments. If actin turns over rapidly in the lamellipodia, there may not be enough time for tropomyosin molecules to redistribute from existing filaments in the stress fibers and bind to newly assembled actin subunits in the lamellipodia.

Isoforms of tropomyosin differ in length and in their affinity for actin filaments [Fine et al., 1973]. Dome et al. [1988] compared the incorporation of fluorescently labeled brain (~30 kD) and gizzard (~42 kD) tropomyosin isoforms into contractile structures in muscle and non-muscle cells. In PtK epithelial cells, both brain and gizzard isoforms incorporate into stress fibers. However, when injected into skeletal myotubes, only the gizzard tropomyosin became incorporated into myofibrils. The simplest explanation is that brain tropomyosin, with its lower affinity for actin filaments, is unable to compete with the endogenous skeletal muscle tropomyosin [Dome et al., 1988]. This also implies that non-muscle tropomyosin may undergo a more rapid exchange among different structures than does muscle tropomyosin.

Microinjection of fluorescently labeled vinculin was performed soon after the identification of the protein [Burridge and Feramisco, 1980]. Although it was uncertain how much its biochemical properties were affected by the fluorescent labeling, the analog did maintain the ability to associate with adhesion plaques. Microinjection has been used to compare the localization of vinculin and meta-vinculin, which is similar to vinculin but is present only in muscle tissues [Feramisco et al., 1982; Siliciano and Craig, 1982; Saga et al., 1985]. A mixture of vinculin and meta-vinculin analogs, labeled with different fluorophores, was microinjected into myotubes and fibroblasts [Saga et al., 1985]. Both analogs became incorporated into periodic, linear structures in myotubes, and into adhesion plaques in fibroblasts, indicating that cellular binding sites cannot discriminate vinculin from meta-vinculin. However, it is still unclear whether or not endogenous vinculin and meta-vinculin assume identical distributions in muscle cells. Furthermore, since vinculin can bind to adhesion plaques through at least two domains [Bendori et al., 1989], it is possible that vinculin and meta-vinculin may be incorporated through different interactions.

Gelsolin is a calcium-activated protein that caps and severs existing actin filaments and nucleates the assembly of new filaments [Yin and Stossel, 1980]. In resting cells, gelsolin is probably inactive, since microinjection of intact gelsolin induces no detectable effect but injection of a calcium-independent fragment causes dramatic disruptions to actin filaments [Cooper et al., 1987]. In order to determine the localization of gelsolin in resting cells, Cooper et al. microinjected a fluorescent analog into fibroblasts [Cooper et al., 1988]. The analog assumed a diffuse distribution without a detectable association with stress fibers, contrary to what was seen by immunofluorescence. However, association of the analog with stress fibers became visible after fixation of microinjected cells. Although alternative explanations are possible, Cooper et al. suggested that gelsolin probably exists predominantly in the soluble form, and makes only transient associations with stress fibers. Chemical fixation may render the association irreversible and create the bright staining of stress fibers. This explanation appears consistent with the photobleaching recovery data, which show gelsolin to be close to 100% mobile, and with the ease of extraction of endogenous gelsolin molecules as shown in the same study.

B. Dynamics of Actin and Myosin as Studied by Microinjected Fluorescent Probes

1. Dynamics of actin in lower eukaryotic cells. Microinjection of fluorescent probes has been applied to study the dynamics of actin-containing structures in *Amoeba proteus* [Taylor et al., 1980; Gawlitta et al., 1980, 1981], and *Physarum polycephalum* [Kukulies and Stockem, 1985; Kukulies et al., 1984, 1985; Brix et al., 1987]. Formation of discrete actin-containing structures was observed during phagocytosis [Stockem et al., 1983], lectin-induced capping [Taylor et al., 1980], and polyamine-induced cleavage [Gawlitta et al., 1981]. The active motility of such organisms makes them attractive targets for examining the organization of actin and myosin. However, detailed analysis has been limited by the irregular shape and the large thickness of the cell. The application of new cell flattening techniques [Fukui et al., 1986], confocal microscopy [Brakenhoff et al., 1989], and ratio imaging [for correcting variations due to the pathlength; Bright et al., 1989] should greatly facilitate the analysis of images from these organisms.

The dynamics of actin and alpha actinin have also been examined in sea urchin eggs [Wang and Taylor, 1979; Hamaguchi and Mabuchi, 1986, 1988]. No distinct structures were detectable in unfertilized eggs with microinjected analogs. However, within 5 min of fertilization, a high concentration of actin and alpha actinin was observed in the membrane cortex and in the fertilization cone, corresponding to the polymerization of actin and elongation of microvilli reported previously [Begg and Rebhun, 1979]. However, results of photobleaching recovery indicate that actin subunits undergo very rapid on-off reactions along the cortex [Hamaguchi and Mabuchi, 1988]. The intensity of cortical fluorescence decreases 5–6 min after fertilization [Wang and Taylor, 1979; Hamaguchi and Mabuchi, 1986, 1988], but appears to increase again before the first mitosis [Wang and Taylor, 1979]. During mitosis, injected actin can be detected within the mitotic spindle, but this is likely due to a large accessible volume in the spindle since injected ovalbumin shows a similar distribution. During cytokinesis, the contractile ring, which is known to contain bundles of actin [Schroeder, 1981], cannot be detected with the fluorescent actin, indicating that although actin filaments are organized, the concentration may not be much higher than that in other regions of the cell. Injection of fluorescent alpha actinin similarly revealed no clear concentration but only a slight thickening of the cortex on the equatorial plane [Mabuchi et al., 1985]. However, the contractile ring has been detected with injected alpha actinin in dividing PtK epithelial cells, which maintain a spread morphology during cytokinesis, facilitating the detection of fine structures [Sanger et al., 1987].

2. Dynamics of stress fibers and developing myofibrils. Much attention has been paid to the stress fibers and adhesion plaques in cultued higher animal cells. Wehland and Weber [1981] microinjected fluorescent phalloidin into rat mammary cells and detected changes in the pattern of stress fibers over time. Direct time-lapse observations of stress fibers were subsequently made by following microinjected actin analogs [Wang, 1984]. Formation of new stress fibers appears to originate at discrete sites near the leading and trailing edges of the cell. The nascent fibers subsequently extend toward the nucleus. As the cell moves and changes shape, stress fibers undergo concomitant changes such as reorientation and fragmentation. Elongation of stress fibers is achieved by the formation of new segments near the edge and subsequent merging with existing fibers, while the distal ends of existing fibers (which are associated with adhesion plaques) remain stationary [Wang, 1984].

Reorganization of stress fibers at the molecular level was examined by photobleaching techniques [Kreis et al., 1982; Wang, 1987]. Fluorescence recovery of stress fibers labeled with either fluorescent actin or alpha actinin shows a half-time of ~10 min, independent of the length of the bleached segment [Kreis et al., 1982; McKenna et al., 1985b]. It was originally proposed that this may reflect an on-off equilibrium of actin subunits and accessory proteins along the length of stress fibers [Kreis et al., 1982]. However, a similar half-time was subsequently obtained with the photobleaching of fluorescent phalloidin, which associates tightly with actin filaments but not monomers [Wang, 1987]. Thus it is possible that an exchange of filamentous actin along stress fibers may be responsible for the fluorescence recovery, and that actin filaments may be mobile in living cells. A second mode of reorganization, which has been detected in stress fibers near the trailing edge, is the centripetal translocation of actin and alpha-actinin along stress fibers. The movement occurs at a rate of 0.2–0.3 μm/min [McKenna and Wang, 1986], and may be due to either an active flux or a passive stretching.

The organization of stress fibers can be disrupted under various conditions, for example, during the initiation of mitosis [Sanger et al., 1987], during oncogenic transformation [Vasiliev, 1985], or following treatments with various agents such as metabolic inhibitors and phorbol esters [Rifkin et al., 1979; Bershadsky et al., 1980]. Microinjection of fluorescent analogs has been used as a powerful tool to determine the pathways of the destruction and recovery of stress fibers [Wehland et al., 1980; Sanger et al., 1980; Meigs and Wang, 1986; Wang, 1986; Sanger et al., 1987; Glascott et al., 1987; DeBiasio et al., 1987; Stickel and Wang, 1988]. There are clearly multiple pathways for the disruption. For example, while phorbol esters and a synthetic peptide that disrupts the association of extracellular matrices induce a dissociation of stress fibers from adhesion plaques [Meigs and Wang, 1986; Stickel and Wang, 1988], metabolic inhibitors induce contractions toward the adhesion sites [Wang, 1986].

The mobility of proteins may also be profoundly affected. For example, although substrate adhesion structures in transformed cells are similar to adhesion plaques in many ways, studies with photobleaching recovery indicate that alpha actinin exchanges at a much faster rate at the adhesion structure in transformed cells than in normal cells [Stickel and Wang, 1987].

Fluorescent analog cytochemistry also has been performed on developing muscle cells to examine the mechanism of myofibrillogenesis. Arrays of punctate structures, which are precursors of Z-lines and are labeled with fluorescent alpha actinin, can be detected near the periphery of developing myotubes and cardiac myocytes [Sanger et al., 1984a; Sanger et al., 1986a]. Sanger et al. [1986a, 1986b] reported that the distance between neighboring punctate structures in developing myotubes increases from 0.9–1.3 μm to 1.6–2.3 μm over 24 h, as the punctate structures develop into Z-lines. They proposed that the increase in distance is a result of direct elongation of nascent sarcomeres. However, similar experiments by McKenna et al. [1986] failed to reveal such an elongation. Instead, adjoining punctate structures appear to coalesce to form fewer, larger punctate structures with a longer distance, while the distance between groups of coalescing punctate structures gradually becomes more regular and approaches the length of sarcomeres in mature myofibrils (~2 μm). Thus the apparent increase in the distance between Z-line precursors may be achieved through coalescence. However, it is difficult to rule out direct growth based on negative results. It would be informative to follow directly the extension of myofibrils at their ends. If direct elongation takes place as proposed by Sanger et al. (1986a), the ends should move by a distance equal to as much as 70% of the starting myofibril length over 24 h.

3. Dynamics of lamellipodia. Lamellipodia, or "ruffles," are the most motile area in cultured fibroblasts and epithelial cells. Previous studies have demonstrated the presence of a high concentration of actin filaments and small filament bundles in this area [Small, 1981]. However, it is not clear how the filaments are related to the motile activity.

In fibroblasts, actin analogs are incorporated into the lamellipodium within 5 min of microinjection [Glacy, 1983b], indicating that actin subunits may undergo a rapid turnover. The pattern of turnover has been studied by bleaching a small spot on the lamellipodia of cells microinjected with a fluorescent actin analog [Wang, 1985]. The bleached spot moves toward the nucleus at a constant rate of ~0.8 μm/min, suggesting that there is a constant flux of actin subunits in the lamellipodia. Confirming results were subsequently obtained with video enhanced DIC optics in 3T3 cell and neuronal growth cones [Forscher and Smith, 1988; Fisher et al., 1988]. The flux of actin subunits is probably generated by a constant assembly of actin at the filament-membrane interface and a disassembly at the opposite end. This process may play a critical role in cytoplasmic protrusion.

Equally important is the distribution of myosin in the lamellipodium region. Early immunofluorescence observations indicated that myosin may be absent from this area. However, observation of injected myosin analogs indicated that myosin is absent only during the earliest stage of lamellipodia formation, but can be detected in diffuse or bead forms after the initial protrusion stops [DeBiasio et al., 1988]. In a recent study with myosin analogs, McKenna et al. [1989b] demonstrated that bead structures in the lamellipodium area appear as a result of de novo assembly, rather than forward movement of existing structures. The assembly of new beads, which probably correspond to myosin filaments, is coupled to a backward movement of the beads at an average rate of 0.18 μm/min and an organization of the beads into linear arrays. The de novo assembly of beads near lamellipodia is likely to be coupled to a disassembly in other regions of the cell, thus myosin molecules may undergo a constant cycle between the backward moving, filamentous form and the forward diffusing, unassembled form. Although the role of myosin in cytoplasmic protrusion has been questioned [De Lozanne and Spudich, 1987; Knecht and Loomis, 1987], it is possible that the flux of myosin may play an important role in the determination of cell polarity. Disruption of cell polarity has indeed been observed following the microinjection of myosin antibodies [Honer et al., 1988].

IV. MICROINJECTION OF PROBES FOR MICROTUBULES

Compared to actin filaments, the larger size and smaller number of microtubules greatly facilitate their detection and analysis in living cells. However, due to the high susceptibility of tubulin to denaturation, fluorescently labeled tubulin was not successfully prepared until 1981 [Keith et al.]. The first analog was prepared with dichlorotriazinyl fluorescein (DTAF), which reacts primarily with amino groups at a neutral pH. conjugates with other reagents, including succinimidyl esters of carboxyfluorescein and carboxyrhodamine, were subsequently reported (Table I). Assays of these labeled preparations indicate a more-or-less normal polymerization behavior, although some appear to have a relatively low labeling stoichiometry and may not show clear differences from unlabeled controls. However, microinjection of different analogs has so far yielded relatively consistent results in both incorporation behavior and patterns of reorganization, as discussed in the following sections. The availability of rhodamine-labeled tubulin marks a significant advancement since earlier studies with fluorescein conjugates have been limited by photobleaching, photodamage, and interference from autofluorescence.

A. Association of Fluorescently Labeled Tubulin and Microtubule-Associated Proteins With Cellular Structures

Fluorescent tubulins become readily incorporated into fibrous structures after microinjection, showing a distribution very similar to that of tubulin immunofluorescence [Saxton et al., 1984]. The fine linear structures in interphase higher animal cells probably represent single microtubules. The distribution of microinjected tubulin is sensitive to agents such as colcemid, nocodazole, and taxol, as expected for cytoplasmic microtubules [Keith et al., 1981; Wadsworth and Sloboda, 1983; Salmon et al., 1984b; Saxton et al., 1984].

Of great interest are the initial sites of incorporation, which most likely correspond to sites of active assembly of tubulin dimers. Observations of living cells soon after microinjection often give an impression that the incorporation takes place along the entire length of microtubules [Hamaguchi et al., 1985]. However, due to the presence of high concentrations of unincorporated molecules, detailed analysis is possible only after extraction of injected cells. Using immunofluorescence against fluorescein to amplify the signal, Soltys and Borisy [1985] identified discrete segments of labeled microtubules near the centrosomes and at the distal, "plus" ends of existing microtubules in interphase fibroblasts 7–14 min after microinjection. Schulze and Kirschner [1986] subsequently made similar observations with biotin-labeled tubulin and demonstrated that the incorporation near centrosomes represents de novo assembly rather than addition of tubulin to the "minus" ends of existing microtubules.

In metaphase cells, assembly also takes place near the centrosomes [Mitchison et al., 1986]. Like interphase microtubules, the kinetochore microtubules incorporate labeled subunits at the plus end [associated with kinetochores, Schulze and Kirschner, 1986], indicating that the association with kinetochores does not inhibit the assembly of microtubules. Alternatively, it is possible that microtubules may undergo constant on-off reactions with the kinetochore [Cassimeris et al., 1988a], and the incorporation of subunits may occur prior to the association with kinetochores.

Microinjection has also been used to study the possible differences between tubulin isoforms and the effects of post-translational modifications. Hamaguchi et al. [1985] reported that fluorescent analogs prepared from brain and sperm flagella tubulin showed a similar distribution in the mitotic spindle of sand dollar eggs, although it was difficult to rule out subtle differences at the resolution of a light microscope. The effects of tubulin tyrosination on incorporation have been studied by microinjecting fluorescently labeled, detyrosinated tubulin into human fibroblasts [Webster et al., 1987]. The analog was rapidly incorporated into microtubules, indistinguishable from the incorporation of tyrosinated tubulin, and then became tyrosinated over 1 h. In a converse experiment biotinylated tubulin was microinjected into human retinoblastoma cells, which contain discrete populations of tyrosinated-nonacetylated and detyrosinated-acetylated microtubules [Schulze et al., 1987]. The tubulin analog was preferentially incorporated into tyrosinated-nonacetylated and detyrosinated-acetylated microtubules [Schulze et al., 1987]. The tubulin analog was preferentially incorporated into tyrosinated-nonacetylated microtubules over a period of 1–2 h, indicating that tubulins in detyrosinated-acetylated microtubules turn over more slowly. Similar observations were made with the neuron-like PC12 cells [Lim et al., 1989]. The turnover of tubulin appears slower along the axon, where acetylated tubulins are preferentially localized, than at the growth cone. However, this correlation does not hold in many other cell types such as chick fibroblasts [Schulze et al., 1987], which often contain detyrosinated-acetylated microtubules that undergo a rapid turnover. Therefore, stable microtubules may simply have a tendency to accumulate certain species of tubulins, but post-translational modification itself appears insufficient to affect the stability of microtubules.

In addition to tubulin analogs, functional analogs of microtubule-associated proteins (MAPs), including MAP2 and MAP4, have been prepared [Scherson et al., 1984; Vandenbunder and Borisy, 1986; Olmsted et al., 1989]. Since MAP2 has been found predominantly in neuronal cells [Vallee et al., 1984], fluorescent MAP2 should be viewed as an affinity probe, rather than an analog of endogenous proteins, when injected into non-neuronal cells. However, following microinjection into fibroblasts and epithelial cells, both MAP2 and MAP4 became rapidly associated with microtubules [Scherson et al., 1984; Olmsted et al., 1989]. Despite the effects of MAPs in stabilizing microtubules in vitro [Murphy et al., 1977], microinjection of even a significant amount of these analogs does not appear to affect the cold or drug-induced reorganization of microtubules, or the time course of mitosis [Vandenbunder and Borisy, 1986; Olmsted et al., 1989].

B. Dynamics of Microtubules as Studied by Microinjected Analogs

Microinjected tubulin analogs remain associated with microtubules for many hours and through multiple cell cycles [Keith et al., 1981; Wadsworth and Sloboda, 1983], thus providing a powerful means for following changes in microtubule distribution. For cells that are difficult to fix, microinjection of fluorescent tubulins has provided useful details about how microtubules reorganize through the cell cycle [Hamaguchi et al., 1985; Kellogg et al., 1988].

1. Dynamics of microtubules in interphase cells. Studies have been performed to determine the rate of the incorporation of tubulin analogs into microtubules in interphase cells. For higher animal cells, the half-time is estimated to be between 10 and 30 min [Saxton et al., 1984; Schulze and Kirschner, 1987]. Schulze and Kirschner [1986] demonstrated that the incorporation into the plus ends (discussed above) of microtubules follows a steady time course, with a rate independent of the concentration of the tubulin analog microinjected. Thus the incorporation probably reflects the normal turnover of microtubules, rather than polymerization driven by the increase in tubulin concentration following microinjection. The average rate of growth, 3–5 μm/min, is fast enough to allow a microtubule to reach from the centrosome to the cell periphery in 5 min. Therefore, in order to maintain a constant total length of microtubules in a cell at steady state, such a fast assembly must be balanced by an equally rapid disassembly.

More detailed information on the dynamics of microtubules was obtained by examining the behavior of individual microtubules. It appears that the incorporation of tubulin analogs into microtubules occurs asynchronously; in some cells there are even "quasi-stable" microtubules that do not bind microinjected tubulin over an extended period of time [Schulze and Kirschner, 1986, 1987]. These results were extended recently by the direct observations of individual microtubules labeled with rhodamine tubulin analogs [Sammak and Borisy, 1988; Schulze and Kirschner, 1988]. Not only do microtubules show a wide range of growth rates, varying both over time and among different microtubules, individual microtubules also show distinct phases of growth, stability, and shortening. Each microtubule spends on the average 50% of the time growing, 32% of the time shortening, and 18% of the time resting [Sammak and Borisy, 1988]. As expected from the frequency, shortening is more rapid than growth, but rarely results in a complete disassembly of microtubules. Instead, after shortening for ~5 μm, a microtubule is as likely to grow as to continue to shorten. These observations, although differing in quantitative details, are similar to those made with uninjected cells using video-enhanced DIC optics [Cassimeris et al., 1988b]. Thus it is unlikely that the rapid shortening is due to the microinjection of tubulin analogs or photodamage. The manifestation of discrete phases of microtubule assembly and disassembly closely parallels the dynamic instability of microtubules as observed in vitro [Mitchison and Kirschner, 1984; Walker et al., 1988].

The results of these elegant studies are complemented by photobleaching studies. In unfertilized sea urchin eggs microinjected with DTAF tubulin, where no microtubules are detectable, the kinetics of fluorescence recovery indicate a single mobile species with an apparent diffusion coefficient of 5×10^{-8} cm^2/s [Salmon et al., 1984a]. This value is close to what one would expect if tubulin moves freely as dimers. In interphase PtK and BSC cells, there is a similar rapid recovery, reflecting the existence of a pool of unpolymerized tubulin dimers [Saxton et al., 1984; Salmon and Wadsworth, 1986]. However, a large portion of fluorescence recovers with a half-time of 4–5 min. This presumably represents exchange of tubulin dimers on and off polymers.

The mechanism of recovery was further studied by analyzing the location of the bleached spot and the pattern of recovery. No longitudinal movement of the bleached spot was detected with microtubules containing either fluorescent tubulin or MAP2 [Saxton et al., 1984; Sammak et al., 1987; Scherson et al., 1984]. In addition, when a wide band was bleached across a group of microtubules, the recovery occurred asynchronously, such that the bleached area gradually contained an increasing number of completely recovered microtubules [Sammak et al., 1987]. The probability of recovery increases with a decreasing distance between the bleached area and the area of the cell. Taken together, these observations suggest that the recovery is a rapid, asynchronous process occurring predominantly at the distal end of the microtubule, consistent with a model involving repeated incomplete depolymerization and repolymerization as suggested by the incorporation experiments. On the other hand, "treadmilling" of tubulin along microtubules, which involves a continuous assembly of tubulin subunits at the distal end, coupled to a centripetal movement of tubulin dimers along microtubules and disassembly at the centrosomes [Margolis and Wilson, 1981], does not appear to play a major role in the turnover of interphase microtubules.

The method of fluorescence recovery after photobleaching was also used to study axonal transport in neuron-like PC12 cells [Keith, 1987]. The cells were microinjected with DTAF tubulin and subsequently photobleached with an unattenuated mercury arc lamp at 450–490 nm for 5 min. The bleached spot split into two spots moving at different rates. The rate of the slower component is similar to that of neurite growth. However, the main concern is whether the prolonged bleaching, which has been found to cause severe damage to microtubules in vitro [Leslie et al., 1984], induces artifacts. Recent studies indicate that quite different results can be obtained depending on the condition of photobleaching [Keith, 1988; Lim et al., 1989]. The correlation between the transport of microtubules and neurite growth therefore awaits further study.

2. Dynamics of microtubules in mitotic cells. Microtubules in prophase and metaphase mitotic spindles appear to be on the average much more dynamic than those in interphase cells [Salmon et al., 1984b; Saxton et al., 1984; Olmsted et al., 1989]. Incorporation of tubulin analogs into the centrosomal region occurs with a half-time of 5–40 s. Studies with photobleaching recovery of fluorescent tubulin similarly yielded a very short half time, on the order of 10–20 s in cultured higher animal cells and sea urchin eggs. It was originally proposed that such a rapid exchange of subunits may be a result of tubulin dimers undergoing association-dissociation all along the length of microtubules [Inoue and Ritter, 1975; Salmon et al., 1984b]. However, since subsequent incorporation experiments with biotinylated tubulin revealed a clear segregation of labeled segments from unlabeled segments [Mitchison et al., 1986], an end-dependent process is more likely.

The increase in exchange when cells enter prophase may be due to a decrease in the average length of microtubules. This will be coupled to an increase in the number of ends and in the net rate of turnover. In addition, if microtubules undergo cycles of incomplete disassembly-reassembly at the plus ends as in interphase cells, a shorter average length will decrease the average distance of tubulin dimers from the plus ends and increase the population of tubulin dimers undergoing active turnover. However, stimulation of turnover could also be induced by increases in the rate constants of tubulin association and dissociation. Although there is no definitive evidence for such modulation, dramatic effects on the rate of tubulin turnover have been observed following treatment of cells with metabolic inhibitors [Wadsworth and Salmon, 1988]. Heterogeneous dynamics have also been reported for kinetochore, astral, and interzonal microtubules within the same dividing cell [Mitchison et al., 1986; Saxton and McIntosh, 1987]. Furthermore, fluorescent MAPs also show different rates of on-off reactions along microtubules, both between microtubules in interphase and in mitotic cells, and among different subpopulations of microtubules within the same mitotic cell [Olmsted et al., 1989]. These observations are consistent with the existence of a mechanism that can modulate the stability of individual microtubules.

The incorporation of tubulin analogs into the plus ends of kinetochore microtubules appears much slower than the incorporation into non-kinetochore microtubules or into microtubules in interphase cells [Mitchison et al., 1986; Schulze and Kirschner, 1986], possibly due to the association of microtubules with the kinetochore. However, recent photobleaching studies by Gorbsky and Borisy [1989b] indicated that about 50% of microtubules that associate with kinetochores recover rapidly, with a half-time $\sim$70 s. Although there is no definitive explanation, this apparent discrepancy may be explained if kinetochores can release at least a fraction of the associated microtubules and catch microtubules that undergo rapid tubulin exchange. Thus the mechanism responsible for short-term ($\sim$10 s) incorporation at the kinetochore end may be different from that for the long-term (several minutes) exchange along the length of kinetochore microtubules.

Of particular importance is the possible involvement of microtubule treadmilling in the movement of chromosomes. Early experiments with UV microbeams indicate a poleward movement of UV-marked microtubules (shown as areas of reduced birefringence) along metaphase microtubules [Forer, 1965]. This was later cited as a support for the treadmilling model. Treadmilling also is implied by the apparently slow but constant assembly of tubulin at kinetochore during metaphase [Mitchison et al., 1986]. However, photobleaching techniques, which should show a transport of bleached spots toward the poles if treadmilling occurs, have failed to yield definitive evidence. For example, although Hamaguchi et al. [1987] reported indications that spots bleached in the metaphase spindle move poleward, in most studies such movement is undetectable [Salmon et al., 1984b; Saxton et al., 1984; Wadsworth and Salmon, 1986b; Cassimeris et al., 1988a; Gorbsky and Borisy, 1989b]. In addition, the recovery of fluorescence follows an exponential time course, at a rate independent of the size of the bleached spot [Wadsworth and Salmon, 1986b], inconsistent with a flux being the mechanism of recovery. However, Wadsworth and Salmon cautioned that it is difficult to rule out the treadmilling of a subpopulation of microtubules, such as the kinetochore microtubules. The major problem with the photobleaching studies is the difficulty in identifying the bleached spots over an extended period of time due to the recovery of fluorescence. Recent application of a tubulin analog labeled with a photoactivatable probe, indeed, has yielded the first clear evidence of poleward movement [Mitchison, 1989]. The rate, 0.5–0.6 μm/min, is close to the previous estimation based on the rate of incorporation [Mitchison et al., 1986]. However, the percentage of microtubules undergoing such movement appears low, and decreases continuously during observation.

During anaphase, disassembly of microtubules appears to occur at the kinetochore, as suggested by the loss of previously labeled segments [Mitchison et al., 1986]. More definitive evidence for the disassembly at kinetochores is provided by photobleaching studies, which show little or no movement of bleached spots relative to the pole while the distance between the spots and the chromosomes decreases [Gorbsky et al., 1987, 1988; Gorbsky and Borisy, 1989b]. These observations suggest an important role of microtubule disassembly at the kinetochore in chromosomal movement, but are clearly inconsistent with treadmilling being the primary mechanism. Available data also indicate a considerably slower exchange of tubulin dimers during anaphase than during metaphase. Incorporation of tubulin analog appears to occur at a much slower rate than during metaphase [Wadsworth and Sloboda, 1983; Mitchison et al., 1986]. Bleached spots on kinetochore microtubules also recover much more slowly [Gorbsky and Borisy, 1989b].

Experiments have been performed to study the role of interzonal microtubules, those located between the separating chromosomes, in the elongation of spindles during late anaphase and telophase [Saxton and McIntosh, 1987]. Following photobleaching, the recovery is slower near the equator than near the chromosomes. However, the equatorial region later accumulates a higher level of fluorescence than in surrounding regions. When a series of bars perpendicular to the spindle axis were bleached in the interzonal region, the two groups of bleached lines on the opposite sides of the equator moved apart from each other, but the lines within each group maintained their distances. These results suggest a limited on-off reaction of tubulin dimers near the equator, where the plus ends of many microtubules are located. However, microtubules in the equatorial region probably undergo a net assembly at the plus ends while sliding apart from each other toward the poles.

V. MICROINJECTION OF PROBES FOR INTERMEDIATE FILAMENTS

Among the three cytoskeletal systems, intermediate filaments are the least soluble in vitro and show the least reorganization in vivo. Nevertheless, the filaments must accommodate to such processes as cell spreading, locomotion, and division. The dynamics of intermediate filaments represent an important unknown topic in cell biology.

Only recently have analogs of intermediate filament proteins, including biotinylated vimentin and rhodamine-labeled desmin, been microinjected into living cells [Vikstrom et al., 1989; Mittal et al., 1989]. In vitro, both analogs can form filaments indistinguishable from normal intermediate filaments. Microinjection was performed with a carrier solution of low ionic strength and high pH, in order to maintain solubility. The incorporation of injected vimentin and desmin into cellular structures occurs much more slowly than does injected actin or tubulin, on the order of several hours. However, at steady state, both analogs become colocalized with the endogenous vimentin (but not cytokeratin), and, upon exposing cells to nocodazole or acrylamide, form aggregates as expected for endogenous vimentin filaments [Mittal et al., 1989]. The distribution of fluorescent desmin during cell division also is similar to that revealed by immunofluorescence of vimentin.

These observations are significant since they indicate that, despite the low solubility of vimentin and desmin in the cytoplasm, the analogs can disperse from the site of microinjection and find sites of incorporation. Questions therefore arise as to whether or not endogenous intermediate filaments and the small, but identifiable [Soellner et al., 1985], pool of soluble components undergo any dynamic reorganization. Equally revealing is the pattern of incorporation. The microinjected vimentin appears to move as small aggregates into the perinuclear area before incorporating into the filament network [Vikstrom et al., 1989], suggesting that the perinuclear area may contain factors that facilitate the dissolution of vimentin aggregates. In addition, at steady state, vimentin may undergo a continuous assembly in perinuclear area and diassembly near the periphery of the cell. The observations also suggest that there may be a force-generating mechanism responsible for the directional movement of the vimentin aggregates. These possibilities are difficult to address with biotinylated vimentin, but should be quite approachable with fluorescent analogs and photobleaching techniques.

VI. CONCLUSIONS AND PROSPECTUS

So far the strategy in using the fluorescent probes falls into two main categories: either examining the association of microinjected fluorescent probes with cellular structures, or following the movement of the probe itself or associated structures. The results discussed above illustrate that even such relatively simple approaches can yield a tremendous amount of information which is difficult to obtain otherwise. However, it is also important to remember the basic assumptions of the approach and the complicated, unpredictable nature of living cells. For example, it is difficult to be totally sure that the microinjected analogs are incorporated correctly at a molecular level, that *all* biochemical functions of the analogs are maintained in injected cells, and that the observed processes are unaffected by the microinjection or the observation. At the quantitative level, it is even more difficult to make sure that the measured values, such as diffusion coefficients or assembly rates, represent true physiological values. Therefore, whenever possible, it is important to seek support of the results by applying other techniques. Successful examples include the use of video-enhanced DIC optics to confirm the flux of actin subunits at the lamellipodia and the dynamic instability of microtubules. Such parallel experiments will continue to play an important role in the application of fluorescent probes.

The versatility of fluorescence techniques has been discussed in other chapters of this volume. Many powerful methods, such as polarization, resonance energy transfer, and spectroscopy, remain to be explored. Theoretically, it should be quite feasible to apply these techniques and delineate molecular interactions in living cells. Unlike conventional spectroscopy or flow cytometry, microscopic analyses of single cells will reveal not only what happens in the cell, but also where. If the molecule assumes several discrete conformations, it should also be possible to map the distribution of these different states and to follow the changes.

Fluorescent structural probes can also be combined with fluorescent indicators of cellular parameters. For example, many fluorescent indicators are now available for determining the cellular ionic environment such as Ca, pH, Na, or membrane potential [Tsien, 1989; Haugland, 1989]. A combination of these indicators and cytoskeletal probes should reveal correlations between structural dynamics and ionic conditions [Waggoner et al., 1989]. Furthermore, cytoskeletal probes can be applied in conjunction with caged compounds, to allow observations of structural reorganizations in response to stimuli at specific sites. Finally, the application of confocal microscopy should greatly increase the resolution of fluorescence images, and allow the observation of three-dimensional structures such as cells within matrices or tissues. A combination of these powerful techniques will turn living cells into a biochemical laboratory for testing various hypotheses, rather than simply beautiful objects to watch.

ACKNOWLEDGMENTS

The authors wish to thank Drs. E. Luna, E. Salmon, and G. Sluder for reading the manuscript and offering valuable comments. Our research is supported by NIH grant GM-32476.

VII. REFERENCES

Amato PA, Taylor DL (1986) Probing the mechanism of incorporation of fluorescently labeled actin into stress fibers. J Cell Biol 102:1074–1084.

Amato PA, Unanue ER, Taylor DL (1983) Distribution of actin in spreading macrophages: a comparative study on living and fixed cells. J Cell Biol 96:750–761.

Arndt-Jovin DJ, Robert-Nicoud M, Kaufman SJ, Jovin TM (1985) Fluorescence digital imaging microscopy in cell biology. Science 230:247–256.

Axelrod D (1989) Fluorescence polarization microscopy. In Taylor DL, Wang Y-L (eds): Methods in Cell Biology, Volume 30. San Diego: Academic Press Inc., pp 333–352.

Begg DA, Rebhun LI (1979) pH regulates polymerization of actin in the sea urchin egg cortex. J Cell Biol 83:241–248.

Bendori R, Salomon D, Geiger B (1989) Identification of two distinct functional domains on vinculin involved in its association with focal contacts. J Cell Biol 108:2383–2393.

Bershadsky AD, Gelfand VI, Svitkina TM, Tint IS (1980) Destruction of microfilament bundles in mouse embryo fibroblasts treated with inhibitors of energy metabolism. Exp Cell Res 127:421–429.

Brakenkoff GJ, van Spronsen EA, van der Voort HTM, Nanninga N (1989) Three dimensional confocal fluorescence microscopy. In Taylor DL, Wang Y-L (eds): Methods in Cell Biology, Volume 30. San Diego: Academic Press Inc., pp 379–398.

Bright GR, Fischer GW, Zogowska, Taylor DL (1989) Fluorescence ratio imaging microscopy. In Taylor DL, Wang Y-L (eds): Methods in Cell Biology, Volume 30. San Diego: Academic Press Inc., pp 157–190.

Brix K, Stockem W, Kukulies J (1987) Chemically induced changes in the morphology, dynamic activity and cytoskeletal organization of Physarum cell fragments. Cell Biol Int Rep 11:803–311.

Burke M, Sivaramakrishnan M (1981) Subunit interactions of skeletal muscle myosin and myosin subfragment 1. Formation and properties of thermal hybrids. Biochemistry 20:5908–5913.

Burridge K, Feramisco JR (1981) Alpha-actinin and vinculin from nonmuscle cells: Calcium-sensitive interactions with actin. Cold Spring Harbor Symp Quant Biol 46:587–597.

Burridge K, Feramisco JR (1980) Microinjection and localization of a 130K protein in living fibroblasts: A relationship to actin and fibronectin. Cell 79:587–595.

Cassimeris L, Inoue S, Salmon ED (1988a) Microtubule dynamics in the chromosomal spindle fiber: Analysis by fluorescence and high-resolution polarization microscopy. Cell Motil Cytoskel 10:185–196.

Cassimeris L, Pryer NK, Salmon ED (1988b) Real-time observations of microtubule dynamic instability in living cells. J Cell Biol 107:2223–2231.

Chaussepied P, Morales MF (1988) Modifying preselected sites on proteins: The stretch of residues 633-642 of the myosin heavy chain is part of the actin-binding site. Proc Natl Acad Sci 85:7471–7475.

Cooper JA, Bryan J, Schwab III B, Frieden C, Loftus DJ, Elson EL (1987) Microinjection of gelsolin into living cells. J Cell Biol 104:491–501.

Cooper JA, Loftus DJ, Frieden C, Bryan J, Elson EL (1988) Localization and mobility of gelsolin in living cells. J Cell Biol 106:1229–1240.

De Lozanne A, Spudich JA (1987) Disruption of the Dictyostelium myosin heavy chain gene by homologous recombination. Science 236:1086–1091.

DeBiasio R, Bright GR, Ernst LA, Waggoner AS, Taylor DL (1987) Five-parameter fluorescence imaging: Wound healing of living Swiss 3T3 cells. J Cell Biol 105:1613–1622.

DeBiasio RL, Wang L-L, Fisher GW, Taylor DL (1988) The dynamic distribution of fluorescent analogues of actin and myosin in protrusions at the leading edge of migrating Swiss 3T3 fibroblasts. J Cell Biol 107:2631–2645.

DeNofrio D, Hoock TC, Herman IM (1989) Functional sorting of actin isoforms in microvascular pericytes. J Cell Biol 109:191–202.

Detmers P, Weber A, Elzinga M, Stephens RE (1981) 7-Chloro-4-nitrobenzeno-2-oxa-1,3-diazole actin as a probe for actin polymerization. J Biol Chem 256:99–105.

Dome JS, Mittal B, Pochapin MB, Sanger JM, Sanger JW (1988) Incorporation of fluorescently labeled actin and tropomyosin into muscle cells. Cell Differ 23:37–52.

Fallon JR, Nachmias VT (1980) Localization of cytoplasmic and skeletal myosins in developing muscle cells by double-label immunofluorescence. J Cell Biol 87:237–247.

Faulstich H, Zobeley S, Rinnerthaler G, Small JV (1988) Fluorescent phallotoxins as probes for filamentous actin. J Muscle Res Cell Motil 9:370–385.

Feramisco JR, Smart JE, Burridge K, Helfman DM, Thomas GP (1982) Co-existence of vinculin and a vinculin-like protein of higher molecular weight in smooth muscle. J Biol Chem 257:11024.

Feramisco JR, Blose SH (1980) Distribution of fluorescently labeled alpha-actinin in living and fixed fibroblasts. J Cell Biol 86:608–615.

Feramisco JR (1979) Microinjection of fluorescently labeled alpha-actinin into living fibroblasts. Proc Natl Acad Sci USA 76:3967–3971.

Fine RE, Blitz AL, Hitchcock SE, Kaminer B (1973) Tropomyosin in brain and growing neurones. Nature New Biol 245:182–186.

Fisher GW, Conrad PA, DeBiasio RL, Taylor DL (1988) Centripetal transport of cytoplasm, actin, and the cell surface in lamellipodia of fibroblasts. Cell Motil Cytoskel 11:235–247.

Forer A (1965) Local reduction of spindle fiber birefringence in living Nephrotoma suturalis (Loew) spermatocytes induced by ultraviolet microbeam irradiation. J Cell Biol 25:95–117.

Forscher P, Smith SJ (1988) Actions of cytochalasins on the organization of actin filaments and microtubules in a neuronal growth cone. J Cell Biol 107:1505–1516.

Fukui Y, Yumura S, Yumura T, Mori H (1986) Agar overlay method: High-resolution immunofluorescence for the study of the contractile apparatus. Methods Enzymol 134:573–579.

Gawlitta W, Stockem W, Wehland J, Weber K (1980) Organization and spatial arrangement of fluorescein-labeled native actin microinjected into normal locomoting and experimentally influenced Amoeba proteus. Cell Tissue Res 206:181–191.

Gawlitta W, Stockem W, Weber K (1981) Visualization of actin polymerization and depolymerization cycles during polyamine-induced cytokinesis in living Amoeba proteus. Cell Tissue Res 215:249–261.

Glacy SD (1983a) Pattern and time course of rhodamine-actin incorporation in cardiac myocytes. J Cell Biol 96:1164–1167.

Glacy SD (1983b) Subcellular distribution of rhodamine-actin microinjected into living fibroblastic cells. J Cell Biol 97:1207–1213.

Glascott PA, McSorley KM, Mittal B, Sanger JM, Sanger JW (1987) Stress fiber reformation after ATP depletion. Cell Motil Cytoskel 8:118–129.

Gomer RH, Lazarides E (1981) The synthesis and deployment of filamin in chicken skeletal muscle. Cell 23:524–532.

Gorbsky GJ, Sammak PJ, Borisy GG (1987) Chromosomes move poleward in anaphase along stationary microtubules that coordinately disassemble from their kinetochore ends. J Cell Biol 104:9–18.

Gorbsky GJ, Sammak PJ, Borisy GG (1988) Microtubule dynamics and chromosome motion visualized in living anaphase cells. J Cell Biol 106:1185–1192.

Gorbsky GJ, Borisy GG (1989a) Hapten-mediated immunocytochemistry: The use of fluorescent and nonfluorescent haptens for the study of cytoskeletal dynamics in living cells. In Wang Y-L, Taylor DL (eds): Methods in Cell Biology, Volume 29. San Diego: Academic Press Inc., pp 175–193.

Gorbsky GJ, Borisy GG (1989b) Microtubules of kinetochore fiber turn over in metaphase but not in anaphase. J Cell Biol 109:653–662.

Hamaguchi Y, Mabuchi I (1982) Effects of phalloidin microinjection and localization of fluorescein-labeled phalloidin in living sand dollar eggs. Cell Motil 2:103–113.

Hamaguchi Y, Mabuchi I (1986) Alpha-actinin accumulation in the cortex of echinoderm eggs during fertilization. Cell Motil Cytoskel 6:549–559.

Hamaguchi Y, Mabuchi I (1988) Accumulation of fluorescently labeled actin in the cortical layer in sea urchin eggs after fertilization. Cell Motil Cytoskel 9:153–163.

Hamaguchi Y, Toriyama M, Sakai H, Hiramoto Y (1985) Distribution of fluorescently labeled tubulin injected into sand dollar eggs from fertilization through cleavage. J Cell Biol 100:1262–1272.

Hamaguchi Y, Toriyama M, Sakai H, Hiramoto Y (1987) Redistribution of fluorescently labeled tubulin in the mitotic apparatus of sand dollar eggs and the effects of taxol. Cell Struct Function 12:43–52.

Haugland RP (1989) Molecular Probes: Handbook of Fluorescent Probes and Research Chemicals. Eugene: Molecular Probes, Inc.

Herman B (1989) Resonance energy transfer microscopy. In Taylor DL, Wang Y-L (eds): Methods in Cell Biology, Volume 30. San Diego: Academic Press Inc., pp 220–243.

Hiratsuka T, Kato T (1987) A fluorescent analog of colcemid, N-(7-nitrobenz-2-oxa-1,3-diazol-4-yl)-colcemid, as a probe for the colcemid-binding sites of tubulin and microtubules. J Biol Chem 262:6318–6322.

Honer B, Citi S, Kendrick-Jones J, Jockusch BM (1988) Modulation of cellular morphology and locomotory activity by antibodies against myosin. J Cell Biol 107:2181–2189.

Inoue S, Ritter Jr H (1975) Dynamics of mitotic spindle organization and function. In Inoue S, Stephens RE (eds): Molecules and Cell Movements. New York: Raven Press., pp 3–30.

Jacobson K, Ishihara A, Inman R (1987) Lateral diffusion of proteins in membranes. Annu Rev Physiol 49:163–175.

Jockusch BM, Fuchtbauer A, Wiegand C, Honer B (1985) Probing the cytoskeleton by microinjection. In Shay JW (ed): Cell and Molecular Biology of the Cytoskeleton. New York: Plenum Publishing Corp., pp 1–39.

Johnson CS, McKenna NM, Wang Y-L (1988) Association of microinjected myosin and its subfragments with myofibrils in living muscle cells. J Cell Biol 107:2213–2221.

Keith CH (1987) Slow transport of tubulin in the neurites of differentiated PC12 cells. Science 235:337–339.

Keith CH (1988) The recovery of fluorescence after photobleaching of DTAF tubulin-injected PC12 cells differs for cells bleached with a laser or a mercury arc lamp. J Cell Biol 107:28a.

Keith CH, Feramisco JR, Shelanski M (1981) Direct visualization of fluorescein-labeled microtubules in vitro and in microinjected fibroblasts. J Cell Biol 88:234–240.

Kellogg DR, Mitchison TJ, Alberts BM (1988) Behavior of microtubules and actin filaments in living Drosophila embryos. Development 103:675–686.

Knecht DA, Loomis WF (1987) Antisense RNA inactivation of myosin heavy chain gene expression in Dictyostelium discoideum. Science 236:1081–1086.

Kouyama T, Mihashi K (1981) Fluorimetry study of N-(1-pyrenyl) iodoacetamide-labeled F-actin. Eur J Biochem 114:33–38.

Kraft GA, Sutton WR, Cummings RT (1988) Photoactivable fluorophores 3. Synthesis and photoactivation of fluorogenic difunctionalized fluoresceins. J Am Chem Soc 110:301–303.

Kreis TE (1986) Preparation, assay, and microinjection of fluorescently labeled cytoskeletal proteins: Actin, alpha-actinin, and vinculin. Methods Enzymol 134:507–519.

Kreis TE, Birchmeier W (1980) Stress fiber sarcomeres of fibroblasts are contractile. Cell 22:555–561.

Kreis TE, Birchmeier W (1982) Microinjection of fluorescently labeled proteins into living cells with emphasis on cytoskeletal proteins. Int Rev Cytol 75:209–228.

Kreis TE, Winterhalter KH, Birchmeier W (1979) In vivo distribution and turnover of fluorescently labeled actin microinjected into human fibroblasts. Proc Natl Acad Sci USA 76:3814–3818.

Kreis TE, Geiger B, Schlessinger J (1982) Mobility of microinjected rhodamine actin within living chicken gizzard cells determined by fluorescence photobleaching recovery. Cell 29:835–845.

Kukulies J, Stockem W (1985) Function of the microfilament system in living cell fragments of Physarum polycephalum as revealed by microinjection of fluorescent analogs. Cell Tissue Res 242:323–332.

Kukulies J, Stockem W, Achenbach F (1984) distribution and dynamics of fluorochromed actin in living stages of Physarum polycephalum. Eur J Cell Biol 35:235–245.

Kukulies J, Brix K, Stockem W (1985) Fluorescent analog cytochemistry of the actin system and cell surface morphology in Physarum microplasmodia. Eur J Cell Biol 39:62–69.

Leslie RJ, Saxton WM, Mitchison TJ, Neighbors B, Salmon ED, McIntosh JR (1984) Assembly properties of fluorescein-labeled tubulin in vitro before and after fluorescence bleaching. J Cell Biol 99:2146–2156.

Lim SS, Sammak PJ, Borisy GG (1989) Progressive and spatially differentiated stability of microtubules in developing neuronal cells. J Cell Biol 109:253–267.

Lubit BW, Schwartz JW (1980) An antiactin antibody distinguishes between cytoplasmic and skeletal muscle actins. J Cell Biol 86:891–897.

Malm B (1984) chemical modification of cys-374 of actin interferes with the formation of the profilactin complex. FEBS Lett 173:399–402.

Margolis RI, Wilson RL (1981) Microtubule treadmills-possible molecular machinery. Nature 293:705–711.

Mabuchi I, Hamaguchi Y, Kobayashi T, Hosoya H, Tsukita S, Tsukita S (1985) Alpha-actinin from sea urchin eggs: Biochemical properties, interaction with actin, and distribution in the cell during fertilization and cleavage. J Cell Biol 100:375–383.

McKenna NM, Wang Y-L (1986) Possible translocation of actin and alpha-actinin along stress fibers. Exp Cell Res 167:95–105.

McKenna N, Meigs JB, Wang Y-L (1985a) Identical distribution of fluorescently labeled brain and muscle actins in living cardiac fibroblasts and myocytes. J Cell Biol 100:292–296.

McKenna NM, Meigs JB, Wang Y-L (1985b) Exchangeability of alpha-actinin in living cardiac fibroblasts and muscle cells. J Cell Biol 101:2223–2232.

McKenna NM, Johnson CS, Wang Y-L (1986) Formation and alignment of Z-lines in living chick myotubes microinjected with rhodamine-labeled alpha-actinin. J Cell Biol 103:2163–2171.

McKenna NM, Johnson CS, Konkel ME, Wang Y-L (1989a) Organization of myosin in living muscle and non-muscle cells. In Kedes LH, Stockdale FE (eds): Cellular and Molecular Biology of Muscle Development.'' New York: Alan R. Liss, Inc., pp 237–246.

McKenna NM, Wang Y-L, Konkel ME (1989b) Formation and movement of myosin-containing structures in living fibroblasts. J Cell Biol, in press.

Meigs JB, Wang Y-L (1986) Reorganization of alpha-actinin and vinculin induced by a phorbol ester in living cells. J Cell Biol 102:1430–1438.

Mitchison TJ (1989) Polewards microtubule flux in the mitotic spindle: evidence from photoactivation of fluorescence. J Cell Biol 109:637–652.

Mitchison T, Kirschner M (1984) Dynamic instability of microtubule growth. Nature 312:237–242.

Mitchison T, Evans L, Schulze E, Kirschner M (1986) Sites of microtubule assembly and disassembly in the mitotic spindle. Cell 45:515–527.

Mittal B, Sanger JM, Sanger JW (1987a) Visualization of myosin in living cells. J Cell Biol 105:1753–1760.

Mittal B, Sanger JM, Sanger JW (1987b) Binding and distribution of fluorescently labeled filamin in permeabilized and living cells. Cell Motil Cytoskel 8:345–359.

Mittal B, Sanger JM, Sanger JW (1989) Visualization of intermediate filaments in living cells using fluorescently labeled desmin. Cell Motil Cytoskel 12:127–138.

Moll E, Manz B, Mocikat S, Zimmermann H-P (1982) Fluorescent Deacetylcolchicine. New aspects of its activity and localization in PtK-1 cells. Exp Cell Res 141:211–220.

Murphy DB, Johnson KA, Borisy GG (1977) Role of tubulin-associated proteins in microtubule nucleation and elongation. J Mol Biol 117:33–52.

Olmsted JB, Stemple DL, Saxton WM, Neighbors BW, McIntosh JR (1989) Cell cycle-dependent changes in the dynamics of MAP2 and MAP4 in cultured cells. J Cell Biol 109:211–223.

Otey CA, Kalnoski MH, Bulinski JC (1988) Immunolocalization of muscle and nonmuscle isoforms of actin in myogenic cells and adult skeletal muscle. Cell Motil Cytoskel 9:337–348.

Pardo JV, Pittenger MF, Craig SW (1983) Subcellular sorting of isoactins: Selective association of gamma actin with skeletal muscle mitochondria. Cell 32:1093–1103.

Pollard TD, Cooper JA (1986) Actin and actin-binding proteins. A critical evaluation of mechanisms and function. Ann Rev Biochem 55:987–1035.

Reisler E (1982) Sulfhydryl modification and labeling of myosin. Methods Enzymol 85:84–93.

Rifkin DB, Crowe RM, Pollack R (1979) Tumor promoters induces changes in chick embryo fibroblast cytoskeleton. Cell 18:361–368.

Saga S, Hamaguchi M, Hoshino M, Kojima K (1985) Expression of meta-vinculin associated with differentiation of chicken embryonal muscle cells. Exp Cell Res 156:45–56.

Salmon ED, Wadsworth P (1986) Fluorescence studies of tubulin and microtubule dynamics in living cells. In Taylor DL, Waggoner AS, Murphy RF, Lanni F, Birge RR (eds): Applications of Fluorescence in the Biomedical Sciences. New York: Alan R. Liss, Inc., pp 377–403.

Salmon ED, Saxton WM, Leslie RJ, Karow ML, McIntosh JR (1984a) Diffusion coefficient of fluorescein-labeled tubulin in the cytoplasm of embryonic cells of a sea urchin: Video image analysis of fluorescence redistribution after photobleaching. J Cell Biol 99:2157–2164.

Salmon ED, Leslie, JR, Saxton WM, Karow ML, McIntosh JR (1984b) Spindle microtubule dynamics in sea urchin embryos: analysis using a fluorescein-labeled tubulin and measurements of fluorescence redistribution after laser photobleaching. J Cell biol 99:2165–2174.

Sammak PJ, Borisy GG (1988) Direct observation of microtubule dynamics in living cells. Nature 332:724–726.

Sammak PJ, Gorbsky GJ, Borisy GG (1987) Microtubule dynamics in vivo: A test of mechanisms of turnover. J Cell Biol 104:395–405.

Sanger, JW (1975) Intracellular localization of actin with fluorescently labeled heavy meromyosin. Cell Tissue Res 161:431–444.

Sanger JW, Sanger JM, Kreis TE, Jockusch BM (1980) Reversible translocation of cytoplasmic actin into the nucleus caused by dimethylsulfoxide. Proc Natl Acad Sci USA 77:5268–5272.

Sanger JW, Mittal B, Sanger JM (1984a) Formation of myofibrils in spreading chick cardiac myocytes. Cell Motil 4:405–416.

Sanger JW, Mittal B, Sanger JM (1984b) Analysis of myofibrillar structure and assembly using fluorescently labeled contractile proteins. J Cell Biol 98:825–833.

Sanger JW, Mittal B, Sanger JM (1984c) Interaction of fluorescently-labeled contractile proteins with the cytoskeleton in cell models. J Cell Biol 99:918–928.

Sanger JM, Mittal B, Pochapin MB, Sanger JW (1986a) Myofibrillogenesis in living cells microinjected with fluorescently labeled alpha-actinin. J Cell Biol 102:2053–2066.

Sanger JM, Mittal B, Pochapin MB, Sanger JW (1986b) Observations of microfilament bundles in living cells microinjected with fluorescently labeled contractile proteins. J Cell Sci Suppl 5:17–44.

Sanger JM, Mittal B, Pochapin MB, Sanger JW (1987) Stress fiber and cleavage furrow formation in living cells microinjected with fluorescently labeled alpha-actinin. Cell Motil Cytoskel 7:209–220.

Saxton WM, McIntosh JR (1987) Interzone microtubule behavior in late anaphase and telophase spindles. J Cell Biol 105:875–886.

Saxton WM, Stemple DL, Leslie RJ, Salmon ED, Zavortink M, McIntosh JR (1984) Tubulin dynamics in cultured mammalian cells. J Cell Biol 99:2175–2186.

Scherson T, Kreis TE, Schlessinger J, Littauer UZ, Borisy, GG, Geiger B (1984) Dynamic interactions of fluorescently labeled microtubule-associated proteins in living cells. J Cell Biol 99:425–434.

Scholey JM, Smith RC, Drenkhahn D, Groschel-Stewart U, Kendrick-Jones J (1982) Thymus myosin. Isolation and characterization of myosin from calf thymus and thymic lymphocytes and studies on the effects of phosphorylation of its Mr = 20,000 light chain. J Biol Chem 257:7737–7745.

Schroeder TE (1981) The origin of cleavage forces in dividing eggs. Exp Cell Res 134:231–240.

Schulze E, Kirschner M (1986) Microtubule dynamics in interphase cells. J Cell Biol 102:1020–1031.

Schulze E, Kirschner M (1987) Dynamic and stable populations of microtubules in cells. J Cell Biol 104:227–288.

Schulze E, Kirschner M (1988) New features of microtubule behavior observed in vivo. Nature 334:356–358.

Schulze EM, Asai DJ, Bulinski JC, and Kirschner M (1987) Posttranslational modification and microtubule stability. J Cell Biol 105:2167–2177.

Siliciano JD, Craig SW (1982) Meta-vinculin, a vinculin-related protein with solubilty properties of a membrane protein. Nature 300:533–535.

Simon JR, Taylor DL (1986) Preparation of a fluorescent analog: Acetamidofluoresceinyl-labeled Dictyostelium discoideum alpha-actinin. Methods Enzymol 134:487–507.

Simon JR, Gough A, Urbanik E, Wang F, Ware BR, Lanni F, Taylor DL (1988) Analysis of rhodamine and fluorescein-labeled F-actin diffusion in vitro by fluorescence photobleaching recovery. Biophys J 54:801–815.

Small JV (1981) Organization of actin in the leading edge of cultured cells: Influence of osmium tetroxide and dehydration on the ultrastructure of actin meshworks. J Cell Biol 91:695–705.

Soellner P, Quinlan RA, Franke WW (1985) Identification of a distinct soluble subunit of an intermediate filament protein: tetrameric vimentin from living cells. Proc Natl Acad Sci USA 82:7929–7933.

Soltys BJ, Borisy GG (1985) Polymerization of tubulin in vivo: direct evidence for assembly onto microtubule ends and from centrosomes. J Cell Biol 100:1682–1689.

Spring KR, Lowy JR (1989) Characteristics of low light level television cameras. In Wang Y-L, Taylor DL (eds): Methods in Cell Biology, Volume 29. San Diego: Academic Press Inc., pp 270–289.

Stickel SK, Wang Y-L (1987) Alpha-actinin-containing aggregates in transformed cells are highly dynamic structures. J Cell Biol 104:1521–1526.

Stickel SK, Wang Y-L (1988) Synthetic peptide GRGDS induces dissociation of alpha-actinin and vinculin from the sites of focal contacts. J Cell Biol 107:1231–1239.

Stockem W, Hoffmann H-U, Gruber B (1983) Dynamics of cytoskeleton in Amoeba proteus. Cell Tissue Res 232:79–96.

Stryer L (1978) Fluorescence energy transfer as a spectroscopic ruler. Annu Rev Biochem 47:819–846.

Suzuki H, Onishi H, Takahashi K, Watanabe S (1978) Structure and function of chicken gizzard myosin. J Biochem (Tokyo) 84:1529–1542.

Taylor DL, Wang Y-L (1978) Molecular cytochemistry of fluorescently labeled actin. Proc Natl Acad Sci USA 75:857–861.

Taylor DL, Wang Y-L (1980) Fluorescently labeled molecules as probes of the structure and function of living cells. Nature 284:405–410.

Taylor DL, Wang Y-L, Heiple JM (1980) Contractile basis of amoeboid movement VII. The distribution of fluorescently labeled actin in living Amebas. J Cell Biol 86:590–598.

Taylor DL, Amato PA, Luby-Phelps K, McNeil P (1984) Fluorescence analogue cytochemistry. Trends Biochem Sci 9:88–91.

Tsien RY (1989) Fluorescent indicators of ion concentrations. In Taylor DL, Wang Y-L (eds): Methods in Cell Biology, Volume 30.'' San Diego: Academic Press Inc., pp 127–153.

Uster PS, Pagano RE (1986) Resonance energy transfer microscopy: Observations of membrane-bound fluorescent probes in model membranes and in living cells. J Cell Biol 103:1221–1234.

Vallee RB, Bloom GS, Theurkauf WE (1984) Microtubule-associated proteins: Subunits of the cytomatrix. J Cell Biol 99:38–44s.

Vandenbunder B, Borisy GG (1986) Decoration of microtubules by fluorescently labeled microtubule-associated protein 2 (MAP2) does not interfere with their spatial organization and progress through mitosis in living fibroblasts. Cell Motil Cytoskel 6:570–579.

Vasiliev JM (1985) Spreading of non-transformed and transformed cells. Biochem Biophys Acta 780:21–65.

Vigers GPA, Coue M, McIntosh JR (1988) Fluorescent microtubules break up under illumination. J Cell Biol 107:1011–1024.

Vikstrom KL, Borisy GG, Goldman RD (1989) Dynamic aspects of intermediate filament networks in BHK-21 cells. Proc Natl Acad Sci USA 86:549–553.

Wadsworth P, Salmon ED (1986a) Preparation and characterization of fluorescent analogs of tubulin. Methods Enzymol 134:519–528.

Wadsworth P, Salmon ED (1986b) Analysis of treadmilling model during metaphase of mitosis using fluorescence redistribution after photobleaching. J Cell Biol 102:1032–1038.

Wadsworth P, Salmon ED (1988) Spindle microtubule dynamics: Modulation by metabolic inhibitors. Cell Motil Cytoskel 11:97–105.

Wadsworth P, Sloboda RD (1983) Microinjection of fluorescent tubulin into dividing sea urchin cells. J Cell Biol 97:1249–1254.

Waggoner A, DeBiasio R, Conrad P, Bright GR, Ernst L, Ryan K, Nederlof M, Taylor D (1989) Multiple spectral parameter imaging. In Taylor DL, Wang, Y-L (eds): Methods in Cell Biology, Volume 30. San Diego: Academic Press Inc., pp 449–470.

Walker RA, O'Brien ET, Pryer NK, Soboeiro MF, Voter WA, Erickson HP, Salmon ED (1988) Dynamic instability of individual microtubules analyzed by video light microscopy: Rate constants and transition frequencies. J Cell Biol 107:1437–1448.

Wang Y-L (1984) Reorganization of actin filament bundles in living fibroblasts. J Cell Biol 99:1478–1485.

Wang Y-L (1985) Exchange of actin subunits at the leading edge of living fibroblasts: Possible role of treadmilling. J Cell Biol 101:597–602.

Wang Y-L (1986) Reorganization of alpha-actinin and vinculin in living cells following ATP depletion and replenishment. Exp Cell Res 167:16–28.

Wang Y-L (1987) Mobility of filamentous actin in living cytoplasm. J Cell Biol 105:2811–2816.

Wang Y-L (1989) Fluorescent analog cytochemistry: Tracing functional protein components in living cells. In Taylor DL, Wang Y-L (eds): Methods in Cell Biology, Volume 29. San Diego: Academic Press Inc., pp 1–12.

Wang Y-L, Taylor DL (1979) Distribution of fluorescently labeled actin in living sea urchin eggs during early development. J Cell Biol 82:672–679.

Wang Y-L, Taylor DL (1980) Preparation and characterization of a new molecular cytochemical probe: 5-iodoactamidofluorescein labeled actin. J Histochem Cytochem 28:1198–1206.

Wang Y-L, Heiple JM, Taylor DL (1982a) Fluorescent analog cytochemistry of contractile proteins. In Wilson L (ed): Methods in Cell Biology, Volume 25. San Diego: Academic Press Inc., pp 1–11.

Wang Y-L, Lanni F, McNeil PL, Ware BR, Taylor DL (1982b) Mobility of cytoplasmic and membrane-associated actin in living cells. Proc Natl Acad Sci USA 79:4660–4664.

Ware BR, Brvenik LJ, Cummings RT, Furukawa RH, Krafft GA (1986) Fluorescence photoactivation and dissappation (FPD), In Taylor DL, Waggoner AS, Murphy RF, Lanni F, Birge RR (eds):"Applications of Fluorescence in the Biomedical Sciences." New York: Alan R. Liss Inc., pp 141–157.

Warn RM, Flegg L, Warn A (1987) An investigation of microtubule organization and function in living Drosophila embryos by injection of a fluorescently labeled antibody against tyrosinated alpha-tubulin. J Cell Biol 105:1721–1730.

Warren RH, Gordon E, Azarnia R (1985) Tropomyosin in peripheral ruffles of cultured rat kidney cells. Eur J Cell Biol 38:245–253.

Webster DR, Gundersen GG, Bulinksi JC, Borisy GG (1987) Assembly and turnover of detyrosinated tubulin in vivo. J Cell Biol 105:265–276.

Wegner A (1976) Head to tail polymerization of actin. J Mol Biol 108:139–150.

Wehland J, Weber K (1980) Distribution of fluorescently labeled actin and tropomyosin after microinjection in living tissue culture cells as observed with TV image intensification. Exp Cell Res 127:397–408.

Wehland J, Weber K (1981) Actin rearrangement in living cells revealed by microinjection of a fluorescent phalloidin derivative. Eur J Cell Biol 24:176–183.

Wehland J, Osborn M, Weber K (1977) Phalloidin-induced actin polymerization in the cytoplasm of cultured cells interferes with cell locomotion and growth. Proc Natl Acad Sci USA 74:5613–5617.

Wehland J, Weber K, Osborn M (1980) Translocation of actin from the cytoplasm into the nucleus in mammalian cells exposed to dimethylsulfoxide. Biol Cellulaire 3o 109–111.

Wehland J, Willingham MC, Sandoval IV (1983) A rat monoclonal anti؛ ؛dy reacting specifically with the tyrosylated form of alpha-tubulin. I. Biochemical c ۱aracterization, effects on microtubule polymerization in vitro, and microtubule polyn erization and organization in vivo. J Cell Biol 97:1467–1475.

Wolf DE, Edidin M, Dragsten PR (1980) Effect of bleaching light on measurements of lateral diffusion in cell membranes by the fluorescence photobleaching recovery method. Proc Natl Acad Sci USA 77:2043–2045.

Yin HL, Stossel TP (1980) Control of cytoplasmic actin gel-sol transformation by gelsolin, a calcium-dependent regulatory protein. Nature 281:583–586.

Noninvasive Techniques in Cell Biology: 213–236
© 1990 Wiley-Liss, Inc.

9. Optical Methods for the Study of Metabolism in Intact Cells

Robert S. Balaban and Lazaro J. Mandel

Laboratory of Cardiac Energetics, National Heart, Lung, and Blood Institute, National Institutes of Health, Bethesda, Maryland 20892 (R.S.B.); Division of Physiology, Department of Cell Biology, Duke University Medical Center, Durham, North Carolina 27710 (L.J.M.)

I. INTRODUCTION

Metabolic reactions usually involve multiple steps that require the coordinated action of various enzymes within a cell. Although much information is often available regarding the function of individual enzymes in vitro, enzyme properties may be significantly different within the intact cell. These differences may be due to interactions with other reaction steps, modulation by unknown cytosolic parameters, or cellular compartmentation. Optical approaches have been utilized to monitor enzymatic properties within intact cells, due to their noninvasive nature. Most of these have used intrinsic optical probes relying on enzymes or metabolites which change their optical absorption or fluorescent emission as a function of specific cellular alterations. More recently optically labeled substrates or enzymes, as well as exogenous indicators, have been used to study a variety of metabolic reactions in intact cells.

A. The Mitochondrial Electron Chain

The most common application of optical techniques using intrinsic probes has been for redox state determinations of mitochondrial electron chain components. A schematic representation of the mitochondrial respiratory chain is shown in Figure 1A, and its reduced minus oxidized absorption spectrum is shown in Figure 1B. As may be seen from this figure, most of the components of the respiratory chain absorb light within characteristic bands which can be used to identify them. All the cytochromes and NAD absorb light within these bands only in the reduced state, whereas FAD absorbs in the oxidized state. In addition, reduced NAD (NADH) fluoresces in a broad emission band of 425–500 nm when excited with 310–370 nm light, and the oxidized form of NAD (NAD$^+$) does not fluoresce. Accordingly, the intensity of fluorescence at 450 nm is proportional to the concentration of NADH. These intrinsic optical properties have been utilized to determine the redox state of each of these respiratory chain components as a function of specific mitochondrial conditions. The foundation for these measurements rests on the description by Chance and Williams [1956] of five respiratory states for isolated mitochondria and the conditions required to elicit a change from one state to another. Table I has been modified from this study to show the average steady-state redox levels of the various respiratory enzymes in these five states.

State 1 characterizes the mitochondria when first isolated containing low levels of substrate and ADP and exhibiting a low respiratory rate. State 2 is called "starved" because the mitochondria are provided with sufficient oxygen and ADP but are deprived of metabolic substrates. In the extreme case of total substrate deprivation shown here, there is no flow of reducing equivalents into the mitochondrial chain and, thus, all the respiratory components are oxidized. State 3 is the "active" one because it describes the conditions present when the mitochondria are respiring at their maximal rate in the presence of sufficient substrate, ADP (and P$_i$), and oxygen. Note that the respiratory chain components display a progressively increasing level of reduction starting with cytochrome aa_3, which is almost completely oxidized, to NAD, which is about 50% reduced. State 4 is the "resting" state, achieved by depletion of ADP or P$_i$, and characterized by a low respiratory rate and generally a more reduced condition than state 3. State 5 is achieved upon anoxia and causes all the components to become totally reduced.

Having described these extreme states, it is simple to visualize mitochondrial behavior during transitions between states. For example, the transition between state 4 (resting) and state 3 (active) is usually achieved by the addition of ADP. The respiratory rate accelerates rapidly and the respiratory enzymes

---

Abbreviations used: FAD, flavin adenine dinucleotide; NAD, nicotinamide adenine dinucleotide; NADP, nicotinamide adenine dinucleotide phosphate.

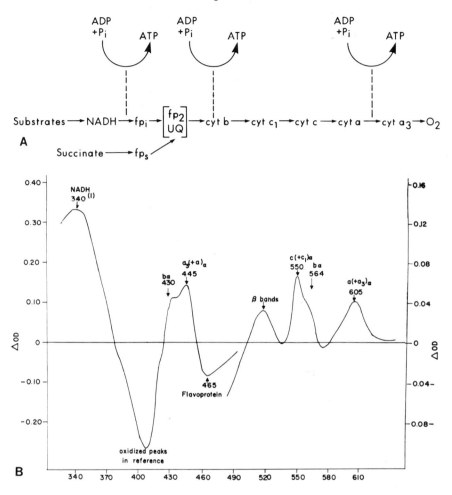

Fig. 1. Optical studies of the mitochondrial electron transport chain. **A**: Schematic diagram of the mitochondrial electron transport chain. **B**: Difference spectrum of the respiratory chain from rat liver mitochondria. The spectrum is the optical difference between a mitochondrial suspension in the presence and absence of oxygen [from Jobsis, 1964].

become more oxidized, with the exception of cytochrome aa_3. The reverse would, of course, happen in a transition from the active to the resting state. The enzyme component that exhibits the most dramatic shifts in redox level is NAD, being 99% reduced in state 4 and 50% in state 3; this is due to its association with the initial step of the mitochondrial respiratory chain. The steady-state redox level of NAD is a function of both its rate of oxidation by the flavoproteins and its rate of reduction by the numerous dehydrogenases of intermediary metabolism. By monitoring the redox state of mitochondrial NAD, the

TABLE I. Average Steady-State Redox Levels of Respiratory Enzymes in Five Respiratory States

State	ADP·P$_i$	Substrate	O$_2$	Q$_{O_2}$	% Reduced NAD	% Reduced Cyt b	% Reduced Cyt c	% Reduced Cyt a, a_3	Rate limitation
1	Low	Low	High	Slow	90	17	7	0	ADP
2	High	0	High	Slow	0	0	0	0	Substrate
3	High	High	High	Fast	53	16	6	4	Respiratory chain
4	Low	High	High	Slow	99	35	14	0	ADP
5	High	High	0	0	100	100	100	100	Oxygen

relative rates of these two categories of redox reactions can be analyzed, thereby providing important information on the respiratory state of the mitochondria.

The transition from state 2 (substrate-free) to state 3 (active) is obtained by the addition of metabolic substrates in the presence of saturating concentrations of ADP and oxygen. All the respiratory chain components become partially reduced, the largest change occurring in NAD. A similar transition can also be obtained from state 2 to state 4 or to an intermediate level between states 3 and 4. In this transition, the redox level of NAD provides direct information regarding the relative ability of individual metabolic substrates to supply reducing equivalents to the respiratory chain.

The transition from either state 3 or 4 to state 5 (anoxia), elicited by removing all the oxygen from the solution, causes the complete reduction of all respiratory enzymes. Under these conditions, cytochrome aa_3 displays the largest change in redox level. In fact, cytochrome aa_3 has such a high affinity for oxygen that this cytochrome remains fully oxidized as long as the local oxygen concentration is above 1 μM [Oshino et al., 1974]. Thus, the reduction of cytochrome aa_3 has been used as an index of anoxia [Chance et al., 1973], although the reduction of cytochrome c [Wilson et al., 1988] or that of NAD have been similarly used for this purpose [Chance, 1976].

B. Cytoplasmic Redox Reactions

In addition to the mitochondrial respiratory chain, many cells contain a second cytochrome-containing respiratory chain associated with the endoplasmic reticulum. This chain is involved in the oxidative metabolism of a diversity of chemicals including steroids, drugs, and numerous toxic substances. As illustrated in Figure 2, both cytosolic NAD and NADP can donate reducing equivalents for the reduction of cytochrome P-450 or b$_5$, while the terminal oxidase is P-450 [Estabrook et al., 1970; Gibson and Skett, 1986]. The redox state of both of these cytochromes can be measured spectrophotometrically, providing important information regarding their content and function in vari-

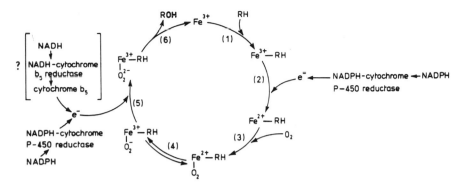

Fig. 2. Catalytic cycle of cytochrome P-450. RH represents the drug substrate, and ROH the corresponding hydroxylated metabolite [from Gibson and Skett, 1986].

ous tissues. NAD and NADP are also involved in numerous other cytosolic redox reactions, especially in glycolysis, the hexose monophosphate shunt, and others. These components can also be measured in the cytoplasm, although this can only be accomplished when the optical contributions from mitochondrial NAD are minimized.

C. Peroxides and Oxygen Free Radicals

The rate of formation of hydrogen peroxide and specific oxygen free radical species may be monitored spectrophotometrically and through chemiluminescent emission, respectively, in response to toxicants or environmental conditions. The former measurement is based on the fact that the fraction of catalase heme present as Compound I depends on the rate of hydrogen peroxide formation [Sies, 1987]. The measurement of oxygen free radicals depends on weak chemiluminescent emission at specific wavelengths which requires the use of single-photon counting and the elimination of possible interfering reactions [Sies, 1987]. Other techniques have also been utilized to perform these measurements with the use of extrinsic probes.

II. APPLICATIONS TO INTACT CELLS AND TISSUES

The application of these optical methods to intact cells and tissues is reviewed next. The subsections are divided according to the enzyme or substrate that is being measured. In most cases, optical techniques are truly noninvasive when the applied optical energy is small as compared with the thermal energy of the sample. However, when larger amounts of energy are required, this could lead to photodynamic damage, which might invalidate the measurement. On the other hand, an optical spectrum may only be obtainable after irreversible reaction with another component, such as carbon monoxide. These limitations need to be kept in mind while considering the ''noninvasive'' character of optical measurements.

A. NAD

As described in the Introduction, NAD is the primary electron source for the electron chain in oxidative phosphorylation and important in the oxidation-reduction occurring in glycolysis as well as numerous redox reactions occurring in the cytosol. For example, in the regulation of mitochondrial oxidative phosphorylation the redox state of mitochondrial NAD may be extremely important in the regulation of hormone action [Denton and McCormack, 1985] or the response to an increase in ATPase hydrolysis in tissue [Balaban, 1990]. By monitoring the mitochondrial NAD redox state several aspects of cellular energetics can also be evaluated, including activation of dehydrogenases [Denton and McCormack, 1985], the concentration of free adenylates [Chance and Connelly, 1957], substrate oxidation kinetic parameters [Balaban and Mandel, 1988], and oxygen delivery [Chance, 1976]. Due to the ubiquitous and important role this metabolite plays in the metabolism of the cell, information on its cellular distribution and metabolism has been a significant goal in the study of cellular metabolism for many years.

Total cellular reduced NAD (NADH) and oxidized NAD have been measured by numerous investigators using enzymatic assays on tissue extracts. Such an approach yields questionable information since no differentiation is possible regarding the various compartments known to contain NAD(H). The determination of NADH and NAD compartmentation has relied on the use of specific redox couples participating in localized enzymatic reactions in the cell which are believed to be at or very close to equilibrium [Williamson, 1976]. For example, lactate dehydrogenase (LDH) has been extensively used in the evaluation of the cytoplasmic NADH redox state, while the glutamate dehydrogenase (GDH) reaction has been used to monitor the mitochondrial matrix spaced. However, this equilibrium method is invasive and has numerous other limitations which include problems in the accurate determination of metabolites in the redox couple, compartmentation of enzymes involved in the redox couple, the compartmentation of the redox couple metabolites within cellular regions where they are not in equilibrium and lack of independent evidence that the reaction is truly at equilibrium. Despite these limitations this approach has provided a wealth of information on this important metabolite [Williamson, 1976; Denton and McCormack, 1985].

For noninvasive studies of the NAD redox state only two methods have been used, optical fluorescence and ^{13}C nuclear magnetic resonance [Unkefer et al., 1983]. As described in the Introduction, NADH absorbs light with a maximum at 340 nm and emits a broad fluorescence band centered at 450 nm. In intact cells this fluorescence is usually detected using the 360 nm line of

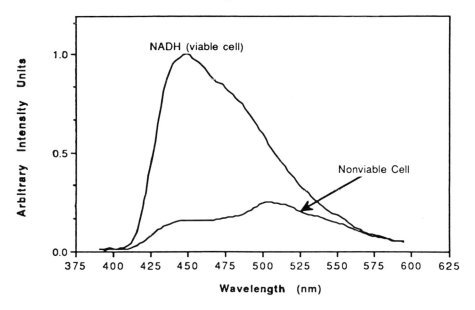

Fig. 3. Autofluorescence emission spectra with 366 nm UV excitation from isolated rat myocytes. Fluorescent spectra from viable and non-viable cells are presented based on the morphology of the cells [from Eng et al., 1989].

a mercury arc lamp as an excitation source,though several groups have also been successful with nitrogen laser lines at 340 nm. In studies of single cell fluorescence, NADH appears to be especially sensitive to photodynamic damage (i.e., bleaching), thus the studies are usually conducted at very low excitation light intensities and limiting exposure to only brief periods [Koke and Williams, 1979; Eng et al., 1989].

The blue-green NADH emission spectrum of a single cardiac myocyte is shown in Figure 3 with a maximum at approximately 447 nm, obtained with a rapid scanning spectophotometer coupled to an inverted microscope [Eng et al., 1989]. This fluorescence spectrum is blue shifted from the free NADH fluorescence signal, suggesting that most of this signal originates from bound NADH [Velick, 1961], as discussed later in this section. In more complex systems such as the perfused heart, other factors can significantly influence the emission spectrum of these probes as well as that of other fluorescent dyes. This is demonstrated in Figure 4 where the emission spectrum of an intact heart is presented for both NADH and the calcium sensitive probe Indo-1. The spectrum of the intact heart is drastically red-shifted with respect to the free NADH fluorescence signal and a specific "dip" at ~415 nm is seen in both spectra. This shoulder and the red shift are due to the absorbance of myoglobin and cytochromes acting as inner filters in this system [Koretsky

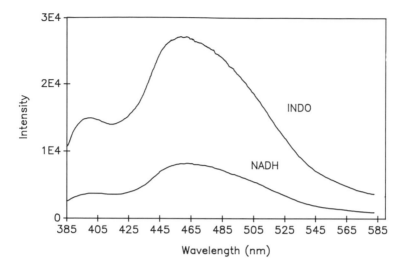

Fig. 4. Autofluorescence and Indo-1 emission spectrum with 340 nm UV excitation from an isolated rabbit heart. Note the extreme dip in the fluorescence spectrum at ~415 nm due to myo-globin. Autofluorescence spectrum was collected from an ischemic heart. Indo-1 data were collected from a heart preloaded with 2 μM Indo-1-am for 30 min at 30°C. Note that the myoglobin effects were nearly identical for both probes. Data was collected as described in Fralix et al. [1990].

et al., 1987]. This aspect of intact organ NADH fluorescence studies or other probes must be corrected to appropriately interpret the emission data since the myoglobin and cytochrome absorbances can change dramatically with the metabolic state of the tissue [Koretsky et al., 1987; Fralix et al., 1990]. Another problem with these intact organ studies is that the penetration of the UV excitation light is extremely limited to only 0.5–1 mm. Thus, the fluorescence measurements are restricted to the surface regions of the tissue and the penetration can also be a function of the concentration of inner filters, such as myoglobin, which also strongly absorbs in the 300 nm region.

This blue-green background fluorescence signal from NADH is usually an annoyance to most investigators attempting to use exogenously added chromophores to the cell, requiring them to use excessive amounts of their probes or expend chemical synthesis efforts in moving the emission of their dyes to regions away from NADH. However, as we hope to point out in this section, this annoying "background" fluorescence can be quite useful.

One of the earliest noninvasive studies of single cell metabolism was conducted by Chance and Thorell in 1959 in grasshopper spermatids. In this study they determined that the majority of the blue NADH fluorescence signal is observed in the mitochondrial rich nebenkern. These studies also revealed that the bulk cytoplasm fluorescence did not respond significantly to anoxia while the nebenkern fluorescence increased as predicted for a mitochondrial NADH pool. Later studies on isolated mitochondria [Avi-Dor et al., 1962; Estabrook, 1962] revealed that the mitochondrial pool fluorescence efficiency was 8–12 times that of free NADH. These results indicate that the majority of the blue fluorescence from tissues with a high density of mitochondria is from the mitochondrial NADH pool. Later functional studies comparing the NADH signal in response to different substrates [Nuutinen, 1984; Balaban and Mandel, 1988] or comparing the changes in cytosolic redox couples and fluorescence during metabolic challenges [O'Connor, 1977], also reached a similar conclusion: most of the blue fluorescence originates from mitochondrial NADH. However, the cytosolic NADH contribution can be quite significant in cells with low mitochondrial density, such as most tissue culture cells [Mandel, 1986], or cells with significant levels of alcohol dehydrogenase which enhances cytosolic NADH fluorescence, such as liver [Balaban and Blum, 1982].

The NADH fluorescence signal has been utilized to monitor mitochondrial metabolic transitions in numerous organ systems both in vitro and in vivo. These studies were conducted on isolated skeletal muscle [Chance and Connelly, 1957], heart [Nuutinen, 1984; Koretsky et al., 1987; Chance et al., 1978; Williamson and Jamieson, 1966], brain [O'Connor, 1977; Dora and Kovach, 1983], and kidney [Franke et al., 1976]. Similar results were obtained in cell preparations and even single cells [Balaban and Mandel, 1988; Balaban et al., 1981; Eng et al., 1989; Doane, 1967; Koke and Williams, 1979]. Data from these preparations revealed that the NADH redox state is an extremely sensitive indicator of numerous aspects of tissue bioenergetics. These aspects include the metabolic state of the mitochondria within the cell relating the substrate availability [Balaban and Mandel, 1988], cytosolic adenylates [Chance and Connelly, 1957], and the presence of functional ischemia or hypoxia [Chance, 1976]. One quantitative example of these approaches is presented in Figure 5. In this study the dose response of two substrates, valerate and citrate, was determined on the NADH fluorescence of isolated renal cortical tubules. The NADH fluorescence responded in a dose-dependent manner to the added substrates, increasing until the reaction pathway was saturated with the substrate. This provided a "net" apparent K_m for the transport and entire oxidation of the substrates. Valerate had a much higher apparent affinity than citrate as well as a higher maximum NADH redox response, indicating a higher relative effectiveness of valerate in providing reducing equivalents to the mitochondrial respiratory chain. Thus, this assay procedure permitted the non-invasive determination of the kinetics of substrate utilization in the intact cell. This type of approach may be extremely valuable in evaluating the specific metabolic pathways in intact cells.

The mitochondrial NADH redox state also provides important information

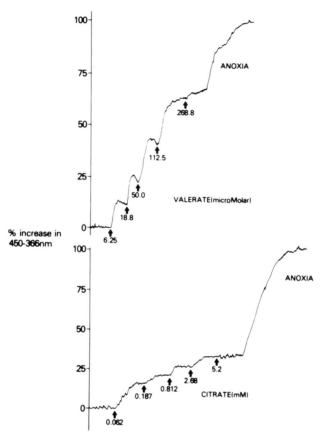

Fig. 5. Dose-response effects of valerate and citrate on NADH fluorescence of rabbit renal cortical tubule suspensions. Anoxia was reached after the suspension had consumed all of the oxygen in the chambers. Note the difference in concentrations of substrates and maximum effect on NADH [from Balaban and Mandel, 1988].

on the metabolic responses to increases in work. The direction of change in the NADH redox state with increases in work reflects the net effect on this "pivot" point in oxidative phosphorylation. If an increase in NADH is observed with work then the net effect of the increase in work was an increase in substrate delivery to the mitochondrial cytochrome chain, assuming oxygen has not become rate limiting (see below). Examples of this occur with the simple addition of substrates to substrate depleted cells, as seen in the renal cell examples above, where the substrate supply has been clearly augmented. However, similar results have been obtained with hormonal stimulation in liver [Balaban and Blum, 1982] as well as with simple work jumps in glucose-perfused hearts

[Katz et al., 1987], indicating that under these conditions the activation of substrate delivery either via dehydrogenase activation or substrate transport has occurred to a greater extent than the demand induced by the increase in work. A decrease in NADH with an increase in work indicates that substrate supply can not keep up with the demand induced by the increase in ATPase turnover or work. This has been the usual response of skeletal muscle [Chance and Connelly, 1957; Jobsis, 1967] as well as the heart [Williamson and Jamieson, 1966] under some experimental conditions. These results have suggested that substrate delivery is not a primary driving force for oxidative phosphorylation in these tissues [Balaban, 1990].

The NADH response to oxygen deprivation has been utilized to map the heterogeneity of tissue metabolic responses to ischemia and hypoxia. Such NADH fluorescence mapping of intact organs has been performed using both flash fluorescence photography as well as video microscopy [Barlow and Chance, 1976; Chance et al., 1978]. These studies demonstrated a very significant heterogeneity of metabolic responses in kidney and heart based on the distribution of NADH fluorescence on the surface of these organs.

As pointed out in the early studies of Chance and Thorell [1959], the signal from NADH is adequate to determine the topological distribution of NADH in single cells. Subsequent to this demonstration, Kohen and coworkers have produced an impressive body of work on the cytosolic localization of reactions involving NADH using aperture limited spectroscopy techniques or scanning procedures [Kohen et al., 1974, 1983, 1985]. These methods have also been coupled to piezoelectric injection techniques to directly assess the metabolism of injected substrates in different regions of the cell, such as the cytosol and nucleus [Kohen et al., 1985].

In more recent studies Eng et al. [1989] have shown that the two dimensional distribution of NADH in mitochondria of single myocytes can be obtained using standard low light microscopy techniques. These studies revealed that the topology of mitochondria could be assessed in the myocyte as well as the topology of metabolic perturbations. An example from this study is presented in Figure 6. This figure represents the digital subtraction of two images: one during a control period, and one after the addition of cyanide to the preparation. Cyanide blocks the respiratory chain of the mitochonria and should nearly maximally reduce mitochondrial NAD. As seen in this difference image the fluorescence changes with cyanide occurred in "strips" in the cytosol consistent with the localization of the mitochondria in these cells. This study demonstrates that the topology of metabolic effects in intact cells is observable using the NADH fluorescence signal from mitochondria.

In summary, NAD plays a pivotal role in the energy metabolism of the cell. The determination of the redox state of this metabolite provides information on the substrate and oxygen delivery to the cell as well as the overall activity

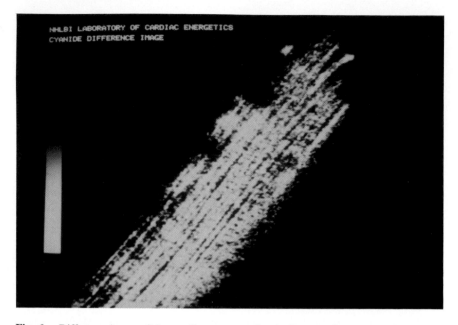

Fig. 6. Difference image of the autofluorescence of a single rat cardiac myocyte in the presence and absence of cyanide. The streaked pattern of fluorescence observed is consistent with the distribution of mitochondria in these cells [from Eng et al., 1989].

status of the cell. With the aid of various localization schemes the topology of these metabolic effects can be evaluated in intact cells using standard low-light videomicroscopy techniques.

B. Oxygen

Oxygen is the terminal electron acceptor in oxidative phosphorylation and in the P450 system. This critical role in the metabolism of the cell makes the determination of oxygen in tissues and cells a major issue in cell biology. This is of particular interest due to the disparity between the O_2 dependence of intact tissues and that of isolated mitochondria. Studies in isolated mitochondria have determined the respiratory and redox changes occur at very low O_2 concentrations, ranging from 0.02 to 0.3 μM [Oshino et al., 1974; Jones, 1986]. In contrast, responses to decreased O_2 occur in intact tissues at arterial pO_2 values of 100 μM and venous values of 25 μM [Jones, 1986]. To determine where the oxygen limitation occurs, numerous attempts have been made to measure oxygen concentration in situ.

Oxygen tension has been determined in intact tissues using a variety of methods [see Jones, 1986]. The noninvasive methods usually involve an optical measurement of an intrinsic or extrinsic chromophore which reacts with oxygen. The major intrinsic probes are the mitochondrial cytochromes and myoglobin absorbance as well as NADH fluorescence which was discussed above. The use of the cytochrome absorbance is based on the optical differences between reduced and oxidized mitochondrial cytochromes as extensively described by various investigators [Chance and Williams, 1956; Wilson and Rumsey, 1987; Jobsis et al., 1977]. Other investigators have taken advantage of alterations in the near infrared absorbance of cytochrome aa_3 with redox state [Jobsis, 1977]. This is the preferred cytochrome to measure oxygen concentration, since it is the final step of the cytochrome chain where oxygen reacts to form water, and its redox state is the least sensitive to other metabolic factors including substrate delivery or activity state. In addition, since this is the site of oxygen reduction in the cytochrome chain, it reports the oxygen tension at the most critical point in the cell, the mitochondrion.

This absorption technique has been applied to numerous isolated organ preparations [Hassinen and Hiltunen, 1975], cell suspensions [Balaban et al., 1980], rapid frozen samples [Gayeski et al., 1987], and in vivo [Lamanna et al., 1977; Balaban and Sylvia, 1981]. These studies have generally been devoted to determining the oxygen tension in these complex systems as well as the investigation of the actual oxidative metabolism process. The major limitations of this approach have been the spectral interference of hemoglobin and myoglobin, light scattering changes in complex tissues, and the low signal-to-noise of these absorption techniques. The interference by hemoglobin has only been a factor in the in vivo preparations and is complicated not only by the oxygenation state of hemoglobin but also by the amount of blood in the tissue which generally changes with the blood flow. Myoglobin is present in most muscle preparations and can significantly interfere with the determination of cytochrome absorbances; indeed, many investigators choose to use myoglobin absorbance characteristics to estimate oxygen [Gayeski et al., 1987]. The low signal to noise ratio for all of these absorbance measurements is the major limitation of performing this type of study in single cells. Indeed, no single cell study has been performed without freezing the sample. For single cell studies fluorescent indicators are necessary and the only strongly fluorescent endogenous oxygen indicator is NADH, as discussed above. However, NADH is not very effective since it is influenced by many factors other than oxygen.

Exogenous fluorescence indicators have also been used to study the distribution of oxygen in cell suspensions, the most common one being pyrene butyric acid, which is selectively quenched by oxygen. Using this dye, Benson et al. [1980] found significant gradients of oxygen content or solubility in single cells dispersed in agar. However, the possibility of gradients in dye would also explain these results. In general, this and other oxygen sensitive probes are hampered by the lack of an internal standard or "isobestic" fluorescence wavelength to correct for dye concentration. This was a similar problem with the earlier pH and Ca dyes, which have now been modified to have the appropriate optical properties in their excitation or emission spectrum to correct for dye concentration or light scattering.

These noninvasive approaches to studying oxygen distribution in intact tissues have provided a very challenging hypothesis that the oxygen concentration may be much lower in tissues than previously believed. This may indicate that oxygen concentrations contribute to the rate limitation of metabolic processes [Balaban, 1990], a very important consequence of this hypothesis. However, the numerous artifacts inherent in the absorption measurements in complex tissues reduces the confidence in many of these measurements. These complications include the influence of light scattering and tissue specific absorbances which modify the frequency dependencies of transmitted or reflected light. The presence of other oxygen sensitive chromophores such as myoglobin and cytoplasmic cytochrome b with overlapping spectral absorbances also interfere with these measurements. In addition, in vivo studies are particularly hampered by the presence of hemoglobin which can severely alter the absorbance or fluorescence characteristics of tissue with changes in oxygen content as well as blood volume. Specific exogenous probes of oxygen tension, properly designed with isobestic points, may help clarify this very important problem in cell biology.

These investigations of tissue oxygenation have been extended to studies in which the oxygen dependence of respiration has been determined in suspensions of isolated cells. This is a controversial area. Studies in hepatocytes by Jones and coworkers [for references see Jones, 1986] found an eightfold higher P_{50} value for oxygen in hepatocytes as compared to isolated mitochondria. These investigators concluded that considerable diffusional gradients existed inside these cells, some of it possibly due to mitochondrial clustering. However, these conclusions have been recently questioned by Robiolo et al. [1990], who utilized the phosphorescence of Pd-coproporphyrin to rapidly measure pO_2 (0.1–0.2 s), providing much better resolution than previously attained by other methods. These investigators found no significant difference between the P_{50} of neuroblastoma cells and that of isolated mitochondria, concluding that no major diffusional gradients for O_2 were present in the cytoplasm.

C. Enzymes and Substrates

Another approach to the study of single cell metabolism is the use of extrinsic metabolic enzymes or substrates which have been labeled with a fluorescent group in a non-active region of the molecule. The study of enzyme distribution and binding may be an important area of metabolic regulation since the organization of enzymes and whole reaction networks may be as important as the concentration and distribution of the substrates for these processes. For example, the cytochrome chain and the glycogen particle are reaction pathways whcrc the geometry and compartmentation of the enzymes involved are critical for this metabolic pathway to function. Thus, enzyme compartmentation or sequestration may be an important, and relatively ignored, aspect of metabolic regulation Pagliaro and Taylor [1988], using rhodamine-labeled aldolase, found that the binding of the enzyme to actin was dependent on its substrate concentrations (i.e., fructose 1, 6 diphosphate). Thus, in the presence of actin, the substrate concentration was influencing the free [E], or V_{max} in Michaelis Menten terms, in addition to its actual participation in the reaction. This study clearly demonstrates that the environment and binding of an enzyme controlled by a substrate may be an important aspect of its regulation other than simple substrate-driven kinetics. Pagliaro and Taylor [1988] also demonstrated that these functional metabolic enzymes labeled with fluorescent groups can be injected into cells and the distribution of these enzymes monitored using video microscopy techniques. The labeled aldolase molecules injected into cells revealed a diffuse distribution of the enzyme throughout the cell. Subsequent photobleaching studies demonstrated, however, that significant fractions of the enzyme were immobilized to varying extents in different regions of the cell. Thus, the *potential* for metabolic regulation by cytoplasmic binding of aldolase exists in the intact cell. These results seem quite promising, but the metabolic and physiological ramifications of these initial observations for aldolase as well as other metabolic enzymes are yet to be established.

Another enzyme which is being studied in a similar manner is hexokinase. In Figure 7 are data from Dr. R. Lynch and F. Fay on vascular smooth muscles in culture. In these images a rhodamine-X-labeled hexokinase isozyme I was microinjected into the cells. The first image (Fig. 7A) is 1 min after the injection, and (Fig. 7B) is taken 9 min later. The localized fluorescence of label in the latter image corresponds to mitochondria within the tissue as shown in the rhodamine 123 fluorescence image in Figure 7C. Thus, these data suggest that hexokinase is localized to the mitochondria in these cells and that the sequestration of this enzyme can occur with the simple injection of the enzyme into the cell. That is, the sequestration is not a consequence of the production of the enzyme in the cell.

This approach may prove to be extremely powerful in mapping the distribution of enzyme units within the cytosol and the potential redistribution which may occur during various stimuli. Coupled to photobleaching methods, the actual bound and free fractions of enzymes in different regions of the cell can also be determined. The major limitations of this system are the generation of functionally active labeled enzymes and the inability to determine the functional aspects of the enzyme in the intact cell. However, the method can accurately provide the location and mobility of the enzymes within the cell, which should significantly help our understanding of the sorting and final distribution of enzymes in intact cells.

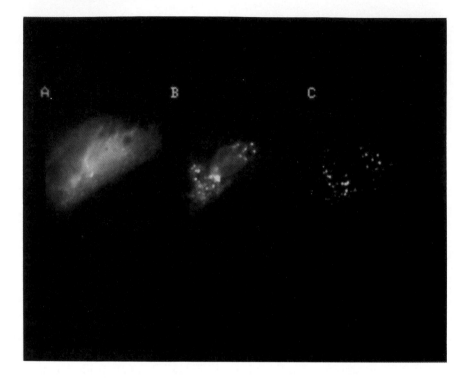

Fig. 7.   Co-localization of hexokinase and mitochondria in vascular smooth muscle cells in culture. A7R5 cells were used in this study. 5 pl of 1 mg/ml rhodamine-X-labeled hexokinase isozyme I was microinjected into the cell. Excitation was at 550 nm and imaging was conducted at 590 nm. **A:** Image taken 1 min after the injection. **B:** 9 min after the injection. After this sequence, rhodamine 123 was introduced to the cells to label the mitochondria. The rhodamine 123 image in (**C**) was excited at 450 nm and light emission was collected at 525 nm. This latter image was digitally filtered using a Gaussian function to maximize the intensity contrast of the image. Note the excellent correlation between the abeled hexokinase distribution (B) and the mitochondrial distribution (C). Data provided by Drs. R. Lynch and F. Fay.

Monitoring the distribution of metabolic events in the intact cell can also be accomplished by using labeled metabolic substrates. The most complete work on the evaluation of exogenously labeled metabolic substrate is the studies by Pagano and Sleight [1985] on the metabolism and distribution of fat and membrane lipid precursors in intact cells. These studies have used a variety of lipid precursors, including phosphatidylcholine, phosphatidylethanolamine, and ceramide. These studies have revealed the mechanisms involved in the transport of lipids through the cell such as monomer insertions, endocytosis, and vesicular transport. In addition, the locations of lipid sorting and metabolism have also been localized for several lipids using this approach [Pagano and Sleight, 1985].

These lipid studies demonstrate the advantages of using fluorescently labeled substrates in the study of cellular metabolism. However, this approach has not been widespread in the study of other metabolic pathways. For energy metabolism, most of the substrates are of low molecular weight (i.e., glucose, lactate short chain fatty acids); thus there is little "room" available on these molecules for a fluorescent probe to be attached without significantly affecting the transport or metabolism of these compounds. This was obviously not the case in the enzyme or lipid studies where relatively large molecules were being studied. Thus, this approach is significantly limited by the construction of fluorescent probes which will mimic the normal transport and metabolic patterns of the native substrate. Potentially the metabolism and binding of hormones or large peptides may be a future target of these types of single cell studies.

D. Cytochrome P-450

The optical measurement of cytochrome P-450 in liver slices has been described in detail by Estabrook et al. [1970]. The slices were mounted in a special cuvette that allowed superfusion and transillumination. Typical spectral changes observed with liver slices from phenobarbital-treated rats are shown in Figure 8. The solid line shows the mitochondrial electron chain spectrum obtained upon substitution of nitrogen for oxygen in the superfusion medium. The cytosolic components become optically visible upon further addition of sodium dithionate (a reducing agent) or carbon monoxide. The latter addition reveals a large peak at 450 nm, which is characteristic of this cytochrome and is the source of its identifying name. From these spectra, it has been determined that in liver, cytochrome P-450 is present in a concentration 3–6 times greater than any of the mitochondrial cytochromes [Estabrook et al., 1970]. Oxygen consumption measurements under conditions that maximized mixed function oxidation through cytochrome P-450 found that as much as 25% of cellular respiration may be due to this pathway in liver slices [Estabrook et al., 1970], and in the perfused liver [Scholz and Thurman, 1969]. The adrenal cortex is the only other tissue with a comparable cytochrome P-450 content to that in the liver, and other tissues such as the kidney cortex and the small intestine have total P-450 contents that are, respectively, one-third and one-twelfth that in the liver [Orrenius et al., 1973].

In intact tissues, care must always be exercized to insure that other components which react with carbon monoxide are first stabilized. This is especially true in the tissues that contain relatively low concentrations of P-450. For example, Jones et al. [1980] found a large optical contribution from indoleamine 2,3,dioxygenase in cells from the rat small intestinal mucosa, which in these cells accounts for about 30% of heme content. In this case, tryptophan was used to enhance the equilibrium binding of CO to this enzyme.

Interactions with substrates produce different types of optical responses in the P-450-CO spectrum [Schenkman et al., 1981]. Type I spectral changes

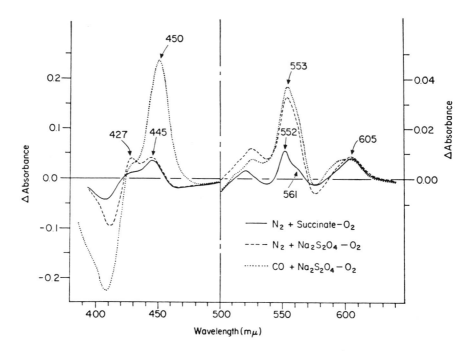

Fig. 8. Spectral changes observed with rat liver slices form phenobarbital treated rats by modifying the composition of the superfusion medium. Difference spectra were collected versus oxygenated Hanks solution [data from Estabrook et al., 1970.]

are produced by a variety of compounds, ranging from simple organic chemicals to drugs (especially barbiturates), detergents, and toxic agents. These compounds cause the appearance of an absorption peak at 385–390 nm and a trough at 420 nm, which are dependent on substrate concentration. The optical changes do not correlate with the total amount of P-450, suggesting that different isoenzymes of this cytochrome are present [Schenkman et al., 1981]. The presence of multiple forms of cytochrome P-450 which preferentially react with different substrates has been confirmed by a variety of techniques. Other compounds, such as aniline, nicotinamide, and pyridine, produce a type II spectral change, characterized by an absorption minimum at 390–405 nm and an absorption peak at 425–435 nm. A third type of spectral change called reverse type I is elicited by compounds such as alcohols and phenacetin, consisting of an absorption maximum at 420 nm and a trough at 388–390 nm.

Some of these spectral changes can be obtained in intact tissues, as demonstrated by Estabrook et al. [1970] in liver slices. However, such a measurement can only be readily made in cells with a high P-450 content as compared to mitochondrial respiratory chain content. Otherwise, optical interference from mitochondrial and other components limit the accuracy of this measurement. Due to these possible complications, most investigators find it more convenient to first isolate a microsomal fraction from the tissue of interest, performing the optical measurements on this fraction.

A different type of optical approach has been utilized by Thurman and coworkers [Ji et al., 1981, 1982; Harris and Thurman, 1986; Arundi et al., 1986]. These investigators developed an optical system of micro-light guides constructed from two strands of 80 μm-diameter glass optical fibers, which separately detect periportal and pericentral events on the surface of the perfused liver [Ji et al., 1982]. Using these fibers and extrinsic fluorescent probes, they measured the relative rates of mixed-function oxidation of several compounds by each of these regions of the liver. For example, the O-deethylation of 7-ethoxycoumarin could be followed fluorometrically since ethoxycoumarin is practically non-fluorescent under the optical conditions employed, but is converted to the highly fluorescent compound 7-hydroxycoumarin by mixed-function oxidation [Ji et al., 1981]. Results show that the rate of O-deethylation is about twice as large in pericentral as in periportal regions of the liver. The same optical approach has been used by this group to study mixed-function oxidation and other metabolic transformations of a variety of compounds in the perfused liver [Harris and Thurman, 1986; Arundi et al., 1986].

E. Peroxides and Oxygen Free Radicals

Hydrogen peroxide is produced in intact cells and tissues in a variety of subcellular localizations [Chance et al., 1979]. Its concentration is about 10^{-8} M in intact bacterial cells and in liver [Sies, 1987]. Quantitation of intracellular hydrogen peroxide production is based on either: a) measuring the steady-state level of the catalase-H_2O_2 intermediate (Compound I), or b) evaluating the flux rate through a peroxidase or catalase [Sies, 1987]. Compound I can be measured optically in intact cells or tissues at the wavelength pair of 660–640 nm. The quantitative measurement is based on the observation that heme occupancy of catalase, i.e., the fraction of catalase heme present as Compound I, depends on the rate of H_2O_2 formation, the total catalase heme concentration, and on the concentration of a hydrogen donor for the peroxidase reaction [Sies, 1987]. Rather than measuring each of these variables, a simplified method was described by Förster et al. [1981] based on the comparison of the response of a test compound or condition with that obtained by external addition of urate, as shown in Figure 9. Such a method was utilized by these investigators to measure peroxisomal fatty acid oxidation, as detected by H_2O_2 production. More recently, Gores et al. [1989] utilized the fluorescent probe 2′,7′-dichlorofluorescin to measure intracellular hydroperoxide production in hepatocytes. The method is based on the entry of the permeable 2′,7′-dichlorofluorescin acetate into the cells, followed by reaction with intracellu-

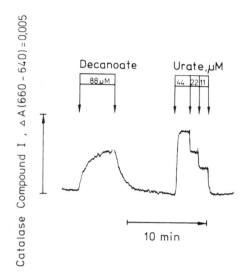

Fig. 9. Quantitation of H_2O_2 production in intact perfused rat liver during decanoate oxidation. Catalase heme occupancy is monitored continuously against time by monitoring the absorbance of Compound I [from Förster et al., 1981.]

lar esterases which trap the membrane-impermeant form of the dye. The basis of the assay is that the non-fluorescent 2′,7′-dichlorofluorescin reacts enzymatically with hydroperoxides to form highly fluorescent 2′,7′-dichlorofluorescein [Gores et al., 1989].

Another optical method utilized to detect hydroperoxide metabolism in intact tissues has been chemiluminescence. Excited oxygen species can emit low-level chemiluminescence from the decay of excited oxygen singlet electrons to the triplet ground state with bands centered at 634 and 703 nm [Sies, 1987]. On the other hand, excited carbonyls emit between 380 and 460 nm. Oxygen singlet electrons tend to occur in low yield and therefore must be considered a side product of major radical producing processes. The weak signal requires photon counting for detection and extreme caution in the interpretation of the data, since there are numerous possible pitfalls as reviewed by Sies [1987]. Chemiluminescence could arise from the breakdown of lipid peroxides, as tested independently by measuring formation of malondialdehyde [Cadenas and Sies, 1984]. On the other hand, chemiluminescence may also arise from other biological reactions, such as photoemission during phagocytosis of leukocytes and macrophages involving NADPH oxidases, oxidation of substrates by xanthine oxidase, and redox cycling of compounds such as menadione and paraquat. In most cases, this photoemission is ascribed to oxygen singlet electrons. Menadione has been found to stimulate oxygen consumption in isolated rat hepatocytes and rat liver mitochondria [DeGroot et al., 1985]. Even though only 10% of this increase is due to redox cycling, this is the critical fraction eliciting the production of oxygen free radical species. The latter appears to be dependent on the steady-state level of the semiquinone.

III. SUMMARY

It is evident that optical spectroscopy is an important tool in the noninvasive study of the metabolic aspects of cellular physiology. This approach provides information on substrate metabolism, oxygen delivery, and the general energetics of the cell as well as various aspects of drug metabolism and toxicology. The major advantage of optical spectroscopy is the direct visualization of metabolic events occuring in their natural environment. This provides the investigator a view of the compartmentation as well as a ''net'' view of the effects of the cytosol on the reaction of interest. The major disadvantage of this approach is that the cytosolic environment can also alter the behavior of the optical probes used, complicating the interpretation of the data. This is true with regard to both intrinsic and extrinsic optical probes. The major limitation for the widespread application of this approach to the study of single cell metabolism is the lack of sensitive fluorescent probes with appropriate characteristics to permit reliable quantitation. Hopefully, with the continued development of these optical probes, this approach will continue to provide unique insights into the complexity and function of the cell.

IV. REFERENCES

Arundi IM, Kauffman FC, E-Mouelli M, Thurman RG (1986) Distribution of sulfatase activity in periportal and pericentral regions of the liver lobule. Mol Pharacol 29:599.

Avi-Dor Y, Olson JM, Doherty MD, Kaplan ND (1962) Fluorescence of pyridine nucleotides in mitochondria. J Biol Chem 237:2377–2383.

Balaban RS (1990) Regulation of oxidative phosphorylation in the mammalian cell. Am J Physiol (in press).

Balban RS, Blum JJ (1982) Hormone induced changes in the NADH fluorescence and O_2 consumption in rat hepatocytes. Am J Physiol 242:C172–177.

Balaban RS, Dennis VW, Mandel LJ (1981) Microfluorimetric monitoring of NAD redox state in isolated perfused tubules. Am J Physiol 240:337–342.

Balaban RS, Soltoff SP, Storey JM, Mandel LJ (1980) Improved renal cortical tubule suspension. Spectrophotometric study of oxygen delivery. Am J Physiol 238:F50–59.

Balaban RS, Mandel LJ (1988) Metabolic substrate utilization by rabbit proximal tubule. An NADH fluorescence study. Am J Physiol 254:F407–F416.

Balaban RS, Sylvia AL (1981) Spectrophotometric monitoring of oxygen delivery to the exposed rat kidney. Am J Physiol 241:F257–F262.

Barlow CH, Chance B (1976) Ischemic areas in perfused rat hearts: Measurements by NADH fluorescence photography. Science 193:990–910.

Benson DM, Knopp JA, Longmuir IS (1980) Intracellular oxygen measurements of mouse liver cells using quantitative fluorescence video microscopy. Biochim Biophys Acta 591:187–197.

Cadenas E, Sies H (1984) Low-level chemiluminescence as an indicator of singlet molecular oxygen in biological systems. In Packer L (ed): Oxygen Radicals in Biological Systems Methods in Enzymology, Vol 105. New York: Academic Press, pp 221–231.

Chance B, Williams CM (1956) The respiratory chain and oxidative phosphorylation. Adv Enzymol 17:65–134.

Chance B, Thorell B (1959) Localization and kinetics of reduced pyridine nucleotide in living cells by microfluorometry. J Biol Chem 234:3044–3050.

Chance B, Barlow C, Nakase Y, Takeda H, Mayesky A, Fishetti R, Grahm N, Sorge J (1978) Heterogeneity of oxygen delivery in normoxic and hypoxic stress: A fluorometer study. Am J Physiol 235:H809–H820.

Chance B, Oshino N, Sugano T, Mayevsky A (1973) Basic principles of tissue oxygen determination from mitochondrial signals. Adv Exp Med Biol 37A:277–292.

Chance B (1976) Pyridine nucleotide as an indicator of the oxygen requirements for energy-linked functions of mitochondria. Circ Res 38:I31–I38.

Chance B, Connelly CM (1957) A method for estimation of the increase in concentration of adenosine diphosphate in muscle sarcosomes following a contraction. Nature 179:1235–1237.

Chance B, Sies H, Boveris A (1979) Hydroperoxide metabolism in mammalian organs. Physiol Rev 59:527–605.

DeGroot H, Noll T, Sies H (1985) Oxygen dependence and subcellular partioning of hepatic menadione-mediated oxygen uptake. Arch Biochem Biophys 243:556–562.

Denton RM, McCormack JG (1985) Ca transport by mammalian mitochondria and its role in hormone action. Am J Physiol 149:E543–E554.

Doane MG 91967) Fluorometric measurements of pyridine nucleotide reduction in giant axon of squid. J Gen Physiol 50:2603–2632.

Dora E, Kovach AG (1983) Effect of topically administered epinephrine, norepinephrine, and acetylcholine on cerebrocortical circulation and the NAD/NADH redox state. J Cereb Blood Flow Metab 3:161–169.

Eng J, Lynch RM, Balaban RS (1989) Nicotinamide adenine dinucleotide fluorescence spectroscopy and imaging of isolated cardiac myocytes. Biophys J 55:621–630.

Estabrook EW, Shigematsu A, Schenkman JB (1970) The contribution of the microsomal electron transport pathway to the oxidative metabolism of liver. In Weber G (ed): Advances in Enzyme Regulation, Volume 8. Fairview Park, New York: Pergamon Press, pp 121–130.

Estabrook RW (1962) Fluorometric measurement of reduced pyridine nucleotide in cellular and subcellular particles. Arch Biochem 4:231–245.

Förster E-C, Fährenkemper T, Rabe U, Graf P, Sies H (1981) Peroxisomal fatty acid oxidation as detected by H_2O_2 production in intact perfused rat liver. Biochem J 196:705–712.

Fralix TA, Heineman FW, Balaban RS (1990) Effects of tissue absorbance on NAD(P)H and INDO-1 fluorescence from perfused rabbit hearts. FEBS Lett (in press).

Franke H, Barlow CH, Chance B (1976) Oxygen delivery in perfused rat kidney: NADH fluorescence and renal functional state. Am J Physiol 231:1082–1089.

Gayeski TEJ, Connett RJ, Honig CR (1987) Minimum intracellular PO2 for maximum cytochrome turnover in red muscle in situ. Am J Physiol 252:H906–H915.

Gibson GG, Skett P (1986) Introduction to Drug Metabolism. London: Chapman and Hall.

Gores GC, Flarsheim CE, Dawson TL, Nieminen A-L, Herman B, Lemasters JJ (1989) Swelling, stress, and cell death during chemical hypoxia in hepatocytes. Am J Physiol 262:C347–C354.

Harris C, Thurman RG (1986) A new method to study glutathione adduct formation in periportal and pericentral regions of the liver lobule by reflectance spectrophotometry. Mol Pharmacol 29:88.

Hassinen IE, Hiltunen K (1975) Respiratory control in isolated perfused rat heart. Role of the equilibrium relations between the mitochondrial electron carriers and the adenylate system. Biochim Biophys Acta 408:319–330.

Ji S, Lemasters JJ, Thurman RG (1981) A fluorometric method to measure sublobular rates of mixed-function oxidation in the hemoglobin-free perfused rat liver. Mol Pharmacol 19:513–516.

Ji S, Lemasters JJ, Christenson V, Thurman RG (1982) Periportal and pericentral pyridine nucleotide fluorescence from the surface of the perfused liver: Evaluation of the hypothesis that chronic treatment with ethanol produces pericentral hypoxia. Proc Natl Acad Sci USA 79:5415–5419.

Jobsis FF (1964) Basic processes in cellular respiration. In Fenn WO, Rahn H (ed): Handbook of Physiology: Respiration. Washington DC: Am. Physiol. Soc., pp 63–124.

Jobsis FF (1967) Oxidative and glycolytic recovery metabolism in muscle. Fluorometric observations on their relative contributions. J Gen Physiol 50:1009–1047.

Jobsis FF, Keizer JH, Lamanna JC, Rosenthal M (1977) Reflectance spectroscopy of cytochrome aa$_3$ in vivo. J Appl Physiol 43:858–872.

Jobsis FF (1977) Non-invasive infrared monitoring of cerebral and myocardial oxygen sufficiency and circulation parameters. Science 198:1264–1267.

Jones DP, Grafstrom R, Orrenius S (1980) Quantitation of hemoproteins in rat small intestinal, mucosa with identification of mitochondrial cytochrome P-450. J Biol Chem 255(6):2283–2390.

Jones DP (1986) Intracellular diffusion gradients of O2 and ATP. Am J Physiol 250:C663–C675.

Katz LA, Korestsky AP, Balban RS (1987) A mechanism of respiratory control in the heart: A ^{31}P NMR and NADH fluorescence study. FEBS Lett 221:270–276.

Kohen E, Kohen C, Salmon JM, Bengtsson G, Thorell B (1974) Rapid microspectrofluormetry for biochemical and metabolic studies in single cells. Biochim Biophys Acta 263:575–583.

Kohen E, Hirschberg JG, Rabinovitch A (1985) Applications of microspectrofluorometry to metabolic control, cell physiology and pathology. Prog ClinBiol Res 196:45–72.

Kohen E, Kohen C, Hirschberg JG, Wouters AW, Thorell B, Westerhoff HV, Charyula KK (1983) Metabolic control and compartmentation in single living cells. Cell Biochem Funct 1:3–16.

Koke JR, Williams LL (1979) Fluorometric measurements of changes in the NADH/NAD ratio in single isolated adult heart cells: Effects of cyanide and 2,4-dinitrophenol. Microbios Lett 12:15–21.

Koretsky AP, Katz LA, Balaban RS (1987) Determination of pyridine nucleotide fluorescence from the perfused heart using an internal standard. Am J Physiol 253:H856–H862.

Lamanna JC, Jobsis FF, Austin GM, Schuler W (1977) Changes in brain metabolism in the cat in response to multiple brief transient ischemic episodes. Exp Neurol 55:304–317.

Mandel LJ (1986) Energy metabolism of cellular activation, growth, and transformation. In Mandel LJ, Benos DJ (eds): The Role of Membranes in Cell Growth and Differentiation. New York: Academic Press, pp 261–291.

Nuuttinen EM (1984) Subcellular origin of the surface fluorescence of reduced nicotinamide nucleotide in the isolated perfused rat heart. Basic Res Cardiol 79:49–58.

O'Connor MJ (1977) Origin of labile NADH tissue fluorescence. In Jobsis FF (ed): Oxygen and Physiological Function. Dallas, TX: Professional Information Library, pp 90–99.

Orrenius S, Ellin A, Jakobsson SV, Thor H, Cinti DL, Schenkman JB, Estabrook RW (1973) The cytochrome P-450-containing mono-oxygenase system of rat kidney cortex microsomes. Drug Metab Dispos 1(1):350–357.

Oshino N, Sugano T, Oshino R, Chance B (1974) Mitochondrial function under hypoxic conditions: The steady states of cytochrome $a + a_3$ and their relation to mitochondrial energy states. Biochim Biophys Acta 368:298–310.

Pagano RE, Sleight RG (1985) Defining lipid transport pathways in animal cells. Science 229:1051–1057.

Pagliaro L, Taylor DL (1988) Aldolase exists in both the fluid and solid phases of the cytoplasm. J Cell Biol 107:981–991.

Robiolo M, Rumsey WL, Wilson DF (1990) Oxygen diffusion and mitochondrial respiration in neuroblastoma cells. Am J Physiol (in press).

Schenkman JB, Sligar SG, Cinti DL (1981) Substrate interaction with cytochrome P-450. Pharmacol Ther 12:43–71.

Scholz R, Thurman RG (1969) Control of mixed function oxidation in perfused rat liver. Proc 28:3623–3625.

Sies H (1987) Intact organ spectrophotometry and single-photon counting. Arch Toxicol 60:138–143.

Unkefer CJ, Blazer RM, London RE (1983) In vivo determination of the pyridine nucleotide reduction charge by carbon-13 nuclear magnetic resonance spectroscopy. Science 22:62–64.

Velick SF (1961) Spectra and structure in enzyme complexes of pyridine and flavine nucleotide. In McElroy WD, Glass B (eds): Light and Life. Baltimore: Johns Hopkins Press, pp 108–143.

Williamson JR (1976) In Hanson RW, Mehlman MA (eds): Gluconeogensis. NY: Wiley, pp 194–204.

Williamson JR, Jamieson D (1966) Metabolic effects of epinephrine in the perfused rat heart. Comparison of intracellular redox states, tissue PO_2 and force of contraction. Mol Pharmacol 2:191–205.

Wilson DF, Rumsey WL (1987) Factors modulating the oxygen dependence of mitochondrial oxidative phosphorylation. Adv Exp Med Biol 222:121–131.

Wilson DF, Rumsey WL, Green TJ, Vanderkoooi JM (1988) The oxygen dependence of mito-chondrial oxidative phosphorylation measured by a new optical method for measuring oxygen concentration. J Biol Chem 263:2712–2718.

Noninvasive Techniques in Cell Biology: 237–272
© 1990 Wiley-Liss, Inc.

10. Optical Studies of Ion and Water Transport in Single Cells

J. Kevin Foskett

Division of Cell Biology, The Hospital for Sick Children,
Toronto, Ontario M5G 1X8, Canada

I. INTRODUCTION

Studies of the mechanisms of water and solute transport by cells have employed a wide variety of experimental techniques ranging from the single molecule level in patch clamp and reconstitution experiments to studies of whole organs. Recent developments in optical technology and their coupling to the light microscope as well as in fluorescent probe synthesis provide new tools to study ion and water transport processes and their regulation at the level of the single intact cell.

Reductionist approaches have been crucial in elucidating the molecular and biochemical bases for a number of transport pathways. Recent molecular biological approaches have resulted in cloning of the genes for several ion transport proteins [Catterall, 1988]. Information from the primary structures and recombination studies can provide new insights into the distribution, function, and pharmacology of these transport pathways. Patch clamp and reconstitution studies have expanded appreciation of the diversity and biophysical properties of ion channels and have provided new insights into the biochemistry and mechanisms of ion channel regulation. At more macroscopic levels, isotopic flux and optical studies using vesicles provide information concerning the transport pathways in specific membrane fractions. Electrophysiological approaches, including conventional and ion-selective microelectrodes, patch clamp techniques, and short-circuit current measurements of intact epthelia, have been integral in defining the transport pathways involved at the cellular and tissue level.

An important goal of ion transport studies is to understand the biophysics and biochemistry of the intact living cell. Thus, relevance for cellular physiology of data from any methodology must be established. Each of the above approaches and techniques for investigating ion and water transport pathways is associated with some degree of invasiveness which may impede extrapolation of data to the physiology of the intact cell. For example, cellular organization which might be critical for normal ion transport is lost in vesicle studies and compromised in studies of cell suspensions (e.g., loss of polarity, reorganization of cytoskeleton). Patch clamping in the whole cell or excised patch configuration results in washout of normal intracellular constituents, and intracellular microelectrodes may perturb normal membrane permeability. The complication of heterogeneity cannot be addressed with most techniques used to study ion and water transport mechanisms. At the tissue and organ level, transport functions in the various cell types may differ and may be differentially regulated. The coordinated transport activities of the individual cell types might be important in determining the overall transport properties of the tissue. At the cellular level, the existence of a cell-specific transport repertoire might be masked by subtle differences among single cells in their physiological status which are reflected in different expressions of various ion transport pathways. An analysis at the single cell level is required to appreciate the role of specific transport pathways within the context of the physiology of that particular cell. At the single cell level, spatial heterogeneity within the cell may be necessary for proper cell function. Localized distributions of ion channels or pumps or of various signaling pathways which regulate these transporters may exist and be critical for proper cell function.

It has become clear that ion and water transport processes play critial roles in regulating diverse cellular functions, including growth and differentiation, cell activation, and motility. Most techniques for studying ion and water transport preclude spatial and temporal correlation of the activities of the various transport pathways with specific cell parameters, including cell behavior such as motility or exocytosis, biochemistry such as levels of intracellular messengers, or morphology such as state of contraction or cell size. As a result, it has become desirable to supplement existing approaches with somewhat less invasive techniques to study ion and water transport. Important adjunct methodologies which can analyze transport within the context of the single living cell with its intact spatial and biochemical characteristics are necessary to allow insights into the spatial/temporal coordination of specific ion and water transport pathways with cell physiological functions.

During the past several years optical methodologies have been used to study ion and water transport processes in single intact cells and are the focus of this review. Emerging technologies for single cell fluorescence quantitation (photometry and imaging) have been focused to date largely on measurement of intracellular levels and distributions of Ca^{2+} and H^+ due to the importance of these ions in cellular regulation and the availability of useful optical probes. The literature on single cell measurements of $[Ca^{2+}]_i$ and pH_i is a rapidly expanding one. The reader is referred to several recent reviews [Tsien, 1989; Tsien and Poenie, 1986; Malgaroli et al., 1987; Cobbold and Rink, 1987; Maxfield, 1989; Bright et al., 1987] for discussion of how these measurements relate to $[Ca^{2+}]_i$ and pH_i regulation per se. The present review will be focused on the use of optical techniques to determine noninvasively cell membrane water permeabilities and rates and pathways of ion permeation, and on the application of ion indicator dyes to follow dynamic changes in intracellular ion concentrations in single living cells. The reader is referred to recent reviews for discussions of optical determinations of membrane potential of single cells (see chapter by Gross, this volume) [Montana et al., 1989; Salzberg, 1989; Grinvald et al., 1988].

II. DYNAMIC SINGLE CELL VOLUME IMAGING OF WATER AND ION TRANSPORT PATHWAYS

A. Background

The water permeability of plasma membranes of most cells is quite high, generally several orders of magnitude greater than ion permeability. Since water is in thermodynamic equilibrium across the membrane and the cytoplasm contains osmotically active, impermeant, charged macromolecules, cells must employ mechanisms to prevent colloid osmotic swelling due to inward diffusion of ions according to the Gibbs-Donnan equilibrium. For animal cells, this appears to be accomplished by regulation of ion "leaks" and active transport pathways. This regulation confers upon the cell an *effective* ion impermeability which means that the membrane can be regarded as semipermeable. For a perfect osmometer with a semipermeable membrane, from the van't Hoff equation

$$\Pi = RT \sum_i \Phi_i c_i$$

or, by expansion and rearrangement

$$V - b = (RT\sum_i \Phi_i Q_i) / \Pi$$

where Π is the osmotic pressure of the medium, R is the gas constant, T is temperature, Φ is the osmotic coefficient, c_i is the concentration of the ith solute in the cell, V is the cell volume, b is the non-osmotically active cell volume, and Q is the amount of solute. Thus, cell volume depends upon the osmotic pressure of the medium and on the solute content of the cell. If Π is changed instantaneously at the cell membrane by addition or subtraction of an impermeant solute, cell volume will change in an inverse fashion at a rate determined by the water permeability of the membrane. Thus, rapid determination in single cells of volume responses to fast changes of medium osmolarity is a method to measure specific membrane water permeabilities. An assumption in this approach is that water movements are much more rapid than solute fluxes, which appears to be true for most cell membranes. Alternatively, under conditions when Π is not altered, changes in cell volume correspond to parallel changes in cell solute content. Thus, measurements of cell volume combined with specific ion substitutions and pharmacological manipulations can provide details concerning specific ion permeation pathways, in a noninvasive manner in intact cells.

B. Cell Volume Imaging

1. Flat-sheet epithelia. In addition to his now classical work on frog skin which provided the conceptual and technical bases for research on ion transport in epithelia in particular, and in other cells as well, Ussing also pioneered the use of optical techniques to study transport mechanisms [Ussing, 1965; MacRobbie and Ussing, 1961; Ussing et al., 1965]. Following his formulation of the two membrane model for epithelia in which vectorial transport of solutes depends on the polarized distribution of specific ion transport pathways [Koefoed-Johnson and Ussing, 1958], MacRobbie and Ussing [1961] postulated that "a study of the volume changes of the epithelium in response to changes in the composition of the two bathing solutions could provide the desired information about the ionic selectives of the two epithelial boundaries." Their experimental protocol consisted of mounting the abdominal skin of a frog on thin glass wool pad on a shallow Lucite dish on the stage of an upright microscope and observing it through a water immersion lens. The thickness of the epithelial layer was determined by focusing on the outside surface of the skin and then on a melanophore immediately below the epithelial layer, and manually recording the difference reading of the fine focus control. Lateral movements of the skin were absent, so changes in cell volume were reflected in changes in the thickness of the tissue. The skin was exposed to separate solutions on either side which could be exchanged by a continuous flow perfusion system. As discussed later, the chamber also incorporated voltage and current electrodes for measuring the transepithelial voltage and the tissue short-circuit current during the optical measurements. With a combination of ion substitutions and replacements, pharmacologic manipulations, and hormone treatments, analysis of tissue volume transients resulted in a description of the specific ion permeabilities on the apical and basolateral cell membranes. MacRobbie and Ussing were also able to localize the water permeability-enhancing effect of antidiuretic hormone to the apical membrane of the epithelial cells and to derive specific water permeabilities for the apical as well as the basolateral membranes. Furthermore, they were able to demonstrate that the tissue could regulate its volume back towards control levels following an osmotic perturbation.

This work was pioneering since it demonstrated that spatially localized transport pathways for ions and water could be resolved by noninvasive optical techniques; it provided the seminal observation that antidiuretic hormone, which is critical in vertebrate body fluid homeostasis, acts by binding to the basolateral membrane but exerts its effects on the apical membrane, hinting at the existence of intracellular second messengers; and it provided the first demonstration of volume regulation by cells exposed to media of altered tonicity, a property now appreciated to exist in most cells. Interestingly, this work, combined with a subsequent optical study [Ussing et al., 1965], led Ussing to discuss concepts such as stretch-activated channels and cytoplasmic viscosity as important determinants of membrane transport, topics which are currently under active investigation [Sachs, 1987; DiBona, 1983].

About 10 years ago, Spring developed a significantly refined approach for analyzing rapid volume changes in single living cells [Spring and Hope, 1978, 1979]. Prompted by controversies concerning the relevance of intercellular dimensions for fluid transport in epithelia, and by recognition of the need to measure membrane water permeability directly to resolve the pathways for epithelial fluid transport, Spring combined high resolution optical microscopy, novel perfusion chamber design, video techniques, and computers to image the morphology of individual cells under various conditions. It is worthwhile to describe the experimental approaches as they have served as the basis for similar studies since then in a variety of cells. Previous studies had demonstrated that the gallbladder from the mudpuppy *Necturus* behaved in vitro as a typical low resistance (''leaky'') epithelium in which fluid absorption is isotonic, driven by transepithelial NaCl absorption. The gallbladder cells are large, columnar, and lack extensive interdigitations of the lateral cell membranes, making them more amenable to optical analyses. The isolated epithelium was mounted in a specially designed chamber on the stage of a microscope. The chamber had a 50 μm thick bath on the side nearest the objective lens, providing several advantages. First, it permitted the use of high magnification ($\times 100$), high numerical-aperture optics (with their attendent short working distances). Second, the small fluid volume allowed solutions to be completely exchanged quite rapidly, providing, for example, the ability to step change the bath osmolarity, as required for accurate measurements of membrane water permeabilities. Finally, unstirred layers, which had up to this time prevented accurate determinations of epithelial cell membrane hydraulic water permeability [Diamond, 1979], were virtually nonexistent. The cells were visualized in differential interference contrast (DIC) optics with a video camera mounted to the microscope. DIC provides high contrast and resolution, as outlined by Allen et al. [1981] and Inoue [1981]. An advantage of DIC when employed with high numerical apertures is the extremely shallow depth of field, i.e., the ability to resolve optically in the depth dimension. Thus, DIC optics permitted ''optical sectioning'' of the cell. A stepper motor drive was coupled to the microscope fine focus mechanism which allowed the focus to be displaced in 0.75 μm increments at high rates under computer control. Beginning at the apical surface, the focus was rapidly displaced in 3 μm intervals until the bottom of the cell was reached. After each focal displacement, a video image of the optical section was captured on a video disk recorder, also under computer control. Optical records of a typical epithelial cell required 1–2 s and could be obtained at ~7 s intervals. An advantage of recording the images on a video disk recorder is that each optical section could be recalled by specifying a discrete track number and examined and analyzed without the jitter inherent in single still frame playback using a video tape recorder. A video cursor generator inserted a cursor into the recorder images and the perimeter was outlined by the experimenter for each optical section. Cell volume was determined from the calculated areas of each optical section and the known vertical displacement between them.

Using this experimental setup for rapid volume measurements of perfused single living cells, Persson and Spring [1982] determined transepithelial salt transport rates, membrane hydraulic conductivities, and ionic mechanisms involved in volume regulation. Isosmotic removal of NaCl from the mucosal bathing solution caused the cells to shrink by ~20%, presumably due to loss of cell solutes. The observed decrease could be explained if 45 mOsm of osmotically active solute were lost from the cell. The rate of this solute loss was similar to previous determination of transepithelial NaCl transport, leading the authors to conclude that the shrinkage rate was determined by the rate of NaCl extrusion from the cell across the basolateral membrane. When NaCl was returned to the medium bathing the apical membrane, the cells swelled to their original volumes. The rate of swelling, if due to isotonic NaCl influx, indicated an apical NaCl influx rate which was similar to the basolateral NaCl active transport rate. These volume-determined rates and permeabilities were in accord with previous measurements.

The above studies complemented the earlier work by Ussing [MacRobbie and Ussing, 1961; Ussing et al., 1965; Ussing, 1965] and Loeschke et al. [1977] in demonstrating that optical measurements of cell volume provide a powerful tool to determine rates and pathways of ion permeation. Implicit in these studies was the assumption that membrane water permeability was high compared to salt permeabilities and that cell volume was determined by cell solute content. The experimental setup used by spring allowed him to measure membrane water permeability directly. In an elegant study, Persson and Spring [1982] performed rapid (<3s) alterations of bath osmolarity while optically monitoring volume of *Necturus* gallbladder cells. The initial rate of swelling or shrinking was used to calculate the initial rate of fluid flow per unit membrane area (J_v) and, therefore, the hydraulic water permeability (L_p) of the cell membrane since, in the absence of hydrostatic pressure differences,

$$L_p = J_v/\sigma\Delta\pi$$

where σ is the solute reflection coefficient and $\Delta\pi$ is the instantaneous osmotic pressure difference. The determined L_p of the apical cell membrane (1.0×10^{-3} cm/s · osmol) and basolateral membrane (2.2×10^{-3} cm/s · osmol) were sufficiently high to conclude that transepithelial fluid absorption was transcellular, requiring a transepithelial osmotic gradient of only 3.5 mOsm. Furthermore, they were able to utilize this same approach to determine relative reflection coefficients for NaCl and urea at the apical cell membrane.

The high hydraulic conductivity of the cell membranes indicates that cell volume will track changes in intracellular solute content. In their original studies, Persson and Spring [1982] discovered that following the initial osmometric response of cell volume to a change in medium osmolarity, the cells regulated their volumes back towards control levels in the continued presence of the osmotic perturbation. Regulation following osmotic shrinkage required the presence of low concentrations (<10mM) of NaCl in the mucosal medium, indicating that the cells swell due to NaCl uptake across the apica membrane. Subsequent studies indicated that the volume regulatory mechanisms responsible for NaCl uptake were different from that employed for transepithelial NaCl absorption. Thus, regulatory volume increase appears to employ parallel apical Na^+/H^+ and Cl^-/HCO_3^- exchangers, whereas transepithelial transport utilizes a NaCl cotransport mechanism, based on pharmacological, ion substitution, and affinity studies of cell volume [Ericson and Spring, 1982a,b]. It should be pointed out that subsequent studies by others using similar and other techniques indicate that such a clear-cut segregation of these transport systems may not be valid [Davis and Finn, 1985a]. Optical studies also determined that regulatory volume decrease (RVD) following osmotic swelling of gallbladder cells is due to a loop-diuretic-sensitive KCl loss into the basolateral bath [Larson and Spring, 1984], which may be regulated by a Ca^{2+} − and cytoskeleton-dependent process [Foskett and Spring, 1985].

Davis and Finn also developed a similar optical sectioning video microscope to study, as had Ussing, a model Na^+ transporting, "tight" epithelium, in their case the amphibian urinary bladder [Davis and Finn, 1981, 1985b, 1987]. Their experimental setup was largely similar to Spring's except for the inclusion of a video tape recorder as the immediate image storage device. Images were subsequently "recaptured" on a video disk for computer-assisted planimetry. Two major themes were developed in their work. First, they observed that osmotic swelling of the urinary bladder cells results in a compensatory volume decrease. This regulation is blocked by high K^+ or 1 mM Ba^{2+} on the serosal side, by removal of Ca^{2+} from the serosal side and by depleting the cells of Cl^-, indicating that it was the result of KCl loss across the basolateral membrane through Ca^{2+}-regulated conductive pathways. Most interesting were the effects on cell volume of specific inhibitors of transepithelial Na^+ transport. They examined the effects of low concentrations of amiloride applied to the apical surface to block conductive Na^+ channels, as well as serosally applied ouabain, to block the Na^+ pump. The prediction was that amiloride, by blocking Na^+ entry while salt exit pathways were unimpeded, would result in cell shrinkage under isosmotic conditions. Conversely, ouabain would be expected to cause cell swelling by blocking salt exit while entry continued. Neither effect was observed. Amiloride had no effect on cell volume under isosmotic conditions, whereas ouabain caused a substantial cell shrinkage. Furthermore, both agents blocked RVD. These and other results suggested that the apical and basolateral membranes communicate in some fashion so as to adjust their transport rates to ensure constancy of cell volume or cell solute composition. Blocking Na^+ entry with amiloride causes the K^+ permeability of the basolateral membrane to become diminished. As a result, salt cannot continue leaving the cell, and the cell does not shrink as originally expected. Furthermore, since the same maneuver blocks RVD, it appears that the K^+ pathway involved in volume regulation is the same one used for transepithelial Na^+ transport. Ouabain blockage of the Na^+ pump causes cell swelling but only if the serosal bath lacks Ca^{2+} and Na^+ is present in the apical bath. The results are consistent with a model for transepithelial transport regulation in which intracellular $Ca^{2+}[Ca^{2+}]_i$ is regulated by a basolaterally located Na^+/Ca^{2+} exchange.

2. Tubules. Flat sheet epithelia, such as those employed in the studies described thus far, present a convenient model for examination of the polarization of transport pathways in apical and basolateral domains since they can be simply stretched across an aperture of an appropriate perfusion chamber to separate mucosal and serosal bathing media. This geometry has been advantageous not only for optical studies of salt and water transport but for other techniques, e.g., electrophysiological, tracer fluxes, as well. Many important transporting epithelia do not present such a simple geometry. In particular, tubular epithelia, e.g., tubules, ducts, present formidable obstacles because of their complicated geometry and small cell size. The development of the isolated perfused kidney tubule technique by Burg [1972] revolutionized studies of salt and water transport by kidney epithelia. However, perfused tubule studies are still limited by considerations of cellular heterogeneity, unstirred layers, and the small size of the cells. Kirk et al. [1983] have outlined several favorable properties of renal tubules for optical studies, including lack of connective tissue and the ability, by optical sectioning of the tubular midline, to view cells along the axis of their apical/basal polarity. Furthermore, the small lumen ($\sim$30-μm) eliminates unstirred layers within it. In recent years, a number of studies have employed video microscopy of cell volume in isolated kidney tubules. These studies have involved examinations of cell volume regulation, a critically important and physiologically relevant process in the kidney, determination of specific membrane water permeabilities, and examination of transport properties of different cell types in heterogeneous tubule segments [Carpi-Medina et al., 1983, 1984; Carpi-Medina and Whittembury, 1988; Welling et al., 1983, 1987; Gonzalez et al., 1982; Kirk et al., 1983, 1984 a,b, 1985, 1987 a,b; Guggino et al., 1985; Guggino, 1986; Lopes and Guggino, 1987; Lopes et al., 1988; Lohr and Grantham, 1986; Hebert, 1986 a,b; Volkl and Lang, 1988; O'Neil and Hayhurst, 1985; Strange, 1988, 1989; Strange and Spring, 1986, 1987a,b].

Early optical investigations of kidney tubule transport utilized non-perfused, lumen-collapsed isolated tubules. More recently, however, perfused tubule stud-

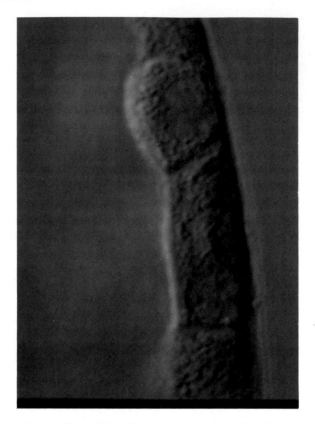

Fig. 1.   Video-enhanced differential interference contrast image of lateral wall of an isolated per-fused cortical collecting tubule from rabbit kidney. The lumen is left of the cells. Upper, bulging cell is an intercalated cell; lower, adjacent cell is a principal cell ($\times$ 2400; courtesy of K. Strange).

ies have provided more informative data (Fig. 1). The tubule-perfusion tech-nique is technically demanding, imposing numerous difficulties which are not encountered in optical studies of flat sheet epithelia or single cells on a cover slip. The tubule must be positioned in the optical path by perfusion- and holding-pipettes mounted to the stage of the microscope. High-resolution optical stud-ies have required that the standard condenser elements be replaced with an objective lens which can fit between the perfusion pipets [Horster and Gundlach, 1979; Kirk et al., 1984 a,b]. The tubule is supported in the bath. Therefore, turbulent flow must be eliminated to prevent tubule movement. On the other hand, bath flow must be high enough to reduce unstirred layers. Switching from one solution to another must be accomplished in such a fashion to pre-vent pressure transients from disrupting the tubule. Strange and Spring [1986] have excellently reviewed the problems and considerations involved in imaging renal tubule cells using video microscopic techniques, and have outlined their elegant solutions to them.

One powerful advantage of imaging techniques is the ability to examine different cell types within a microscopic field simultaneously. Guggino [1986] described two cell types in the early distal tubule of *Amphiuma* based on electrophysiological measurements. The data were corroborated and extended using video microscopy by examining the rate of cell swelling induced by elevations of K^+ in the perfusion bath. Cells identified electrophysiologically as having a high basolateral membrane ionic conductance swelled twice as rapidly as those having a low conductance. The conclusion was that the basolateral membrane of low conductance cells has a significant electroneutral pathway for K^+, probably KCl cotransport, whereas the high conductance cells transport K^+ and Cl^- across the basolateral membrane by conductive pathways.

The cortical collecting duct is one of the major sites of action of antidiuretic hormone (ADH), which dramatically enhances the transepithelial water permeability. This kidney tubule segment is composed of so-called principal and intercalated cells in a ratio of 2:1 (Fig. 1). The apical membrane of principal cells had been believed to be the sole site of action of ADH. However, video microscopic determination of apical and basolateral membrane water permeabilities in the isolated perfused tubules demonstrated that ADH had comparable effects on the apical membrane water permeabilities in both cell types [Strange and Spring, 1987b]. Determination of these water permeabilities was complicated by the small volume of these cells in relation to the length and high water permeability of the basolateral membrane. Volume changes in response to a step change of medium osmolarity were completed in ~ 500 ms. Thus, accurate determinations of specific membrane water permeability required that the opposite side of the cell be bathed in oil to prevent water flux through the cells. Although ADH had similar effects on all intercalated cells, Strange and Spring [1987b] observed two types of intercalated cell morphology. Interestingly, two physiologically distinct groups of intercalated cells were observed. One group failed to regulate their volumes upon exposure to a hyperosmotic stress, whereas the second group showed almost complete regulatory volume increase.

A striking example of heterogeneity of cell types in the kidney is the macula densa, a specialized plaque of cells within the thick ascending limb of the loop of Henle that lies adjacent to the glomerulus. It is believed that these cells ''sense'' the fluid in the tubule lumen and participate in tubuloglomerular feedback by transmitting information to the glomerulus. Kirk et al. [1985] demonstrated that it was possible to visualize these cells in vitro using high-resolution DIC video microscopy. Gonzalez et al. [1988a,b] modified the technique by replacing the video camera with an image-splitting eyepiece connected to a graph recorder. This system is claimed to have a resolution better than the optical system used and has a temporal resolution of 1 s. By combination of ion substitution, pharmacology and rapid alterations of the bathing media, they were able to develop a model for the transport properties of these cells and to conclude that these properties are critical for the tubuloglomerular feedback response.

3. Cell heterogeneity. Optical examinations of other heterogenous epithelia have also been instrumental in defining cell types involved in specific transport functions. Mammalian colon consists of two distinct structural regions: the surface epithelium and the crypts. Welsh et al. [1982] used light microscopy of isolated perfused colonic mucosa to localize cAMP-mediated fluid secretion specifically to the crypts.

An equally complicated geometry exists in the frog and toad skin epithelium. Since the original work of Ussing [Ussing and Zerahn, 1951] the frog skin has been studied as a model Na^+-absorbing epithelium. Although the skin consists of several cell layers, it was established that the apical membrane of the outer cell layer contains the Na^+ permeability pathway, while the basolateral membrane together with the rest of the cells form a functional syncitium. Three pathways for Cl^- permeation across the skin were proposed—a paracellular pathway between cells, and cellular pathways through mitochondria-rich (MR) cells and glands. Electrophysiological evidence suggested a transcellular pathway and variations in skin Cl^- conductance are correlated with the density of MR cells [see Foskett and Ussing, 1986]. However, the effects of MR cells on the transport properties of the frog skin were generally ignored, in spite of the fact that their density can be quite high. This was probably because the distribution and small size of these cells have prevented the use of standard transport techniques. Three studies employing video microscopic optical studies have established that the MR cell is the pathway for conductive Cl^- movement across frog skin [Spring and Ussing 1986; Foskett and Ussing, 1986; Larsen et al., 1987]. These studies employed the "split skin" preparation in which the epithelium is separated from the underlying connective tissue by collagenase treatment. In spite of its several cell layers, the flask-shaped MR cells are easily resolved in high-resolution DIC optics and are therefore amenable to cell volume determinations using optical sectioning techniques (Fig. 2). Spring and Ussing [1986] observed that MR cell volume specifically was influenced by changing the Cl^- concentration or osmolarity of the outside bathing solution. MR cells shrank 20-25% when all Cl^- was removed from the mucosal solution and recovered their volumes when Cl^- was restored. As described below, volume imaging of single MR cells combined with transepithelial voltage clamping techniques demonstrated that this Cl^- permeability is conductive.

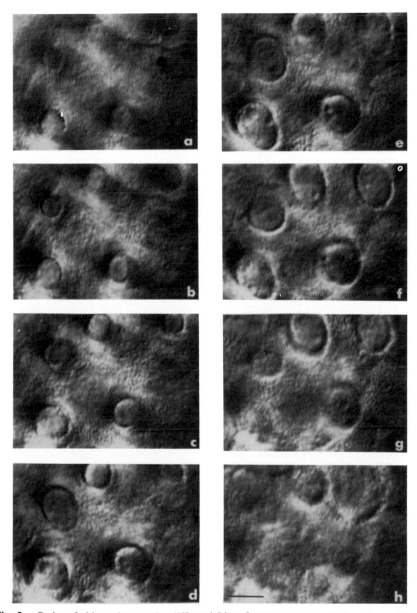

Fig. 2.   Series of video-microscopic, differential interference contrast optical sections of frog skin epithelium perfused in a chamber and voltage clamped to 0 mV. Mitochondria- rich cells are clearly defined against the background epithelium and are round when viewed from above. Optical sections **(a-h)** are displaced by 3 μm. Section a is at the surface. Bar is 10 μm [from Foskett and Ussing, 1986].

C. Volume Imaging of Single Cells Combined With Electrophysiological Approaches

As detailed earlier, cell volume determinations of living cells can provide information concerning specific transport mechanisms when combined with pharmacological and ion substitution strategies. These manipulations change the effective driving forces for ion permeation through specific pathways by changing ion concentrations or the activities of specific mechanisms. Since voltage is an important driving force for ion translocation through conductive pathways, it too can be used in optical determinations of cell volume to dissect specific transport pathways and to localize them to specific cells. MacRobbie and Ussing [1961] first tried this approach in their studies of frog skin. Short-circuiting the skin was associated with variable cell volume changes but both swelling and shrinkage were found. The mammalian kidney cortical collecting duct absorbs Na^+ and secretes K^+ through separate conductive pathways in the apical membrane. However, as previously described, this tubule segment contains at least two cell types—principal and intercalated cells. O'Neil and Hayhurst [1985] injected current via the perfusion pipette across an isolated perfused collecting duct while they observed each cell type in interference contrast optics. They observed the principal cell, but not the intercalated cell, to swell. Swelling was sensitive to luminal amiloride and Ba^{2+}, blockers of Na^+ and K^+ channels, respectively, demonstrating that the principcal cells only possess these transport pathways (as well as a substantial Cl^- conductance in the basolateral membrane).

Similar reasoning led Foskett and Ussing [1986] to examine the cell volume response of individual MR cells in frog skin to transepithelial voltage clamping. If MR cells are localized pathways of conductive movements of Cl^- and have an appreciable K^+ conductance in its basolateral membrane, as do other cells, then voltage clamping the tissue to serosa-positive voltages should drive KCl into the cell, causing cell swelling. Apical Cl^- and voltage-dependent swelling was in fact observed. These studies complemented extracellular current-recording localization of Cl^- current to MR cells [Foskett and Ussing, 1986; Katz et al., 1985]. However, since the electrical determinations only indicated that the conductance was associated with the cell, without allowing localization of that pathway, the volume experiments proved that the conductance must be at least in part in the cellular pathway. Subsequent studies by Larsen et al. [1987] demonstrated the power of volume imaging combined with voltage clamping techniques for analysis of ion transport in a minority cell type which is small and not accessible to microelectrodes, such as the MR cell. They were able to confirm that MR cells are localized sites of Cl^- conductance and that voltage induced swelling is caused by an uptake of Cl^- from the apical bath and K^+ from the serosal bath. They demonstrated that MR cells also have an amiloride-sensitive Na^+ conductance in the apical membrane, an ouabain-sensitive Na^+ pump in the basolateral membrane and a passive Cl^- permeability in both membranes. Furthermore, these optical studies were used to calculate the Na^+ current carried by a single MR cell (~10 pA), the osmotically inactive intracellular volume (21%) and the intracellular steady-state Cl^- concentration (20 mM).

A novel approach for analyzing transport in hepatocytes utilized optical determinations of extracellular volume. Isolated hepatocytes form couplets in which two hepatocytes become joined by tight junctions at their apical poles, retaining their normal polarity. Secretion into the canalicular space between the hepatocytes causes the space to swell. Employing video microscopic optical-sectioning with high-resolution DIC optics, Gautam et al. [1989] have shown that biliary secretion can be quantified by following the change in the volume of the canalicular space. More recently, Weinman et al. [1989] combined this approach with electrophysiological techniques. By optically tracking the size of the canalicular space during intracellular current injection, they determined that taurocholate transport across the apical (canalicular) membrane is electrogenic.

III. SINGLE CELL FLUORESCENCE MEASUREMENTS
A. Background

The preceding sections demonstrate that optical imaging of single cell volume is a powerful tool for analyzing ion and water transport pathways in intact cells. However, many important transport parameters and activities may not be reflected in volume changes. For example, determinations of membrane potential and intracellular ion concentrations, both required to define driving forces for transport, require other methodologies such as isotopic or chemical distributions or microelectrode techniques. However, isotope/chemical equilibrium distribution methods are slow and represent a cell population response, while intracellular microelectrode techniques are technically demanding and not suitable for many cells. Recent developments in computer and video technology and in optical probe synthesis have now made it possible to quantitate with high spatial and temporal resolution fluorescent indicator dyes inside single cells.

The attributes of fluorescence which make it especially suitable for studying single intact cells are its sensitivity and specificity. The sensitivity of fluorescent molecules to their environment may be exploited to derive information about that the intracellular environment(s). The specificity of fluorescence allows detection of a specific parameter within a complex mixture such as the cytosol. Fluorescent probes have been recently developed which can, at least in principle, be used to quantitate specific ions including Ca^{2+}, Cl^-, Na^+, K^+, Mg^{2+}, and Zn^{2+} [reviewed by Haugland and Minta, this volume; and by Tsien, 1989]. In addition, fluorescent probes which are transported across biological membranes by specific ion transport pathways (e.g., NBD-taurine on the Cl^-/HCO_3^- exchanger) and probes whose fluorescence is sensitive to concentration, and therefore useful for tracking volume (intra- and extracellular) (e.g., fluorescein sulfonate), are now available. Given the current state of computer and imaging technology, continued progress in fluorescent probe development will result in a rapidly expanding use of fluorescence techniques to study ion and water transport mechanisms.

B. Methodology

The basic equipment required for quantitative fluorescence measurements at the single cell level are a microscope, a light-source for excitation of fluorescence, and a low-light-sensitive device to detect weak fluorescence emissions. The detector can be a photomultiplier tube, providing a mask can be inserted into an intermediate image plane to ensure light collection from only the cell of interest. The output can be to a chart recorder or, in more sophisticated systems, to a computer through appropriate devices. The advantages of such a system include very low-light sensitivity, high temporal resolution, and manageability of data collection and analysis. The disadvantage is the absence of spatial information. Thus, only one cell can be examined at a time, and spatial distribution of fluorescence within a cell cannot be determined. Imaging systems overcome these disadvantages, but at the expense of temporal resolution, complexity of data storage, and analysis and complexity (and cost) of necessary computer hardware and software.

For either photometry or imaging, fundamental considerations apply regarding microscope design, fluorescence excitation and detection, and data interpretation. In the development of the microscope and digital image processing system used in my laboratory to quantitate fluorescence in single living cells the following considerations were guiding principles:

1. Maximize light through-put in the microscope.
2. Minimize fluorescence excitation light flux.
3. Maximize fluorescence detection.
4. Provide ability to examine relevant spectral properties.
5. Maximize data throughput to improve temporal resolution.
6. Provide an appropriate physiological environment for the cells on the stage of the microscope, with the ability to rapidly alter the environment.
7. Allow-high resolution DIC transmitted light imaging to be performed simultaneously with low light-level fluorescence imaging to permit correlation of fluorescence with cell/tissue structure.

1. Fluorescence light transmission. Most modern microscopes employ epi-illumination of fluorescence. In this method, the objective lens serves the dual function of excitation illumination and emission detection. Excitation and emission beams are spectrally separated by a dichroic mirror, which, when combined with appropriate excitation and emission narrow band-pass interference filters, provides a fluorescence signal with minimal contamination of the excitation light. A high numerical aperture (NA) objective lens is used since the light gathering ability of the lens is proportional to NA. In an epifluorescence microscope, the image brightness is proportional to the fourth power of the NA. In addition to the NA, choice of magnification is important since light brightness is inversely proportional to the square of it. Finally, the lens must be constructed of appropriate glass, cements, and coatings to ensure high light through-put over a wide spectral range with low lens autofluorescence. In our work, we utilize a Zeiss IM-35 inverted microscope equipped with quartz optics and usually employ as the objective lens a Zeiss $63 \times$, NA $= 1.25$ neofluor. A Nikon $40 \times$, NA $= 1.3$ is also employed in some applications. Fluorescence excitation is from a 75 W xenon arc lamp which is aligned for Koehler illumination to maximize illumination uniformity. Normally, low-light-level fluorescence is associated with the elimination of optical elements such as phase rings and polarizers which attenuate fluorescence excitation and emission intensities. As discussed below, however, it is possible to provide DIC imaging in a low-light-level fluorescence microscope without the fluorescence light losses normally associated with this contrast-enhancing technique.

2. Fluorescence excitation level.   Incorporation and illumination of fluorescent probes in living cells may not be without adverse physiological effects. This is especially true in single cell measurements because the same cell is repeatedly illuminated. Excited fluorescent probes (intrinsic and extrinsic) may generate toxic free-radical species, buffer intracellular ion levels, or have other diverse metabolic or toxic effects. Two general strategies are employed to minimize these effects. First, attempts are made to minimize the amount of dye incorporated into the cell. Second, the total excitation light flux is reduced. Since this flux is the integral of the absolute light intensity over time, strategies involve reducing the exposure time by use of intermittent excitation and attenuating the excitation intensity.

Shuttering devices, such as those from Vincent Associates, placed between the arc lamp and the microscope, allow continuously variable excitation times down to ~ 5 ms. Filter wheels or solenoid-activated filters switchers used for changing excitation wavelengths may also be used as shuttering devices, as discussed below. Shorter pulses can be achieved using pulsed laser excitation or acousto-optic modulation [Spring and Smith, 1987], but the arc lamp/shutter combination is the most practical solution for most applications. Attenuation of excitation is most easily achieved by introduction of commonly available neutral-density filters between the lamp and the preparation.

3. Fluorescence detection.   Because fluorescence intensity is proportional to the amount of probe as well as the level of excitation illumination, minimization of these requires enhanced light-detection systems. For non-imaging applications a photomultiplier tube is used [Wampler and Kutz, 1989]. For imaging, a wide variety of technologies have been developed. The reader is referred to several excellent recent reviews [Wampler and Kutz, 1989; Aikens et al., 1989; Spring and Lowy, 1989]. Choice of the appropriate detector involves considerations of its sensitivty to very low light levels, speed, and resolution. Low-light-level images are generally of poor quality due to insufficient photon flux. Signal to noise can be improved by integrating the signal at the expense of temporal resolution. Imaging rapid transients of fluorescence thus becomes a trade-off between light level, signal/noise, and speed. I employ a two-stage inverted type image intensifier (Videoscope) coupled by relay lenses to a newvicon camera (models 65 and 70, Dage MTI or KS-2000N, Videoscope) or more recently, to a video-rate charge-coupled device (CCD) (model 72, Dage MTI) as described by Spring and Smith [1987]. Automatic gain and black level are disabled. The intensifier approaches the theoretical limit for photon detection and resolution and has a fast (<1 ms) response time. The newvicon tube has moderate light sensitivity and response times with high resolution (~700 lines) at high light levels.

   Various aspects of the camera performance should be tested for suitability for quantitative fluorescence measurements. Tsay et al. [1990] have recently analyzed various performance characteristics for several low-light cameras. Of critical importance for quantitative work is response linearity. To test for linearity, a convenient, accurate and reproducible way to modulate the light level to the camera is to vary the angle (θ) between crossed polarizers. Light intensity is proportional to $\sin^2\theta$. Thus, a plot of camera output (determined with an oscilloscope or digital image processor) vs. $\sin^2 \theta$ over the range of light intensities encountered during experiments should be linear for the device to be suitable for quantitation. The response speed of the detector is determined by the lag, which represents the amount of time required for the persistence of the previous video field, or the building up of the new video field, to achieve a particular level. Lag is generally available in camera specifications as a time corresponding to 67% of the final value or the amount of persistence at the end of 50 ms (lag operates for intensity increases as well as decreases). However, in my experience lag depends on the operating characteristics of the camera as well on its previous history requiring that it be empirically determined under the conditions of the experiments. This is accomplished by rapidly shuttering on or off input light and recording the images in real video time on a video device or recording the video output on a storage oscilloscope. In our measurements 95% response was generally achieved within 100–130 ms. Similar values were obtained by Tsay et al. [1990] for this camera (53% complete response in 33 ms, 87% in 100 ms). The implication of these measurements is that temporal resolution of the camera is limited to these frequencies unless the images can be deconvoluted to remove the signal remaining from the camera's history. Recent descriptions of video imaging systems which acquire alternative video frames at different excitation wavelengths have not adequately addressed this problem. Probably the best solution to this problem is to couple a video-rate CCD to an image intensifier such as that described above. Tsay et al. [1990] report that the response of the Dage 72 CCD coupled to the KS1381 image intensifier is 96% complete within 1 video frame time (33 ms). Another parameter which determines the frequency response of the imaging system is signal/noise, since this parameter must be increased in low-light applications by temporal integration of the signal. Integration of sequential video frames improves signal/noise by the square root of the number of frames. As a compromise between improved signal/noise and temporal resolution, we typically integrate 8 frames ($\sim .25$ s) during ion-indicator dye imaging experiments.

4. Spectral information. Many fluorescent probes have single excitation and emission peaks. Measurement requires a single set of optical filters. Meaningful quantitation of such probes is difficult, especially in imaging systems, since fluorescence of an indicator dye is determined specifically by the level of the sensitive parameter (i.e., Ca^{2+} for fura-2) and nonspecifically by optical path length, probe concentration, excitation intensity, and detector sensitivity. Thus, even in spatially unresolved measurements of single cells, fluorescence emission intensity will depend nonspecifically upon the degree of dye loading, which may be time-dependent due to bleaching and leakage, cell thickness, and position of the cell in the illumination (if there exists illumination heterogeneity) and the detector fields. Spatially resolved determinations are further complicated by such dependencies within the cell.

Some fluorescence indicator dyes overcome these technical difficulties as a result of spectral shifts due to interaction with the parameter of interest, or due to existence of another wavelength which is independent of that parameter (isoexcitation point). The ratio of the fluorescence intensity at two appropriate excitation (excitation dyes) or emission (emission dyes) wavelengths normalizes nonspecific parameters (above) such that the computed ratio intensity becomes, to a first approximation, a measure of the parameter of interest. Excitation ratio-type dyes are available for determinations of Na^+ and K^+. Excitation and emission dyes are available for determination of Ca^{2+} and H^+ (see chapter by Haugland and Minta, this volume). Thus, the microscope must have the capability to illuminate the preparation and/or record the emission at >1 wavelength. The acceptable amount of time between acquisition of the two signals (images) will depend on the speed at which the parameter of interest is changing and the degree and speed at which a parameter-insensitive wavelength is changing (if an isosbestic point is used). For most applications, it is desirable to alternate between the two wavelengths as quickly as possible. A number of systems are used to achieve this. Emission-type dyes can provide simultaneous determinations at two appropriate wavelengths. In photometry systems, a dichroic mirror separates the two wavelengths to separate photomultiplier tubes. This is considerably more difficult in imaging systems because of the amount of information on the emission side. Image processing to register the images from the two cameras is necessary. Continued development of video image processing hardware and software combined with faster computers should result in increasing use of this approach. Another approach, recently described by Balaban et al. [1986], involves digital video imaging of the first-order diffraction from a slit in the image plane. The image contains spatial information along the vertical (= slit) axis, while each horizontal line represents the fluorescence emission spectrum. Thus, spatially and spectrally resolved processes can be resolved at video framing rates (33 ms).

At present, most imaging syystems use excitation-type ratio dyes. Kurtz et al. [1987] has developed a system to excite at two wavelengths simultaneously using two lamps, each with a rapid chopping system. Fluorescence is demodulated using a pair of lock-in amplifiers, providing a high signal/noise response. The system is only suitable for photometry, however. Imaging systems generally employ a single lamp with a stepper-motor controlled filter wheel or solenoid-activated filter holder, or dual lamps (coupled to dual monochromators, or separate interference filters) and a chopper device positioned between the arc lamps and the microscope. Acousto-optic modulators allow wavelength switching and intensity modulation of laser excitation in the μsec domain. Recently Kurtz [1987] described a similar device for use with arc lamps. When commercially available, this should be the method of choice for excitation wavelength modulation. Filter switching devices rapidly insert interference filters into the excitation light path in <2 video frame times [Foskett, 1988]. Systems to chop between two lamps are limited only by the speed of the chopper. Because of the camera lag in video imaging systems, it is desirable to execute a single switch from one wavelength to another as quickly as possible, to allow the camera to *begin* to settle in as quickly as possible. However, time spent at the new wavelength must be sufficient to account for lag and frame integration. For intensified video-rate CCDs, lag is <33 ms.   An 8-frame integration requires 266 ms, for a total of ~ 300 ms. If filter switching requires 2 video frames, effective continuous acquisition of pairs of images requires:

process:	$\Delta\lambda$	lag	integrate	$\Delta\lambda$	lag	integrate	$\simeq 730$ ms
time (msec):	66	33	266	66	33	266	

Even if switching was essentially instantaneous (acousto-optic modulation), temporal resolution is still only ~600 ms.

It is clear that reduction of frame integration time is required to improve temporal resolution of the system. Thus, increased temporal resolution requires acceptance of noisier images and/or higher excitation light intensities. Tolerance of either must be determined for each type of experiment, dye, and cell type. During continuous switching, the effective temporal resolution can be improved to ~350 ms mathematically by using λ1 on either side of λ2 in ratio calculations if the elapsed time between acquisition of λ1 and λ2 is the same as between λ2 and the next λ1. However, as discussed below, realizing such temporal resolution may be limited by the data throughput of the system.

5. Data throughput. Video images contain a tremendous amount of information. Typical video image processing systems digitize each image to 512 × 480 × 8 bits, or ~250 KB. Capture of a video image results in corresponding occupation of digital video memory. When video memory becomes filled, images must be transferred to another storage area, which requires time, decreasing temporal resolution. Image capture to a real-time digital disk avoids this problem, but the expense (although considerably less in recent years) is generally prohibitive. Another solution is to digitize after the experiment from a primary recording device such as a laser disc recorder, although this adds another noise component to the data. Our experience thus far has been to digitize and store images during the experiments. In our system, writing images to hard disk becomes rate limiting. To improve digital image throughput, we have used several complementary strategies. The first involves sacrificing some spatial resolution. Images are digitized at 256 × 240 resolution instead of 512 × 480, reducing the amount of data by 75% and increasing throughput correspondingly. The second strategy involves creation of virtual frame buffers in computer random access memory (RAM). Writing images out of the frame buffers to RAM is considerably faster than to hard disc. Depending upon the amount of RAM which can be dedicated to this function, "bursts" of high temporal-resolution imaging can be performed during an experiment. At less critical times during the experiment, the data are written out of RAM to hard disk, freeing up RAM for another burst. One MB RAM can store sixteen 256 × 240 images, or 8 pairs. We dedicate 4 MB RAM as virtual frame buffers, or enough space for 32 pairs of images, or 64 ratios (as described above). Another strategy involves identifying regions-of-interest (ROI) for storage. In many experiments, only discrete areas of the entire image contain useful information (e.g., 1 cell in the middle of the field). Elimination of uninteresting regions of the image will speed data throughput.

6. Physiological environment. The cells on the stage of the microscope should be perfused with physiologically appropriate salt solution. Thus, provisions for CO_2 gassing and temperature regulation should be incorporated. Further, many experiments require the system to be perturbed. Thus, perfusion switching devices (e.g., pinch valves) should be incorporated to change rapidly perfusion fluids such that the new solution reaches the cell of interest at a known time.

7. Correlation of fluorescence with cell structure. Full advantage of the high spatial and temporal resolution provided by low-light level video imaging of fluorescence is realized only when the cell structure can also be imaged with high contrast and resolution and with equivalent temporal resolution. In that way, a high correlation between cell structure and fluorescence can be achieved. DIC imaging permits resolution of structural detail that surpasses other light microscopic methods. The use of video for DIC imaging has greatly extended the use of resolution and contrast provided by the light microscope [Allen et al., 1981; Inoue, 1981). Thus, a combination of low-light-level fluorescence with DIC imaging provides the best means to correlate fluorescence distribution with cell structure in living cells. However, low-light-level fluorescence is associated with elimination of contrast-enhancing optical elements such as polarizers. For example, a polarizer generally transmits only 30–40% of the incoming light. In a microscope equipped for DIC, the fluorescence emission will pass through a Wollaston prism and a polarizer. If the excitation light also passes through the same elements, the resultant fluorescence intensity will be <10% compared with that of the same microscope without these elements. The loss of transmitted light image resolution and contrast resulting from the elimination of these optical elements in low-light-fluorescence imaging is further compounded by another necessity of video imaging. Because the video detector is operated in a low-light-level mode for the fluorescence imaging, the transmitted light intensity used for brightfield imaging must be greatly attenuated. This results in an image with poor signal characteristics that is degraded by noise in the camera, even in digital image processing systems where it is possible to integrate a number of sequential images. As a result of the combined effects of eliminating contrast-enhancing optical elements and imaging under photon-poor conditions, low-light-level fluorescence is assocaited with poor transmitted-light imaging, making correlation of fluorescence with simultaneous cell structure and function difficult. Recently, however, two systems have been developed which permit DIC imaging in a low-light-level fluorescence microscope.

a. System I. Developed by Foskett [1988], this technique requires two video detectors, the means to separate spectrally the two images to the respective cameras, and a judicious placement of the polarizers in the optical path (Fig. 3a). The transmitted light optical path incorporates the elements required for DIC imaging. The light from the halogen lamp is polarized by a calcite prism and passes through a Wollaston prism before illuminating the preparation. Fluorescence excitation from the xenon arc is directed to the preparation by a dichroic mirror. Fluorescence emission as well as transmitted light that passes through the specimen pass through the objective lens and a second Wollaston

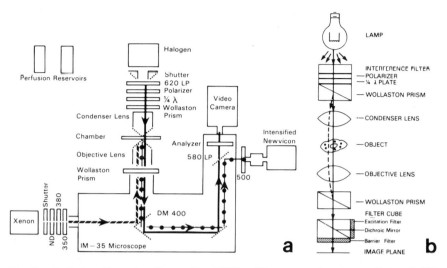

Fig. 3. Optical configurations for combined DIC and fluorescence imaging. **a:** System I. LP, long pass filter; DM; dichroic mirror; 350, 380, 500, band-pass interference filters; ND, neutral density filter; fluorescence excitation, ---; fluorescence emission, —•••—; transmitted light, ——. See text for details [from Foskett, 1988]. **b:** System II. See text for details [from Spring, 1990].

prism associated with it. In the standard Zeiss IM-35 configuration, the light would next pass through a second polarizer (analyzer). However, because this polarizer would be in the fluorescence emission light path, it was repositioned more distally in the optic path, as described later. In the standard microscope configuration, the light proceeds through the microscope to a set of prisms designed to either allow the light to continue to the oculars or to direct it at right angles through a side port to a video camera. With the use of a trinocular, a second camera is mounted above the oculars and adjusted to be parfocal with the side port camera. To direct fluorescence emission to one camera and transmitted light to the other, the prisms are replaced with a dichroic mirror mounted at 45° to the incident light path. The dichroic mirror can separate light based on spectral properties; thus, it becomes possible to image the transmitted light and fluorescence separately if they are spectrally distinct. The halogen light is initially filtered to long enough wavelengths to pass through this dichroic. The dichroic mirror that reflects the fluorescence excitation to the preparation passes these long wavelengths without interference. The dichroic mirror that replaces the prisms to direct the light to the appropriate cameras is chosen to be able to pass the long wavelength halogen light to the overhead camera for transmitted light imaging and reflect the fluorescence emission out the side port through an interference filter for fluorescence imaging. The analyzer (calcite prism) is positioned between this dichroic mirror and the

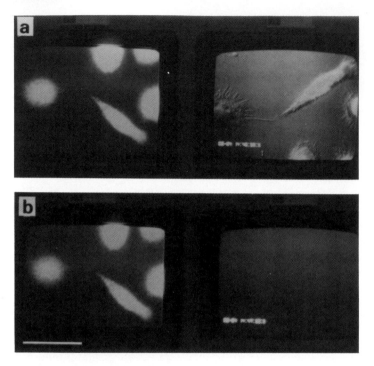

Fig. 4. Simultaneous DIC and fluorescence imaging. Human monocytes, grown in culture, perfused on stage of microscope and loaded with Ca^{2+}-sensitive dye, fura-2. **Left monitor:** Fluorescence emission. **Right monitor:** Simultaneous DIC image. **b:** Transmitted light shuttered off demonstrating that the two images are not spectrally overlapping. See text for details. Bar is 50 μm [from Foskett, 1988].

transmitted light camera, completing the DIC light path. Thus, the overhead camera views the preparation in high light, contrast, and resolution DIC optics, whereas the side-port camera views the low-light-level fluorescence (Fig. 4). The only DIC optical element in the fluorescence light path is a Wollaston prism. My measurements indicate that this component results in a fluorescence diminution of ~2–4% for most wavelengths.

   b. System II. Another method, recently described by Spring [1990], takes advantage of the light polarizing properties of dichroic mirrors (Fig. 3b). The dichroic mirror which reflects fluorescence excitation into the preparation produces two separate transmission curves depending on the incident beam polarization. Therefore, the dichroic can act as the analyzer in DIC imaging if the wavelength of the transmitted light is properly selected to fall in the narrow spectral region where the p-plane is transmitted whereas the s-plane is not. Absence of an analyzer eliminates the need for a second detector. However, the gain of the intensified camera must be reduced during DIC imaging, and gated on during fluorescence imaging. Thus, high-resolution DIC and low-light-level fluorescence imaging cannot be performed simultaneously. However, if the gain of the intensifier is under computer control, it will be possible rapidly to image sequentially in DIC and low-light-level fluorescence. A significant advantage of this approach is that a single detector results in precise correlation between the two images.

The advantage of technique I is that DIC imaging and low-light-level fluorescence can be performed simultaneously (Fig. 4). A correlate of this is that the intensified camera can be replaced with a photomultiplier tube, allowing photometry to be performed simultaneously with DIC imaging. Thus, micromanipulation of cells, including electrophysiology and microinjection, can be performed using high-resolution optics while the cell fluorescence is monitored with high temporal resolution [Foskett et al., 1989]. The system can also perform dual-emission fluorescence imaging by incorporation of an intensifier in front of the transmitted light camera. The disadvantages include the expense and complication of operating two cameras and the lack of perfect registration between the DIC and fluorescence images, rendering ambiguous precise correlations between the two images.

Although I have not yet implemented it, both approaches could be combined to provide DIC imaging and dual emission fluorescence in a single microscope. Thus, the camera with gatable-gain of system II is placed at the side port of system I. Another intensified camera sits atop the trinocular. The side-port camera would view the DIC image generated by the method of system II, as well as shorter-wavelength fluorescence emission. The top camera would view long-wavelength emission.

C. Specific Approaches

1. Simultaneous determinations of $[Ca^{2+}]_i$ and cell volume in single cells. The development of systems which permit both DIC and fluorescence measurements allows the power inherent in cell volume determinations of single living cells for analysis of ion and transport pathways, discussed earlier, to be combined with the power of quantitative fluorescence imaging of ion- and volume-indicator dyes. This was recently demonstrated in a study in which DIC microscopy and digital imaging of the calcium indicator dye fura-2 were performed simultaneously in single rat salivary gland acinar cells to examine the effects of agonist stimulation on cell volume and $[Ca^{2+}]_i$ [Foskett and Melvin, 1989]. $[Ca^{2+}]_i$ was determined by analysis of ratio images of fura-2 constructed from pairs of images obtained at 340 nm and 390 nm excitation. Cell volume was determined by planimetry of single optical sections of images of acinar cells stored on a video disk recorder. In combined DIC/fluorescence microscopy, the data load is exacerbated by the requirement to obtain DIC images in addition to the pair of fluorescence images. While reduction of dig-

ital resolution to 256×240 pixels for fluorescence images is tolerable, loss
of DIC resolution is considerable and not acceptable. In lieu of reducing temporal resolution as a result of the necessity to digitally store full-resolution
DIC images, the DIC images are captured directly to disc recorder under computer control. Thus, addition of DIC image aquisition to the low light level
fluorescence microscope does not result in an appreciable loss of temporal
resolution.

$[Ca^{2+}]_i$ is believed to be the primary regulator of salivary fluid secretion
by activating K^+ and Cl^- permeabilities. Osmotic consequences of KCl fluxes
would have important implications because cell water content affects ion activities, thereby influencing the driving forces for secretion, and rapid alterations
of cell size may modulate specific ion channel activities via cell volume-sensing
mechanisms. By simultaneous optical determinations of cell volume and
$[Ca^{2+}]_i$ during stimulation of single salivary gland acinar cells, it was demonstrated that agonist stimulation of fluid secretion is initially associated with
a rapid tenfold increase in $[Ca^{2+}]_i$ which precedes a rapid, substantial cell
shrinkage (Fig. 5) [Foskett and Melvin 1989]. In spite of considerable heterogeneity among individual cells, subsequent changes of cell volume in the continued presence of agonist are tightly coupled to dynamic levels of $[Ca^{2+}]_i$,
even during $[Ca^{2+}]_i$ oscillations (Fig. 6). Experiments with Ca^{2+} chelators
and ionophores demonstrated that elevation of $[Ca^{2+}]_i$ is necessary and sufficient to cause changes in cell volume. It was proposed that agonist-induced

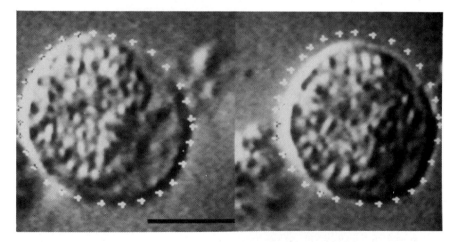

Fig. 5. DIC image of a single rat salivary gland acinar cell perfused on the stage of the microscope. **Left:** Control, 1 s before application of 10 μM carbachol. **Right:** 10 s following exposure to carbachol. Cursor marks defined by the perimeter of the control cell have been transposed
to the stimulated cell image, demonstrating cell shrinkage associated with cell activation. $[Ca^{2+}]_i$,
determined simultaneously, is 50 nM (left) and 700 nM (right). Bar is 10 μM.

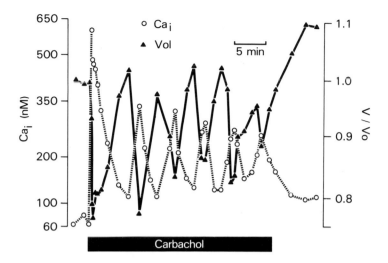

Fig. 6. Carbachol (10 μM)-induced oscillations of single salivary gland acinar cell $[Ca^{2+}]_i$ (○) and volume (▲). (Reproduced from Foskett and Melvin [1989] from *Science* 244:1582–1585; copyright by the AAAS.)

changes in $[Ca^{2+}]_i$, by modulating specific ion permeabilities, result in solute movement into or out of the cell causing volume to change accordingly. Simultaneous measurement of intracellular Cl^- activity with a Cl^- sensitive fluorescent dye support this conclusion (below).

2. Fluorescence determinations of cell volume and fluid transport. Determination of single cell volumes by planimetry is tedious and laborious. One solution to this problem is to teach a computer to recognize cell edges [Marsh et al., 1985]. More recent novel optical approaches which should permit real time tracking of cell volume and fluid transport employ fluorescent dyes. Verkman [Kuwahara et al., 1988; Verkman, 1989; Kuwahara and Verkman, 1988; Fushioni et al., 1989] has determined the diffusional and osmotic water permeabilities of isolated perfused kidney collecting tubules by imaging and photometry of amino-naphthalenetrisulfonic acid, a H_2O/D_2O sensitive fluorophore, and fluorescein sulfonate, a self-quenching dye, respectively, in the tubule lumen. This method may be suitable for single cell volume studies by using the permeant diacetate derivative of fluorescein sulfonate, which becomes trapped in the cell by esterase cleavage.

3. Chloride transport. Cells utilize a variety of mechanisms to transport Cl^-, including anion exchange, channels (conductive), and cation-coupled cotransport. Fluorescent probes are available to measure $[Cl^-]$ and to directly or indirectly examine the activity of Cl^-/base exchangers.

a. Cl^-/HCO_3^- exchange. NBD-taurine is a fluorescent substrate for the

Cl^-/HCO_3^- exchanger which has been used in red blood cells to analyze the kinetics of the anion exchanger [Eidelman et al., 1981]. Foskett [1985] measured the efflux of NBD-taurine from a field of gallbladder epithelial cells which had been loaded with the dye by a 1–3 h incubation. Exposure of the tissue to a hypertonic solution enhanced a disulfonic stilbene-sensitive component of the washout curve, corroborating previous data suggesting that cell shrinkage activates apical membrane Cl^-/HCO_3^- exchange. The activity of Cl^-/base exchangers can also be assayed using fluorescent pH-indicator dyes. Paradiso et al. [1987] imaged 2',7'-bis(carboxyethyl)-5(6)-carboxyfluorescein (BCECF) fluorescence in intact gastric glands. BCECF emission intensity is sensitive to pH when excited at 490–500 nm, but insensitive at 440–450 nm excitation. Excitation ratio-imaging (490/450) was used to spatially resolve intracellular pH (pH_i) in isolated gastric glands which are composed of a number of different cell types. Removal of extracellular Cl^- rapidly and reversibly caused the oxyntic cells to become strongly alkaline. The data were most consistent with the activity of a Cl^-/HCO_3^- exchanger, indicating that Cl^-/HCO_3^- exchange was a dominant transport pathway in the basolateral membrane of oxyntic cells. Since oxyntic cells are responsible for gastric acid secretion, this finding indicated that Cl^-/HCO_3^- exchange could serve two functions necessary for acid secretion—provide Cl^- for HCl secretion and provide a pathway for HCO_3^- out of the cell. A similar illustration of the use of pH_i measurement to analyze Cl^- transport led to the identification of and suggested a role for Cl^-/base (OH^-, HCO_3^-, formate) exchange in NaCl absorption by the perfused proximal tubule of the kidney (Alpern, 1987).

b. Intracellular Cl⁻. 6-Methoxy-N-[3-sulfopropyl] quinolinium (SPQ) is a water soluble dye which interacts with halides by a collisional quenching mechanism. It is particularly useful for biological studies because of its insensitivity to anions of biological relevance other than Cl^- or to ionic strength and pH [Illsley and Verkman, 1987]. However, its sensitivity to bromide, iodide, and thiocyanate can be exploited since these anions are transported by some Cl^- transport pathways. SPQ is zwitterionic with a negatively charged sulfonic acid group at neutral pH. Therefore, it is not expected to be membrane permeant. In spite of this, it has proved possible to load cells by soaking them in 10–20 mM SPQ for 10 min–2 h or by transiently exposing the cells to a hypotonic medium [Dho and Foskett, unpublished]. With regard to the latter strategy, we have found that many, but not all, cells respond to a 50% dilution of the medium by loading SPQ. Use of this protocol will require appropriate controls to ensure that hypotonic shock does not disrupt normal physiological function. For example, exposure of bovine chromaffin cells to a 50% dilution appears to inhibit exocytosis in isotonic medium in response to depolarizing agents (Duarte and Foskett, unpublished). In these cells, as well as in some epithelial cells (cultured sweat gland and pancreas), only a very modest (10%)

dilution of the medium greatly facilitates loading with SPQ. A significant disadvantage of this dye is that its spectral characteristics are not modified by Cl^-. Thus, Cl^- cannot be spatially revolved using ratio techniques. In spite of this, Krapf et al. [1988] were able to quantitate $[Cl^-]_i$ in isolated perfused kidney proximal tubules. SPQ was loaded into the cell by a 10 min luminal exposure with 20 mM SPQ. Half-time for leakage was 8–9 min. Calibration of SPQ fluorescence using the ionophores nigericin and tributyl-tin (described below) indicated that $[Cl^-]_i$ was ~28 mM. The quenching constant of the dye in the cell was determined to be 12 M^{-1}, compared with 118 M^{-1} in free solution. The reduced sensitivity of SPQ to intracellular Cl^- may be due to the viscosity of the cytoplasm [Krapf et al., 1988].

Rat parotid acinar cells can also be loaded with SPQ by simple incubation with the dye for ~45 min [Foskett, 1990]. In spite of the fact that it diffuses into the cell, leakage is minimal (Fig. 7). SPQ fluorescence imaging was performed simultaneously with DIC imaging of single cells, as described earlier. Calibration of a separate group of cells used nigericin (a K^+/H^+ exchanger) and tributyl-tin (a Cl^-/OH^- exchanger) with $[K^+]_o = [K^+]_i$ (150 mM) at pH = 7.3 with varying $[Cl^-]_o$. This method clamps $[Cl^-]_i$ to $[Cl^-]_o$. The calibration procedure indicated that SPQ fluorescence is related to $[Cl^-]_i$ according to a Stern-Volmer relation with an intracellular quenching constant of 17 M^{-1}. Resting $[Cl^-]_i = 60$ mM. Stimulation with an agonist was asso-

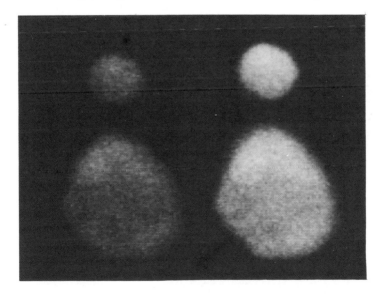

Fig. 7. SPQ fluorescence in rat salivary gland acinus *(bottom)* and single acinar cell *(top)* before *(left)* and 10 s following exposure to 10 μM carbachol *(right)*. Enhanced fluorescence during stimulation indicates that $[Cl^-]$ has fallen.

ciated with a parallel (temporally and quantitatively) cell shrinkage and decrease of $[Cl^-]_i$ to 29 mM, representing a 62% loss of cell Cl^- content when cell volume changes were taken into account. Subsequent cell volume changes were tightly coupled to $[Cl^-]_i$. The changes in $[Cl^-]_i$ detected with SPQ are nearly exactly as expected based on the assumption that cell volume changes are due to changes in cell solute content. Taken together with similar experiments measuring $[Ca^{2+}]_i$ (described above), that data indicate that dynamic changes of cell volume during fluid secretion provide an accurate indicator of cell solute content, $[Ca^{2+}]_i$ and $[Cl^-]_i$.

4. Sodium transport. Sodium transport pathways include energy requiring pumps (e.g., Na/K-ATPase), channels, cotransporters (e.g. NaCl, NaK2Cl, $NaHCO_3$, Na^+ glucose) and exchangers (e.g., Na^+/H^+).

a. Na^+ pathways. The numerous studies of the roles and consequences of the Na^+/H^+ exchanger in pH_i regulation indirectly probe Na^+ transport although most of these studies have not been concerned primarily with Na transport per se. However, some epithelial cells employ Na^+/H^+ and Na^+/HCO_3^- exchange as mechanisms for transepithelial Na^+ transport. In such cells, pH_i measurements have been used to study these mechanisms. In gastric mucosa, Cl^- secretion is thought to be driven by parallel Na^+/H^+ and Cl^-/HCO_3^- exchangers in the basolateral membrane. Using imaging of BCECF fluorescence in isolated gastric glands, Paradiso et al. [1987b] demonstrated that oxyntic cells possess Na^+/H^+ exchangers in the basolateral membrane. Krapf [1988] used BCECF fluorescence determination of pH_i in perfused thick ascending limbs to demonstrate the presence of electrogenic Na^+/HCO_3^- cotransport. This kidney tubule segment is important for reabsorption of much of the HCO_3^- delivered out of the proximal tubule. The demonstration of Na^+/HCO_3^- cotransport in parallel with Na^+/H^+ exchange on the basolateral membrane provides a mechanism for transepithelial HCO_3^- transport.

b. Na activity. BCECF fluorescence has been used to measure cytosolic Na^+ activity ($[Na^+]_i$) in intact synaptosomes using null point methods [Kongsamut and Nachshen, 1988]. After determination of pH_i with BCECF, the external medium pH (pH_o) is adjusted so $pH_o = pH_i$. Using monensin, a Na^+/H^+ exchanger, $[Na^+]_o$ is adjusted to produce no pH_i change. Under these conditions, $[Na^+]_i = [Na^+]_o$. This technique should be applicable to intact cells. However, this method requires monensin in the membrane and it cannot be used to follow changes of $[Na^+]_i$.

Recently, Minta and Tsien [1989] developed fluorescent Na^+-indicator dyes which have shown promise in initial studies. Sodium-binding benzofuran isophthalate (SBFI) binds Na^+ with a dissociation constant of 17–19 mM, which causes a spectral shift somewhat similar to the way fura-2 responds to Ca^{2+}. Thus, excitation ratio measurements allow quantitation of $[Na^+]_i$ in cells. Harootunian et al. [1989] loaded SBFI into lymphocytes and fibroblasts by incubation with the acetoxymethyl ester of the dye. The dye was found to partition into acidic compartments as well as the cytoplasm. Intracellular dye was calibrated by clamping $[Na^+]_i = [Na^+]_o$ using pore-forming antibiotics (gramicidin) and ionophores (nigericin, monensin). Intracellular calibration would appear to be necessary since the excitation spectra inside the cells differed from the spectra in buffer solutions. In fibroblasts, changes of $[Na^+]_i$ $\geqslant$ 1 mM could be resolved. Mitogenic stimuli, the Ca^{2+}-ionophore ionomycin, and ouabain each increased $[Na^+]_i$.

Negulescu et al. [1990] have digitally imaged SBFI in isolated gastric glands. In these cells, dye loading appeared to be uniform and 2 mM changes of $[Na^+]_i$ could be resolved. Basal $[Na^+]_i$ was 8-9 mM. Estimates of Na^+ influx and passive and active efflux were determined by measuring $[Na^+]_i$ during modulation of the activity of the Na^+/K^+-ATPase. In an elegant series of experiments, they were able to determine cytosolic Na^+-buffering (none) and the Na^+ activation curve for the Na^+/K^+-ATPase. Muscarinic agonists stimulated Na^+ influx without effect on passive efflux. cAMP, on the other hand, although more potent in stimulating acid secretion, had no effect on Na^+ metabolism. These as well as preliminary studies in smooth muscle cells [Moore et al., 1989] indicate that the dye will find wide applicability in defining the roles of $[Na^+]_i$ in cell physiological processes and for exploring the activities of various Na^+ transport pathways, e.g., Na^+/K^+-ATPase and Na^+/H^+ exchanger.

5. Potassium transport. Karpen et al. [1986] have used Cs^+ quenching of anthracene-1,5-dicarboxylic acid to monitor cation permeability. PC-12 cells were loaded with the dye by incubation in 3 mM for 1 h at 37°C at pH 6.8. Because of the high pKa of the carboxylic acid group (~4), a low pH is necessary to protonate enough dye to permeate into the cell. Cs^+ quenches the dye according to a Stern-Volmer relation. However, the quenching constant is only 1.4 M^{-1}.

Minta and Tsien [1989] have described the synthesis of a K^+-indicator dye SBFI. The effective K_d is ~70 mM when $[K^+] + [Na^+] = 135$ mM. The dye does not discriminate against Na^+ very well (1.5 fold), although $[K^+]_i$ is usually much higher than $[Na^+]_i$. However, at appropriate $[K^+]$ for cytoplasm (90–140 mM), the fluorescence of the dye is relatively insensitive to changes in $[K^+]$. No biological applications have yet been reported.

6. Ammonia transport. Cell membranes are generally believed to be highly permeable to lipophilic and small, uncharged molecules. In an elegant study, Kikeri et al. [1989] utilized BCECF fluorescence to demonstrate that this is not true for the apical membrane of renal tubule cells of the medullary thick ascending limb (TAL) of Henle. Exposure of most cells to NH_4^+ usually leads to cellular alkalinization due to rapid NH_3 entry. However, in perfused TAL, exposure on the apical side to NH_4^+ leads instead to a rapid cellular acidification, reflecting a high NH_4^+ permeability. K^+ channel and $Na^+/K^+/2Cl^-$ cotransporter inhibitors blocked the acidification, demonstrating that NH_4^+ permeates through these pathways and, most importantly, that NH_3 permeability is exceedingly low. It is interesting that the apical membrane is highly impermeable to H_2O as well, suggesting that this membrane has a unique lipid composition/structure which confers these impermeabilities.

IV. CONCLUSIONS

The studies summarized in this review demonstrate the power of optical techniques for noninvasive studies of specific water and ion transport pathways in intact cells. There is no question that these approaches will become increasingly employed in such studies. In the future, new dye chemistry, in particular, and new imaging/photometry/computer hardware will increase the applicability, sensitivity, and resolution (temporal and spatial) of these techniques. In particular, fast-responding dyes which exhibit a large change in fluorescence in response to membrane voltage changes will provide the ability, when combined with ion-sensitive dyes, to define completely driving forces for ion translocation in single intact cells. The combined use of membrane potential-, ion-, and volume-sensitive dyes with appropriate spectral separations will allow cellular morphology and behavior to be precisely correlated with the driving forces for transport and the levels of intracellular regulatory molecules, e.g., Ca^{2+}, H^+. Continued development of light-sensitive molecular chelators will permit not only the measurement of these parameters, but their control as well. It seems realistic to expect, therefore, that optical studies of intact cells will continue to grow in importance as complementary techniques to more standard electrophysiological approaches to understanding mechanisms of ion and water transport in cells.

V. REFERENCES

Aikens RS, Agard DA, Sedat JW (1989) Solid-state imagers for microscopy. In Wang YL, Taylor DL (eds): Methods in Cell Biology, Volume 29. Boston: Academic Press, Inc., pp 291–313.

Allen RD, Allen NS, Travis JL (1981) Video enhanced contrast, differential interference contrast (AVEC-DIC) microscopy: A new method capable of analyzing microtubule-related motility in the reticulopodial network of Allegromia laticollaris. Cell Motil 1:291–302.

Alpern RJ (1987) Apical membrane chloride/base exchange in rat proximal convoluted tubule. J Clin Invest 79:1026–1030.

Balaban RS, Kurtz I, Cascio HE, Smith PD (1986) Microscopic spectral imaging using a video camera. J Microsc 141:31–39.

Bright GR, Fisher GW, Rogowska J, Taylor DL (1987) Fluorescence ratio imaging microscopy: temporal and spatial measurement of cytoplasmic pH. J Cell Biol 104:1019–1033.

Burg MB (1972) Perfusion of isolated renal tubules. Yale J Biol Med 45:321–326.

Carpi-Medina P, Gonzalez E, Whittembury G. (1983) Cell osmotic water permeability of isolated rabbit proximal convoluted tubules. Am J Physiol 244:F554–F567.

Carpi-Medina P, Lindemann B, Gonzalez E, Whittembury G (1984) The continuous measurement of tubular volume changes in response to step changes in contraluminal osmolality. Pflugers Arch 400:343–348.

Carpi-Medina P, Whittembury G (1988) Comparison of transcellular and transepithelial water osmotic permeabilities (Pos) in the isolated proximal straight tubule (PST) of rabbit kidney. Pflugers Arch 412:66–74.

Catterall WA (1988) Structure and function of voltage-sensitive ion channels. Science 242:50–61.

Cobbold PH, Rink TJ (1987) Fluorescence and bioluminescence measurement of cytoplasmic free calcium. Biochem J 248:313–328.

Davis CW, Finn AL (1981) Regulation of cell volume of frog urinary bladder. In Dinno MA, Callahan AB (eds): Membrane Biophysics: Structure and Function in Epithelia. New York: Alan R Liss, Inc., pp 25–36.

Davis CW, Finn AL (1985a) Effects of mucosal sodium removal on cell volume in Necturus gallbladder epithelium. Am J Physiol 249:C304–C312.

Davis CW, Finn AL (1985b) Cell volume regulation in frog urinary bladder. Fed Proc 44:2520–2525.

Davis CW, Finn AL (1987) Interactions of sodium transport, cell volume, and calcium in frog urinary bladder. J Gen Physiol 89:687–702.

Diamond JM (1979) Osmotic water flow in leaky epithelia. J Membr Biol 51:195–216.

DiBona DR (1983) Cytoplasmic involvement in ADH-mediated osmosis across toad urinary bladder. Am J Physiol 245:C297–C307.

Eidelman O, Zamgvill M, Razin M, Ginsburg H, Cabantchik ZI (1981) The anion-transfer system of crythrocyte membranes. N-(7-nitrobenzofura-zan-4-yl)taurine, a fluorescent substrate-analogue of the system. Biochem J 195:503–513.

Ericson A-C, Spring KR (1982a) Coupled NaCl entry into Necturus gallbladder epithelial cells. Am J Physiol 243:C140–C145.

Ericson A-C, Spring KR (1982b) Volume regulation by Necturus gallbladder: Apical Na^+/H^+ and Cl^-/HCO_3^- exchange. Am J Physiol 243:C146–C150.

Foskett JK (1985) NBD-taurine as a probe of anion exchange in gallbladder epithelium. Am J Physiol 249:C56–C62.

Foskett JK (1988) Simultaneous Nomarski and fluorescence imaging during video microscopy of cells. Am J Physiol 255:C566–C571.

Foskett JK (1990) $[Ca^{2+}]_i$ modulation of Cl^- content controls cell volume in single salivary acinar cells during fluid secretion (submitted).

Foskett JK, Gunter-Smith PJ, Melvin JE, Turner RJ (1989) Physiological localization of an agonist-sensitive pool of Ca^{2+} in parotid acinar cells. Proc Natl Acad Sci USA 86:167–171.

Foskett JK, Melvin JE (1989) Activation of salivary secretion: coupling of cell volume and $[Ca^{2+}]_i$ in single cells. Science 244:1582–1585.

Foskett JK, Spring KR (1985) Involvement of calcium and cytoskeleton in gallbladder epithelial cell volume regulation. Am J Physiol 248:C27–C36.

Foskett JK, Ussing HH (1986) Localization of chloride conductance to mitochondria-rich cells in frog skin epithelium. J Membr Biol 91:251–258.

Fushioni K, Dix JA, Verkman AS (1989) Real time measurement of osmotic and diffusional water transport in the vasopressin-sensitive kidney collecting tubule measured by quantitative ratio imaging. Biophys J 55:158a.

Gautam A, Ng OC, Strazzabosco J, Boyer JL (1989) Quantitative assessment of canibular bile formation in isolation hepatocyte couplets using microscopic optical planimetry. J Clin Invest 83:565–573.

Gonzalez E, Carpi-Medina P, Whittembury G (1982) Cell osmotic water permeability of isolated rabbit proximal straight tubules. Am J Physiol 232:F321–F330.

Gonzalez E, Salomonsson M, Muller-Suur C, Persson AEG (1988a) Measurements of macula densa cell volume changes in isolated and perfused rabbit cortical thick ascending limb. I. Isoosmotic and anisosmotic cell volume changes. Acta Physiol Scand 133:149–157.

Gonzalez E, Salmonsson M, Muller-Suur C, Persson AEG (1988b) Measurements of macula densa volume changes in isolated and perfused rabbit cortical thick ascending limb. II. Apical and basolateral cell osmotic water permeabilities. Acta Physiol Scand 133:159–166.

Grinvald A, Frostig RD, Lieke E, Hildesheim R (1988) Optical imaging of neuronal activity. Physiol Rev 68:1285–1366.

Gross DJ (1990) Quantitative single cell fluorescence imaging of indicator dyes. In Foskett JK, Grinstein S (eds): Noninvasive Techniques in Cell Biology. New York: Wiley-Liss, Inc., pp 21–51.

Guggino WB (1986) Functional heterogeneity in the early distal tubule of the Amphiuma kidney: Evidence for two modes of Cl^- and K^+ transport across the basolateral cell membrane. Am J Physiol 250:F430–F440.

Guggino WB, Oberleithner H, Giebisch G (1985) Relationship between cell volume and ion transport in the early distal tubule of the Amphiuma kidney. J Gen Physiol 86:31–58.

Harootunian AT, Kao JPY, Eckert BK, Tsien RY (1989) Fluorescence ratio imaging of cytosolic free Na^+ in individual fibroblasts and lymphocytes. J Biol Chem 264:19458–19467.

Haugland RP, Minta A (1990) Design and application of indicator dyes. In Foskett JK, Grinstein S (eds): Noninvasive Techniques in Cell Biology. New York: Wiley-Liss, Inc. pp 1–20.

Hebert SC (1986a) Hypertonic cell volume regulation in mouse thick limbs. I. ADH dependency and nephron heterogeneity. Am J Physiol 250:C920–C931.

Horster M, Gundlach H (1979) Application of differential interference contrast with inverted microscopes to the in vitro perfused nephron. J Microsc 117:375–379.

Illsley NP, Verkman AS (1987) Membrane chloride transport measured using a chloride-sensitive fluorescent probe. Biochemistry 26:1215–1219.

Inoue S (1981) Video image processing greatly enhances contrast, quality and speed in polarization-based microscopy. J Cell Biol 89:346–356.

Karpen JW, Sachs F, Pasquale EB, Hess G (1986) Spectrophotometric detection of monovalent cation flux in cells: Fluorescent microscope measurement of acetylcholine receptor-mediator ion flux in PC-12 cells. Anal Biochem 157:353–359.

Katz U, Van Driessche W, Scheffey C (1985) The role of mitochondria-rich cells in the chloride current conductance across toad skin. Biol Cell 55:245–250.

Kikeri D, Sun A, Zeidel ML, Hebert SC (1989) Cell membrane impermeable to NH_3. Nature 339:478–480.

Kirk KL, Bell D, Barfuss DW, Ribadeneira M (1985) Direct visualization of the isolated and perfused macula densa. Am J Physiol 248:F890–894.

Kirk KL, DiBona DR, Schafer JA (1984a) Morphologic response of the rabbit cortical collecting tubule to peritubular hypotonicity: quantitative examination with differential interference contrast microscopy. J Membr Biol 79:53–64.

Kirk KL, DiBona DR, Schafer JA (1987a) Regulatory volume decrease in perfused proximal nephron: evidence for a dumping of cell K^+. Am J Physiol 252:F933–F942.

Kirk KL, Schafer JA, DiBona DR (1983) Microscopic methods for analysis of function in isolated renal tubules. In Dinno M, Callahan AB, Rozzell TC (eds): Membrane Biophysics II: Physical Methods in the Study of Epithelia. New York: Alan R. Liss, Inc., pp 21–36.

Kirk KL, Schafer JA, DiBona DR (1984b) Quantitative analysis of the structural events associated with antidiuretic hormone-induced volume reabsorption in the rabbit cortical collecting tubule. J Membr Biol 79:65–74.

Kirk KL, Schafer JA, DiBona DR (1987a) Cell volume regulation in rabbit proximal straight tubule perfused in vitro. Am J Physiol 252:F922–F932.

Koefoed-Johnsen V, Ussing HH (1958) The nature of the frog skin potential. Acta Physiol Scand 42:298–308.

Kongsamut S, Nachshen DA (1988) Measurement of cytosolic sodium ion concentration in rat brain synaptosomes by a fluorescence method. Biochim Biophys Acta 940:241–246.

Krapf R (1988) Basolateral membrane H/OH/HCO_3 transport in the rat cortical thick ascending limb. Evidence for an electrogenic Na/HCO_3 cotransporter in parallel with a Na/H antiporter. J Clin Invest 82:234–241.

Krapf R, Berry CA, Verkman AS (1988) Estimation of intracellular chloride activity in isolated perfused rabbit proximal convoluted tubules using a fluorescent indicator. Biophys J 53:955–962.

Kurtz I (1987) Apical Na^+/H^+ antiporter and glycolysis-dependent H^+-ATPase regulate intracellular pH in the rabbit S3 proximal tubule. J Clin Invest 80:928–935.

Kurtz I, Dwelle R, Katzka P (1987) Rapid scanning fluorescence spectroscopy using an acousto optic tunable filter. Rev Sci Instrum 58:1996–2003.

Kuwahara M, Barry CA, Verkman AS (1988) Measurement of osmotic water transport in perfused cortical collecting tubules by fluorescence microscopy. Biophys J 54:595–602.

Kuwahara M, Verkman AS (1988) Direct fluorescence measurement of diffusional water permeability in the vasopressin-sensitive kidney collecting tubule. Biophys J 54:587–593.

Larsen EH, Ussing HH, Spring KR (1987) Ion transport by mitochondria-rich cells in toad skin. J Membr Biol 99:25–40.

Larson M, Spring KR (1984) Volume regulation by Necturus gallbladder: Basolateral KCl exit. J Membr Biol 81:219–232.

Loeschke K, Eisenbach GM, Bentzel CJ (1977) Water flow across Necturus gallbladder and small intestine. In Kramer M, Lauterbach F (eds): Intestinal Permeation. Amsterdam: Excepta Medica, pp 406–412.

Lohr JW, Grantham JJ (1986) Isovolumetric regulation of isolated S2 proximal tubules in anisotonic media. J Clin Invest 78:1165–1172.

Lopes AG, Amzel LM, Markakis D, Guggino WB (1988) Cell volume regulation by the thin descending limb of Henle's loop. Proc Natl Acad Sci USA 85:2873–2877.

Lopes AG, Guggino WB (1987) Volume regulation in the early proximal tubule of the Necturus kidney. J Membr Biol 97:117–125.

MacRobbie EAC, Ussing HH (1961) Osmotic behavior of the epithelial cells of frog skin. Acta Physiol Scand 53:348–365.

Malgaroli A, Milani D, Meldolesi J, Pozzan T (1987) Fura-2 measurement of cytosolic free Ca^{2+} in monolayers and suspensions of various types of animal cells. J Cell Biol 105:2145–2155.

Marsh DJ, Jensen PK, Spring KR (1985) Computer-based determination of epithelial cell size and shape. J Microsc 137:281–292.

Maxfield FR (1989) Measurement of vacuolar pH and cytoplasmic calcium in living cells using fluorescence microscopy. In Fleischer S, Fleischer B (eds): Methods in Enzymology, Volume 173. San Diego: Academic Press, Inc., pp 745–771.

Minta A, Tsien RY (1989) Fluorescent indicators for cytosolic sodium. J Biol Chem 264:19449–19457.

Montana V, Farkas DL, Loew LM (1989) Dual-wavelength ratiometric fluorescence measurements of membrane potential. Biochemistry 28:4536–4539.

Moore EDW, Minta A, Tsien RY, Fay FS (1989) Measurement of intracellular $[Na^+]_i$ with SBFI, and Na^+ regulation in smooth muscle cells. Biophys J 55:471a.

Negulescu PA, Harootunian A, Tsien RY, Machen TE (1990) Fluorescent measurements of cytosolic free Na in gastric cells (submitted).

O'Neil RG, Hayhurst RA (1985) Functional differentiation of cell types of cortical collecting duct. Am J Physiol 248:F449–F453.

Paradiso AM, Tsien RY, Machen TE (1987) Digital image processing of intracellular pH in gastric oxyntic and chief cells. Nature 325:447–450.

Persson B-E, Spring KR (1982) Gallbladder epithelial cell hydraulic water permeability and volume regulation. J Gen Physiol 79:481–505.

Sachs F (1987) Baroreceptor mechanisms at the cellular level. Fed Proc 46:12–16.

Salzberg BM (1989) Optical recording of voltage changes in nerve terminals and in fine neuronal processes. Annu Rev Physiol 51:507–526.

Spring KR (1990): Differential interference contrast and fluorescence microscopy without significant light loss. In Herman B, Jacobsen K (eds): Optical Microscopy in Biology. New York: Alan R. Liss, Inc. (in press).

Spring KR, Hope A (1978) Size and shape of the lateral intercellular spaces in a living epithelium. Science 200:54–58.

Spring KR, Hope A (1979) Fluid transport and dimensions of cells and interspaces of living Necturus gallbladder. J Gen Physiol 73:287–305.

Spring KR, Lowy RJ (1989) Characteristics of low light level television cameras. In Wang Y-L, Taylor DL (eds): Methods in Cell Biology, Volume 29. Boston: Academic Press Inc., pp 269–289.

Spring KR, Smith PD (1987) Illumination and detection systems for quantitative fluorescence microscopy. J Microsc 147:265–278.

Spring KR, Ussing HH (1986) The volume of mitochondria-rich cells of frog skin epithelium. J Membr Biol 92:21–26.

Strange K (1988) RVD in principal and intercalated cells of rabbit cortical collecting tubule. Am J Physiol 255:C612–C621.

Strange K, (1989) Ouabain-induced cell swelling in rabbit cortical collecting tubule: NaCl transport by principal cells. J Membr Biol 107:249–261.

Strange K, Spring KR (1986) Methods for imaging renal tubule cells. Kidney Int 30:192–200.

Strange K, Spring KR (1987a) Cell membrane water permeability of rabbit cortical collecting duct. J Membr Biol 96:27–43.

Strange K, Spring KR (1987b) Absence of significant cellular dilution during ADH-stimulated water reabsorption. Science 235:1068–1070.

Tsay T-T, Inman R, Wray B, Herman B, Jacobson K (1990) Characterization of low light level camera for digitized video microscopy. In Herman B, Jacobson K (eds): Optical Microscopy in Biology. New York: Alan R. Liss, Inc. (in press).

Tsien RY, Poenie M (1986) Fluorescence ratio imaging: a new window into intracellular ionic signalling. Trends Biochem Sci 11:450–455.

Tsien TY (1989) Fluorescent indicators of ion concentrations. In Taylor DL, Wang Y-L, (eds): Methods in Cell Biology, Volume 30B. San Diego: Academic Press, Inc., pp 127–156.

Ussing HH (1965) Relationship between osmotic reactions and active sodium transport in the frog skin epithelium. Acta Physiol Scand 63:141–155.

Ussing HH, Biber TUL, Bricker NS (1965) Exposure of the isolated frog skin to high potassium concentrations at the internal surface. II. Changes in epithelial cell volume, resistance and response to antidiuretic hormone. J Gen Physiol 48:425–433.

Ussing HH, Zerahn K (1951) Active transport of sodium as the source of electric current in the short-circuited isolated frog skin. Acta Physiol Scand 23:109–127.

Verkman AS (1989) Mechanisms and regulation of water permeability in renal epithelia. Am J Physiol 257:C837–C850.

Volkl H, Lang F (1988) Ionic requirement for regulatory cell volume decrease in renal straight proximal tubules. Pflugers Arch 412:1–6.

Wampler JE, Kutz K (1989) Quantitative fluorescence microscopy using photomultiplier tubes and imaging detectors. In Wang Y-L, Taylor DL (eds): Methods in Cell Biology, Volume 29. Boston: Academic Press, Inc., pp 239–267.

Weinman SA, Graff J, Boyer JL (1989) Voltage-driven, taurocholate-dependent secretion in isolated hepatocyte couplets. Am J Physiol 256:G826–G832.

Welling LW, Welling DJ, Ochs TJ (1983) Video measurement of basolateral membrane hydraulic conductivity in the proximal tubule. Am J Physiol 245:F123–F129.

Welling LW, Welling DJ, OChs TJ (1987) Video measurement of basolateral NaCl reflection coefficient in proximal tubule. Am J Physiol 253:F290–F298.

Welsh MJ, Smith PL, Fromm M, Frizzell RA (1982) Crypts as the site of intestinal fluid and electrolyte secretion. Science 218:1219–1221.

Noninvasive Techniques in Cell Biology: 273–310
© 1990 Wiley-Liss, Inc.

11. Vibrating Probe Technique for Studies of Ion Transport

Richard Nuccitelli

Zoology Department, University of California, Davis, California 95616

I. INTRODUCTION

All cells are surrounded by a plasma membrane, a lipid bilayer which generally has a relatively low permeability to charged molecules including ions. Among the most common proteins found in all plasma membranes are ion-transporting ATPases that are responsible for the generation of ion concentration gradients across the membrane. In animal cells the most common form of this ion pump is an Na^+/K^+-ATPase, and in plant cells and prokaryotes, the most common form is an H^+-ATPase. It is this ion concentration gradient that drives the diffusion of ions through specialized protein channels in the membrane to separate charge across the plasma membrane, resulting in a membrane potential which is inside negative with a magnitude of roughly 70 mV in animal cells and often over 200 mV in plant cells where electrogenic ion pumps also contribute to the potential. Thus, in all cells there is a built-in "battery" that is generated by the ion pumps and charge separation that can drive ionic currents through the cell. These currents or fluxes are involved in a wide variety of cellular mechanisms because the energy stored in this membrane potential can be utilized for the transport of solutes into the cell, for signal transduction and even high-energy chemical bond formation. Thus, it should not surprise anyone that extracellular ion currents are commonly found around cells and can be associated with many different activities as I will review here. One of the few exceptions to this is the cyanobacterial filament [Jaffe and Walsby, 1985].

Transcellular ionic currents are by no means restricted to single cells. Multillular systems are also found to generate ionic currents, mainly due to the nature of their outer layer which is generally formed by an epithelial sheet of cells that are usually electrically coupled and pump ions across the sheet in a polarized manner. Thus, this sheet of cells is to the organ or organism which it encloses as the plasma membrane is to the cytoplasm it encloses. One main difference is that most epithelia pump cations in an apical-basal direction, generating an inside-positive potential in contrast to the inside-negative potential generated by the plasma membrane in a single cell. Thus, the natural voltage across an epithelium will drive positive ions out in regions of low resistance, whereas the plasma membrane's voltage will drive positive ions in in regions of low resistance.

These electrical properties of cells and tissues naturally lead to the prediction that ion fluxes should be a common property of biological systems and this has been confirmed with radioactive ion tracer work. When $^{45}Ca^{2+}$ fluxes across the egg of the brown alga, *Pelvetia,* were measured using the ingenious technique of separating the rhizoid and thallus ends of the egg with a nickel screen, a total Ca^{2+} current of 2 pA was found to enter the rhizoid and leave the thallus end [Robinson and Jaffe, 1975]. This net influx would generate a surface current density at the rhizoid tip of 0.05 $\mu A/cm^2$ while the Cl^- efflux in the same region is 0.2–1.0 $\mu A/cm^2$. Thus, in some cases one must be able to detect current densities in the 0.1 $\mu A/cm^2$ range in order to determine the steady current pattern around single cells. In epithelial systems, the extracellular currents are generated by many cells working in parallel. Moreover, they are mature and specialized so they are often much larger in magnitude. The vibrating probe technique allows for the noninvasive measurement of these very small endogenous currents, and both this technique and the major results from its application to both plant and animal cell types will be covered here.

My goal in this review is to discuss many of the results that have come from vibrating probe studies of both single cell and multicellular systems in both plant and animals. I have attempted to include all of the papers using this technique over the past 15 years in the reference list. However, I certainly won't be able to discuss all of these here. Instead, I will attempt to at least list most of these studies in a tabular format and will concentrate on a few that have provided some insight into the role of these ionic currents in cellular functions. Other reviews that might be of interest include those by Jaffe and Nuccitelli [1977], Weisenseel and Kicherer [1981], Borgens [1982, 1989a,b], De Loof [1986], Harold [1986a, chapter 14; 1986b], Jaffe [1979, 1981a,b, 1986], Robinson [1985, 1989], Nuccitelli [1983a,b, 1984, 1988a,b], Borgens and McCaig [1989], McGinnis [1989], and Vanable [1989]. Also of interest are the proceedings of two international meetings on ionic currents in development [Nuccitelli, 1986; *Biological Bulletin,* April 1989, Supplemental Volume 176].

II. THE VIBRATING PROBE TECHNIQUE

In order to detect these small, steady ionic currents, one must use a measuring system that has very low noise. One can calculate this from Ohm's law in a continuous medium,

$$I = E / \rho = -(1/\rho)(\Delta V/\Delta r) \qquad (1)$$

where I is current density, E is the electric field strength, ρ is the resistivity of the medium, and ΔV is the voltage difference measured over the small distance, Δr. The most challenging measurement here is the voltage gradient. For $I = 0.1 \ \mu A/cm^2$, and $\rho = 25 \ \Omega\text{-cm}$ in seawater, this voltage gradient will only be 6 nV over a 20 μm displacement, 25 μm outside the cell's plasma membrane. Standard 3 M KCl-filled microelectrodes have resistances on the order of $10^6 \ \Omega$ and their noise places a limit of 10 μV on their resolution. They are therefore not sufficiently sensitive to detect the required 6 nV potential. Lionel Jaffe and I developed a new technique 15 years ago called the vibrating probe that has much greater sensitivity and can detect such small voltages [Jaffe and Nuccitelli, 1974]. A similar technique had been described in 1950 for measurement of plant surface potentials [Blüh and Scott, 1950]. The main idea is to lower the electrode's noise about 1,000-fold both by reducing its resistance and by signal averaging over time using a phase-sensitive lock-in amplifier. The electrode's resistance was reduced by replacing the fluid-filled pipette with a metal sphere with a relatively large capacitance. By vibrating this capacitor between two points at different potentials, the charge on the capacitor fluctuates and the voltage is detected through these charge variations. Thus, we essentially "capacitatively couple" to the external field and the effective resistance is now an impedance that will be reduced as the vibration frequency and capacitance are increased. This impedance can be reduced to levels near the limiting access resistance of the electrode. A lock-in amplifier is used to detect the voltage change and amplifies only those signals exhibiting the vibration frequency, eliminating interference from other common sources of noise such as the 60 Hz line voltage.

This vibrating probe technique has been used in about 30 laboratories around the world and a National Vibrating Probe Center is now sponsored by the NIH at the Marine Biological Lab at Woods Hole, Massachusetts. Largely due to research at this National Center, several improvements in the technique have occurred over recent years [Dorn and Weisenseel, 1982; Nawata, 1984; Scheffey, 1986a, 1988; Jaffe and Levy, 1987]. Wire probes illustrated in Figure 1 have replaced the original solder-filled glass capillary design and two-dimensional, computer-controlled vibrating probes are now available. In order to generate two-dimensional data, the probe must be moved in two dimensions at the same time. An electromagnetic coil-driven design [Freeman et al., 1986] parallelogram bender [Nuccitelli, 1986], and a Pi bender design [Scheffey, 1988] have all been used to accomplish this successfully. One such two-dimensional current pattern generated in my laboratory illustrates the steady currents around a 7.5-day-old mouse embryo (Fig. 2).

III. ADVANTAGES AND DISADVANTAGES OF THIS TECHNIQUE

The main advantage of the vibrating probe is that it provides a noninvasive means of detecting the spatial distribution of ion transporters in the plasma membrane. The probe itself never comes in contact with the cell or tissue under study, so it disturbs the system less than other techniques for detecting ion channel distributions, such as the patch clamp. This means that the vibrating probe can be used over extended time periods to follow any changes in the extracellular current pattern that occur and it can be continuously moved to various positions around the cell or tissue so that the entire external field pattern can be obtained. This is another advantage over the patch clamp technique, where only a very small proportion of the surface is studied in a given experiment.

The disadvantages of the technique are few, but there are some technical points to beware of. In order to maintain the maximum sensitivity, a low tip impedance is required. The main factor influencing this impedance is the capacitance of the electrode tip, which is dependent on surface area. As an experiment progresses, the tip will often become coated with organic matter suspended in the medium, lowering the effective surface area and increasing impedance. Thus, for very small current density measurements, one must clean the probe tip or even apply a fresh coat of platinum at regular intervals (typically once

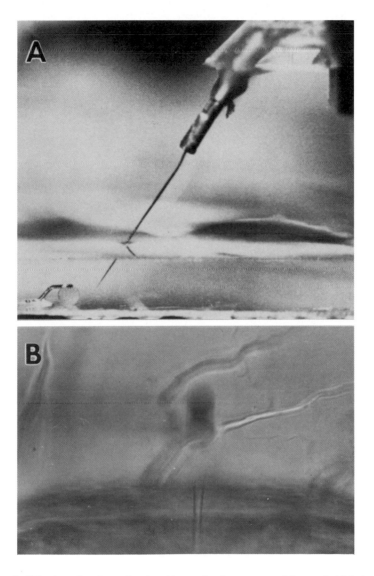

Fig. 1. **A:** Side view of a wire probe placed into a chamber near a medaka egg that is being held down by a gold loop. The egg's diameter is 1.1 mm. **B:** View of the probe vibrating near the micropyle of the egg as seen from below through the inverted microscope. Probe diameter, 20 μm.

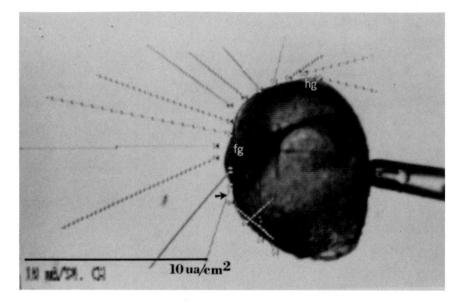

Fig. 2. Parasaggital view of a 7.5 day mouse embryo in the recording chamber with the computer vectors superimposed over the video image. The mouse embryo is held from the right with a fire-polished holding pipette. The magnitude of the current is represented by the length of the vector, while the direction of the vector shows the current direction. (Courtesy of Winkel and Nuccitelli [1989].)

or twice a day). A second concern is that the vibration of the probe stirs the medium quite actively, resulting in both the disruption of any extracellular concentration gradients and the potential to disturb fragile cellular structures such as growth cones. This disruption of extracellular concentration gradients greatly reduces the likelihood that Donnan potentials generated by secreted polyelectrolytes would interfere with the measurement of extracellular currents [Jaffe and Nuccitelli, 1974]. This is a benefit of the stirring action, but there may be one disadvantage when one is studying proton currents. Theory suggests that if the current-carrier has a much greater mobility than other ions in the external medium (i.e., if H^+ is the main current carrier) such stirring can result in an overestimate of the actual current density [Ferrier and Lucas, 1986a,b]. However, there is as yet no direct experimental test of this hypothesis.

IV. APPLICATIONS OF THE VIBRATING PROBE TO STUDY ION TRANSPORT

A. Ionic Currents Around Single Plant Cells

Single cells from more than 16 different plant species have been studied with this technique thus far (Table I). Plants are particularly vigorous current

TABLE I. Ionic Currents Around Single Plant Cells

Species	Cell type	Current magnitude and pattern	Reference*
Blastocladiella emersonii	Water mold spore	1 μA/cm^2 enters rhizoid	1
Allomyces macrogynus	Water mold hypha	0.5 μA/cm^2 enters rhizoid and leaves vegetative hyphae	2,3
Achlya bisexualis	Water mold hypha	0.2 μA/cm^2 enters apical region	3–13
Neurospora crassa	Fungal hypha	0.2 μA/cm^2 enters hyphal tip	3, 14–16
Trichoderma harzianum	Soil fungal hypha	0.2 μA/cm^2 enters hyphal tip	17
Chara corallina	Algal internodal cell	Alternating bands of inward and outward current up to 70 μA/cm^2	18–21a
Nitella flexilis	Algal internodal cell	Alternating bands of inward and outward current up to 25 μA/cm^2	22
Bryopsis plumosa	Regenerating segment of unicellular marine alga	Up to 10 μA/cm^2 enters regenerating ends and leaves central region	23
Acetabularia crenulata	Marine alga	500 μA/cm^2 enters basal region	24, 25
Micrasterias rotata	Algal desmid	1–5 μA/cm^2 enters expanding daughter half cell, concentrated at lobe tips	25a, 26
Pelvetia fastigiata	Algal egg	1 μA/cm^2 enters prospective rhizoid; rate of elongation proportional to current	27–31
Griffithsia pacifica	Algal rhizoidal cell	8 μA/cm^2 enters growing tip but is not necessary for elongation	32
Vaucheria terrestris	Algal filament	0.5 μA/cm^2 leaves region of light-stimulated chloroplast aggregation and 5 μA/cm^2 enters growing apex	33,34
Funaria hygrometrica	Moss caulonema tip cell	0.5 μA/cm^2 enters the nuclear zone and within a few min of cytokinin addition, current decreases at the nuclear zone and enters the distal end, preceding and predicting the presumptive division site	35
Onoclea sensibilis	Fern gametophyte	0.07 μA/cm^2 leaves tip of apical cell	36,37
Lilium longiflorum	Pollen grain	0.5 μA/cm^2 enters prospective germination region and then along pollen tube	38,39

*1: Stump et al. [1980]. 2: Youatt et al. [1988]. 3: Gow [1989]. 4: Armbruster and Weisenseel [1983]. 5: Kropf et al. [1983]. 6: Kropf et al. [1984]. 7: Gow [1984]. 8: Gow et al. [1984]. 9: Harold et al. [1985a,b]. 10: Kropf [1986a,b]. 11: Gow and McGillivray [1986]. 12: Money and Brownlee [1987]. 13: Thiel et al. [1988]. 14: Schreurs and Harold [1988]. 14a: Slayman and Slayman [1962]. 15: McGillviray and Gow, [1987]. 16: Takeuchi et al. [1988]. 17: Horwitz et al. [1984]. 18: Lucas and Nuccitelli [1980]. 18a: Lucas [1982]. 19: Lucas et al. [1984]. 20: Lucas and Sanders [1989]. 21: Fisahn et al. [1989]. 21a: Fisahn and Lucas [1990]. 22: Dorn and Weisenseel [1984]. 23: Nawata et al. [1989]. 24: Bowles and Allen [1984, 1986]. 25: Dazy et al. [1981, 1986]. 25a: Troxell et al. [1986]. 26: Troxell [1989]. 27: Nuccitelli [1978]. 28: Nuccitelli and Jaffe [1974]. 29: Nuccitelli and Jaffe [1975]. 30: Nuccitelli and Jaffe [1976a,b]. 31: Brawley and Robinson [1985]. 32: Waaland and Lucas [1984]. 33: Blatt et al. [1981]. 34: Kataoka and Weisenseel [1988]. 35: Saunders [1986a,b]. 36: Racusen et al. [1986]. 37: Racusen et al. [1988]. 38: Weisenseel et al. [1975]. 39: Weisenseel and Jaffe [1976].

generators due to the fact that the H^+-ATPase is highly electrogenic and contributes substantially more to the membrane voltage in plants than does the Na^+/K^+-ATPase in animal systems. Moreover, growth is often driven by turgor pressure which is generated by ion transport and the water movements that follow, so plants tend to have more active ion pumps than animal cells. In general, the ionic currents found in single plant cells exhibit a spatial pattern that is highly correlated with the polarity of the cell. Thus, in a variety of tip-growing cells from water mold spores and fungal hyphae to pollen grains, positive current enters the localized region of secretion and growth and leaves over a much larger neighboring region. In some cases such as the algal egg, *Pelvetia,* the current is directly linked to the growth, but in others such as *Griffithsia* there is no clear relationship between current density or polarity and growth [Waaland and Lucas, 1984]. The key to this variability probably lies in the ionic components of the current. Some ions, such as Ca^{2+}, can have profound effects on local vesicle secretion and growth, while others, such as Cl^- or K^+ have no known local effect but can influence the more global turgor pressure. Ionic currents in the fungus, *Achlya,* illustrate this point well because they do not exhibit a tight correlation with tip growth.

1. Ionic currents in the fungal hyphae of *Achlya.* *Achlya* normally exhibits an inward current at the tip shown in Figure 3 carried by H^+ symport with amino acids (see Table I for references). However, when a branch forms nearby, the inward current in this new growth region dominates and actually drives current out of the original tip. I favor the hypothesis that this is because the relative permeability of the tip region is greater than other parts of the cell and the outward current which must flow due to the new branch inward current is focused at that region. The reversed current at the new tip means only that there is now more passive cation efflux (perhaps K^+) flowing out in that region than there is H^+ flowing in. The local controls on growth in this system are not sensitive to the current direction because the tip can grow at similar rates with either inward or outward current, but this growth might be sensitive to the ion concentration gradients generated by the current. I would predict that the local pH_i gradients expected to result from the current would be more closely correlated with growth rate and hope that pH_i imaging of growing hyphae is done soon to test this prediction.

2. Ionic currents in the polarizing *Pelvetia* egg. The best case thus far for a causal role of transcellular currents in influencing growth and polarity is actually the first cell ever studied with the vibrating probe, the egg of the brown alga, *Pelvetia.* These eggs have no predetermined axis of polarity and a rhizoid outgrowth can germinate at any point on their surface. This process involves a localized secretion of wall-softening enzymes and an increase in turgor pressure to five atmospheres, accomplished by pumping K^+ and Cl^- into the cell. The cell then bulges or germinates at the weakest region where wall softening has

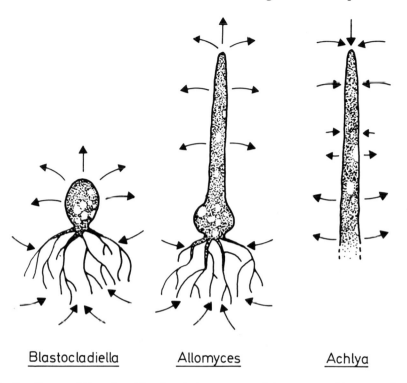

Blastocladiella   Allomyces   Achlya

Fig. 3.   Pattern of flow of positive electrical current around three water molds measured with the vibrating probe. Currents enter the rhizoids of *Blastocladiella emersonii* [Stump et al., 1980] and *Allomyces macrogynus* and the hyphal apex of *Achlya bisexualis*. It is suggested that these three sites are locations of ion-coupled nutrient transport. (Courtesy of Youatt et al. [1988].)

occurred, and the axis of polarity is established. The establishment of this axis of secretion can be influenced by a variety of environmental vectors including light and temperature, and, since the light receptors are located in the plasma membrane [Jaffe, 1958], it is not surprising that this membrane plays a key role in the signal transduction. The investigations using the vibrating probe, stimulated by results from earlier population measurements [Jaffe, 1966], indicated that ionic currents are already present around the egg as early as measurement is possible (30 min post-fertilization) and tend to enter on the side at which germination occurs some 8 h later [Nuccitelli, 1978]. The early current pattern is unstable and shifts position, often with more than one inward current region. However, the current enters mainly on the side where germination will occur and is usually largest at the prospective cortical clearing region where the rhizoid The current pattern observed during the 2 h period prior to germination is more stable and the site of inward current always predicts the germination site, even when the axis is reversed by light direction reversal (Fig. 4).

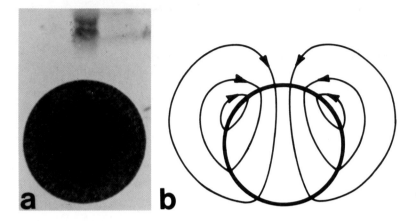

Fig. 4.   Current measurements around the *Pelvetia* egg. **a:** Photomicrograph of a 3-h-old *Pelvetia* egg with a probe vibrating perpendicular to the plasma membrane in the horizontal plane. The egg's diameter is 100 μm. **b:** The electrical current pattern generated by the germinating *Pelvetia* egg just before germination, which will occur at the top of the figure. (Courtesy of Nuccitelli [1983a].)

In order to determine how this transcellular current might be influencing the polarization process, it is important to determine which ions are carrying the current. By varying the composition of the artificial seawaters in which these measurements were made, we determined that Ca^{2+} influx was responsible for a small fraction of the current and Cl^- efflux carried the bulk of it. This Ca^{2+} component was elegantly demonstrated by Robinson and Jaffe [1975] when they measured $^{45}Ca^{2+}$ influx and efflux at either end of polarized eggs by tightly fitting them into holes in a nickel screen so that one end at a time could be exposed to $^{45}Ca^{2+}$. They directly measured 2 pA of Ca^{2+} current traversing each egg. Therefore, one consequence of the transcellular ionic current in this egg is the generation of a Ca^{2+} flux through the cell which has recently been shown to result in a gradient in the intracellular concentration of Ca^{2+} ($[Ca^{2+}]_i$) [Brownlee and Wood, 1986; Brownlee and Pulsford, 1988; Brownlee, 1989]. It appears to be this gradient which plays the primary role in polarizing the egg.

We now know that a central component of the polarization mechanism in the egg of the brown alga involves a cytoplasmic gradient in $[Ca^{2+}]_i$. This hypothesis is supported by the three criteria required for physiological significance:

1. There indeed exists an endogenous $[Ca^{2+}]_i$ gradient in the egg (although measured long after polarization) [Brownlee and Wood, 1986; Brownlee and Pulsford, 1988; Brownlee, 1989];

2. Modifying the direction of this gradient during polarization changes the orientation of the axis of polarization [Robinson and Cone, 1980];

3. Inhibiting the gradient prevents polarization [Speksnijder et al., 1989].

The gradient was detected in germinated zygotes by Brownlee and Wood using Ca^{2+}-sensitive microelectrodes and fura-2 fluorescence, and 10-fold higher Ca^{2+} was detected in the tip region than in the sub-tip region of the cell. Robinson and Cone imposed a gradient of Ca^{2+} on unpolarized eggs by placing them near a fixed source of Ca^{2+} ionophore in the dark. A majority of the eggs then germinated on the hemisphere facing the ionophore where the Ca^{2+} influx is expected to be the greatest. Finally, the magnitude of any natural gradients can be greatly reduced by injecting into the cell a molecule that can rapidly shuttle Ca^{2+} from regions of high concentration to regions of low concentration. Speksnijder et al. [1989] recently used various BAPTA Ca^{2+} buffers to do that and found that tip growth could be completely blocked for weeks while the cell remained viable and even divided. This is the strongest evidence yet that a gradient in $[Ca^{2+}]_i$ is required for cell polarization.

What cellular processes is this gradient influencing to generate the polarization? In this system vesicle secretion determines where the wall-softening enzymes will be released to generate the rhizoid outgrowth as the turgor pressure increases. Therefore, the gradient must influence the process of vesicle localization and secretion. Actin filaments have been implicated in this process because zygotes cannot be polarized when incubated with cytochalasin D [Quatrano, 1973; Brawley and Quatrano, 1979]. Brawley and Robinson [1985] made a connection between this cytochalasin effect and the transcellular current by demonstrating that the cytochalasin D significantly reduces the inward current at the rhizoid pole after 2.5 h of incubation, and also changes the distribution of actin filaments in the growing tip. Thus, they hypothesized that the inward current, carried in part by Ca^{2+}, stimulates F-actin localization at the rhizoid pole and causes more Golgi vesicle transport to that region. They further speculated that the membranes surrounding these vesicles might contain Ca^{2+} channels so that upon fusion an increase of Ca^{2+} influx and inward current into the rhizoid tip could result. This would explain why the addition of cytochalasin D reduces the current. This is a very plausible hypothesis that comes closest to integrating the observed transcellular ionic currents with the overall mechanism of polarization in the fucoid egg.

B. Ionic Currents Around Multicellular Plant Tissues

1. Carrot and tobacco embryos. The number of applications of the vibrating probe technique to the study of multicellular plant tissues is much smaller than for single cells and is restricted to embryos, roots and leaf epidermis (Table II). In both the slower developing cell line from carrot (RCC 27) and tobacco embryos, the current pattern in similar (Fig. 5): about 1 $\mu A/cm^2$ enters the cotyledons or presumptive cotyledon and leaves the radicle [Brawley et al.,1984; Gorst et al., 1987; Rathore et al., 1988; Rathore and Robinson, 1989; Overall and Wernicke, 1986]. However, in faster developing (RCC48) line of

TABLE II. Ionic Currents Around Multicellular Plant Tissues

Species	Tissue type	Current magnitude and pattern	Reference*
Daucus carota	Carrot embryos	Up to 2 $\mu A/cm^2$ enters the cotyledons and leaves the radicle	1–4
Nicotiana tabacum	Haploid embryos from pollen	1 $\mu A/cm^2$ enters that end that will become cotyledon and leaves presumptive radicle	5
Hordeum vulgare *Trifolium repens* *Nicotiana tabacum* *Pisum sativum* *Zea mays*	Roots and root hairs	2 $\mu A/cm^2$ enters elongating root zone and growing tips of root hairs	6–11
Lepidium sativum	Root	During gravistimulation, current begins to leave upper side of root cap	12
Zea mays *Lepidium sativum*	Root	Current leaves most of elongation region and enters root cap in roots grown on moist paper; during gravi-stimulation, current leaves upper surface of root cap and enters lower surface root cap	13–15
Commelina communis	Leaves and leaf epidermis	Up to 4 $\mu A/cm^2$ leaves leaf surface when stomata are open; when fully closed a small inward current is detected	16

*1: Brawley et al. [1984]. 2: Gorst et al. [1987]. 3: Rathore et al. [1988]. 4: Rathore and Robinson [1989]. 5: Overall and Wernicke [1986]. 6: Weisenseel et al. [1979]. 7: Miller et al. [1986]. 8: Miller et al. [1988]. 9: Hush and Overall [1989]. 10: Miller [1989]. 11: Lucas and Kochian [1987]. 12: Behrens et al. [1982]. 13: Bjorkman and Leopold [1987a,b]. 14: Bjorkman [1989]. 15: Iwabuchi et al. [1989]. 16: Bowling et al. [1986].

Daucus carota inward current enters both the cotyledon and the radicle and leaves in the middle of the embryo. Rathore and Robinson [1989] speculate that somatic embryos frequently undergo precocious differentiation so the differences in current pattern might reflect differences in the extent of differentiation in somatic embryos from the two culture lines. One popular hypothesis that correlates these embryonic currents with growth was put forth by Raven [1979] who suggested that the electrical field generated by these currents might result in the polar transport of growth hormones such as indole-3-acetic acid (IAA). As a result of this polar transport, the basal cells would be exposed to higher levels of IAA and would be expected to have higher levels of electrogenic H^+ extrusion. The external pH gradient observed along the embryonic axis in somatic embryos is in support of this hypothesis and it would be most

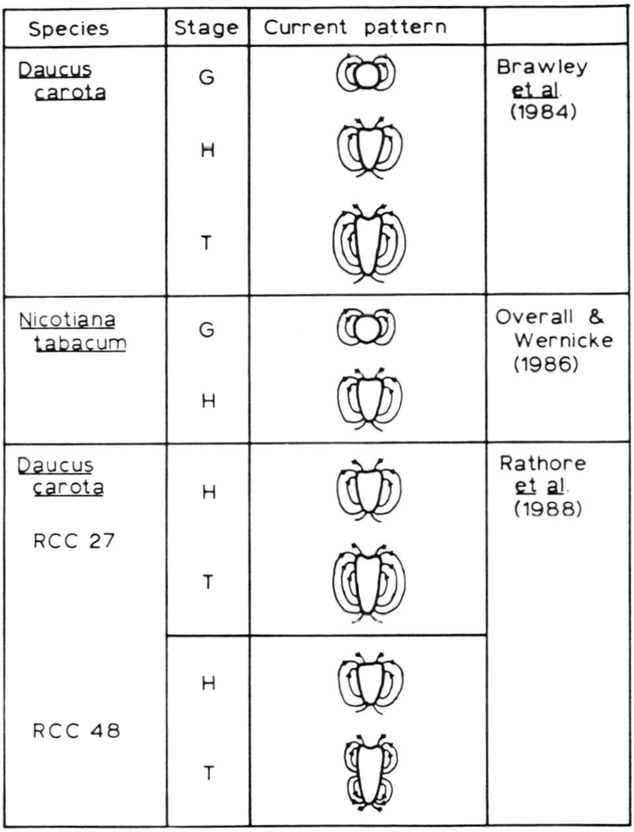

Species	Stage	Current pattern	
Daucus carota	G		Brawley et al. (1984)
	H		
	T		
Nicotiana tabacum	G		Overall & Wernicke (1986)
	H		
Daucus carota RCC 27	H		Rathore et al. (1988)
	T		
RCC 48	H		
	T		

Fig. 5.   Patterns of electrical currents measured around developing embryos in culture at globular (G), heart (H), and torpedo (T) stages. (Courtesy of Rathore and Robinson [1989].)

interesting to study the spatial pH$_i$ profile in these embryos to determine if the current might be influencing this important parameter as well.

2. Roots and root hairs. Studies of extracellular currents around growing roots and root hairs make up the bulk of the work on multicellular plant tissues. No fewer than seven plant species have been investigated and most exhibit a similar pattern of about 1 μA/cm^2 entering the meristematic and elongating cells at the tip of primary roots and growing tips of elongating hairs while exiting the more mature tissue (Fig. 6) [Weisenseel et al., 1979; Behrens et al., 1982; Miller et al., 1986, 1988; Hush and Overall, 1989; Miller, 1989; Rathore and Robinson, 1989]. However, there appears to be some variability in the extent of the inward current region in the elongation zone because in

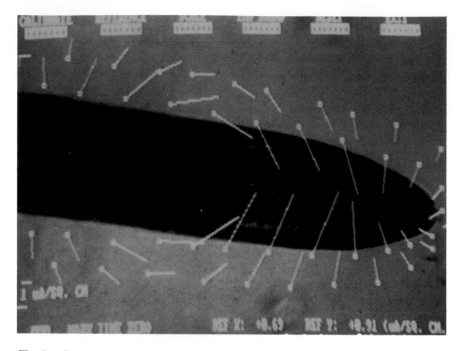

Fig. 6.   Current pattern around a lateral root of radish developing in a IAA-free medium. (Courtesy of Rathore et al. [1990].)

both *Lepidium* and *Zea mays* current actually exits along most of this zone and only enters near the tip [Behrens et al., 1982; Bjorkman and Leopold, 1987a,b; Bjorkman, 1989; Iwabuchi et al., 1989].

  The most interesting aspect of this root current is the change in pattern associated with gravity sensing. When roots are gravistimulated by placing them in a horizontal position, the outward current on the upper side of the cap lateral to the columella increases by 0.4 μA/cm^2 abruptly at 4 min after gravistimulation in *Zea mays* [Bjorkman and Leopold, 1987a,b] and 6 min after gravistimulation in *Lepidium* [Behrens et al., 1982]. In *Lepidium* there is a tendency for an inward current to appear along the lower surface of the root in the bending region [Iwabuchi et al., 1989]. This correlates quite well with the presentation time in *Zea mays*, so there may be a correlation between this change in current density and the modification of the local growth pattern leading to bending. It is also worth noting that the spatial pattern of electric potential has been mapped along tha azuki bean root with a linear array of electrodes, and this pattern correlates well with a flow of current into the elongation zone [Toko et al., 1987].

C. Ionic Currents Around Single Animal Cells

1. Insect follicles. Single cells from 16 different species have been investigated with the vibrating probe technique thus far, resulting in 30 papers. By far the most intensely studied system has been the insect follicle on which more than one-third of these papers have concentrated. This lively interest is largely due to the pioneering studies of Woodruff and Telfer [1973] who showed that there is standing voltage gradient between nurse cells and oocyte in the silk moth follicle and provided good evidence that this voltage difference is involved in the unidirectional transport of protein and nucleic acids from nurse cells to oocyte within the follicle syncytium [Woodruff and Telfer, 1980]. The only way that such a voltage gradient can be maintained in the conductive cytoplasm is by a steady current flow. The extracellular current pattern around such follicles was first measured by Jaffe and Woodruff [1979] in cecropia, but has subsequently been studied in many other species and is generally found to enter the nurse cell end and leave the oocyte (see Table III for references). However, when the epithelium surrounding the nurse cells is removed, an apparent outward current is observed at the nurse cell surface, indicating that the epithelium may be contributing to the current pattern described above [Woodruff et al., 1986b].

Whatever the true extracellular current pattern, in order to generate the observed voltage gradient across the cytoplasmic bridge, positive current must flow along the bridge from the oocyte into the nurse cells. Jaffe and Woodruff proposed that the bridge current would flow in this direction if both the nurse cells and the oocyte exhibited a polarized distribution of inward and outward current regions so that current would be expected to exit the vegetal end of the nurse cells and enter the animal end of the oocyte. Since these two regions are tightly apposed in the follicle, they could act as a battery that would drive current across the apposed membranes and this current would then leak back into the nurse cells along the pathway of lowest resistance through the cytoplasmic bridge (Fig. 7A). This bridge current would generate a voltage of the observed polarity across the cytoplasmic bridge. Another model of current flow has recently been suggested by Verachtert and De Loof [1989]. Their model places the main current pumps in the plasma membranes of the nurse cells and oocyte as well, but rather than being concentrated in the apposed membranes, they suggest that the pumps are concentrated at the poles with current entering the oocyte and leaving the nurse cell end (Fig. 7B). They propose that the extracellular current measured with the vibrating probe is opposite to this because the follicle cells are generating that current. One advantage of this model is that it explains the current reversal observed when the follicle layer is partially removed. However, two properties of the follicle cell layer appear to be inconsistent with this model. First, there are often gap junc-

TABLE III. Ionic Currents Around Single Animal Cells

Species	Cell type	Current magnitude and pattern	Reference*
Physarum polycephalum	Slime mold plasmodia	15 μA/cm^2 enters numerous protrusions	1
Chaos chaos	Crawling amoebae	0.1 μA/cm^2 enters rear quarter of amoeba and leaves psuedopods	2
Hyalophora cecropia	Silk worm polytrophic ovariole	Up to 20 μA/cm^2 enters nurse cell end and leaves oocyte	3,4
Sarcophaga bullata	Fleshfly polytrophic ovariole	Up to 40 μA/cm^2 enters nurse cell end and leaves oocyte	5,6
Drosophila melanogaster	Fruitfly polytrophic ovariole	10 μA/cm^2 enters nurse cells and leaves near oocyte end of follicle	7–9
Locusta migratoria	Locust panoistic ovariole	Up to 75 μA/cm^2 enters posterior end terminal follicle with outward current over most of surface	5
Blattella germanica	Cockroach panoistic oocyte	3 μA/cm^2 enters ventral side oocyte	10–11a
Dysdercus intermedius Rhodnius prolixus	Telotrophic ovariole	30 μA/cm^2 enters germarium and leaves syncytial tropharium	12–14
Lymnaea stagnalis	Snail egg	0.5 μA/cm^2 enters animal hemisphere and leaves vegetal hemisphere	15
Limulus polyphemus	Ventral photoreceptor	1,000 μA/cm^2 pulse enters distal end of isolated photoreceptor upon saturating light stimulation	16
Xenopus laevis	Frog oocyte	1 μA/cm^2 enters animal hemisphere and leaves vegetal hemisphere prior to maturation	17
Xenopus laevis Oryzias latipes	Activating frog and fish egg	A pulse of up to 30 μA/cm^2 enters the site of sperm-egg fusion and spreads over the egg	18–21
Goldfish	Retinal ganglion growth cones	Up to 0.1 μA/cm^2 enters tips of filopodia and leaves at juncture of filopodia and growth cone	22
Discoglossus pictus	Activating frog egg	A pulse of up to 60 μA/cm^2 enters only the animal dimple of the egg	23,24
Xenopus laevis Lymnaea stagnalis	Dividing egg	0.4 μA/cm^2 leaves the region of new membrane insertion at the furrow and leaves neighboring pigmented membrane; for *lymnaea*, outward current is vegetal	25,26
Mouse	Enlarged 8-cell stage blastomere	1 μA/cm^2 enters apical end of polarized blastomere	27,28

(continued)

TABLE III. Ionic Currents Around Single Animal Cells *(continued)*

Species	Cell type	Current magnitude and pattern	Reference*
Mouse	Fibroblast L cells	Current enters leading edge of migrating cells as detected using a "current focusing" technique on wounded monolayers	29

*1: Achenbach and Weisenseel [1981]. 2: Nuccitelli et al. [1977]. 3: Jaffe and Woodruff [1979]. 4: Woodruff et al. [1986a,b]. 5: Verachtert and De Loof [1986]. 6: Verachtert and De Loof [1988, 1989]. 7: Overall and Jaffe [1985]. 8: Sun and Wyman [1989]. 9: Bohrmann et al. [1984, 1986]. 10: Kunkel [1986]. 11: Kunkel and Bowdan [1989]. 11a: Bowdan and Kunkel [1990]. 12: Dittmann et al [1981]. 13: Huebner and Sigurdson [1986]. 14: Diehl-Jones and Huebner [1989]. 15: Zivkovic and Dohmen [1989]. 16: Payne and Fein [1986]. 17: Robinson [1979]. 18: Kline and Nuccitelli [1985]. 19: Kline [1986]. 20: Jaffe et al. [1985]. 21: Nuccitelli [1987]. 22: Freeman et al. [1985]. 23: Nuccitelli et al. [1988]. 24: Campanella et al. [1988]. 25: Kline et al. [1983]. 26: Dohmen et al. [1986]. 27: Nuccitelli and Wiley [1985]. 28: Wiley and Nuccitelli [1986]. 29: Oiki et al. [1989].

tions between the follicle cells and nurse cells. This would make it impossible for current to enter the follicle cells at the nurse cell end without entering the nurse cells at the same time. Second, the low resistance that is often found between follicle cells at the oocyte end should allow current from the oocyte to pass right through the follicle layer with the same polarity rather than with

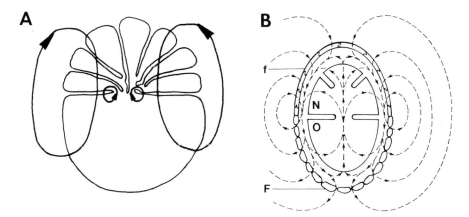

Fig. 7. Models for the ionic current patterns of a vitellogenic polytrophic insect follicle. **A:** Model proposed by Jaffe and Woodruff [1979] in which current enters the animal half of plasma membranes of both the nurse cells and oocyte while leaving the vegetal regions. This drives current back across the low resistance of the cytoplasmic bridge to generate the required electrical potential across the bridge. **B:** Model proposed by Verachtert and De Loof [1989] incorporating the follicle cell layer into the pattern. O, oocyte; N, nurse cells; f, squamous follicle cells; F, cylindrical follicle cells. The follicle cells are represented separated from the oocyte and nurse cells for clarity, but they are often electrically coupled to them. (Courtesy of Verachtert and De Loof [1989].)

the reversed polarity envisioned in their model. Whatever the exact mechanism generating the current, the laws of physics dictate that there must be some steady current traversing the cytoplasmic bridge if a voltage difference is to be maintained in this conducting medium. And it is the evidence for this voltage difference that has stimulated so much interest in this system in the first place.

The evidence for a voltage gradient between nurse cells and oocyte in insect follicles comes from two types of experiments. The first type is the direct measurement of membrane potentials with intracellular microelectrodes and the second is the study of the movement of charged proteins across the bridge. As mentioned above, the electrical measurements were first done by Woodruff and Telfer [1973] on cecropia follicles, but have also been carried out in *Drosophila* by several groups [Bohrmann et al., 1986b; Woodruff, 1989] (Sun and Wyman, pers. commun.). There is considerable controversy over the potential measurements in *Drosophila* follicles because two laboratories have not detected any significant differences in the potential between the oocyte and nurse cells [Bohrmann et al., 1986b] (Sun and Wyman, pers. commun.), while one laboratory has reported a significant voltage gradient [Woodruff, 1989]. These are very difficult measurements that are heavily dependent on the medium used and experimental technique, so this controversy is hard to judge. However, Woodruff's recent paper reports on 159 paired measurements in a very positive persuasive manner. He found that the nurse cells are about 5 mV more negative than the oocyte in 80% of the cases. This important work clearly merits further study so that the conflicting reports can be resolved.

The second type of evidence in support of electrophoretic transport involves the study of the movement of labeled proteins injected into either the nurse cells or oocyte. It has long been known that the transport of cytoplasmic components across the bridge is polarized, flowing only in the direction of nurse cell to oocyte [see review by Gutzeit, 1986]. The polarity of protein transport across the intercellular bridge connecting these two cell types can be reversed by reversing the endogenous electric field within this bridge [Woodruff and Telfer, 1973]. While it is perhaps not too surprising that proteins could be electrophoresed along an intercellular bridge by an imposed field, Woodruff and Telfer [1980] then discovered even more compelling evidence that electrophoresis is involved in the transport of proteins across the bridge. They have shown that the polarity of movement of a given protein across the bridge is dependent only on its net electric charge. Lysozyme is a basic protein with an isoelectric point of 11.5 and, when injected into the follicle, will only move from the oocyte to the nurse cell. However, when the net charge was reversed on this same protein by methylcarboxylation of its ϵ-amino groups, it reversed its transport direction and was only observed to move from the nurse cell to the oocyte. In further support for the electrophoresis mechanism, neutral proteins with isoelectric points near 7 moved in both directions across the bridge. Thus in the cecropia follicle, the polarity of protein transport, which is a critical component of the oocyte's polarity, is determined by the electrical field across the intercellular bridge. The cecropia follicle is still the best example in which transcellular ionic currents play an active, causal role in polarized transport.

One of the other polytrophic ovarioles in which these protein injection studies have been carried out is that of *Drosophila*. Again there is a controversy with these studies because Bohrmann and Gutzeit [1987] and Sun and Wyman (pers. commun.) failed to obtain charge-dependent migration, whereas Woodruff [1989; Woodruff et al., 1988] did. These groups are currently trying to determine why they obtained contradictory results, however, the difficulty in proving a negative result combined with the convincing nature of Woodruff's data lead me to believe that electrophoretic transport occurs in *Drosophila* as well.

Three telotrophic ovarioles have also been investigated using this technique. In this type of ovariole the oocytes are linked to a single, shared group of nurse cells in a tropharium that is much farther from the oocyte than in polytrophic ovaries. Charge-dependent translocation of microinjected proteins has been demonstrated in ovarioles of *Rhodnius prolixus* and *Oncopeltus fasciatus*, but this was restricted to the tropharium [Telfer et al., 1981; Woodruff and Anderson, 1984]. However, in the most recent work using the ovariole from *Dysdercus intermedius*, negatively charged proteins were found to migrate according to the voltage gradient from the tropharium into the oocytes via the trophic cords, while positively charged proteins remained in the tropharium [Munz and Dittmann, 1987]. The effectiveness of this polarized transport is greatest on the tropharium side, since both negatively and positively charged proteins injected into the previtellogenic oocytes moved into the trophic cords. Thus, intercellular electrophoresis is used by a number of insect species as part of the mechanism of polarized transport. Moreover, a recent study of the nurse cell-oocyte complex of a polychaete indicates that steep voltage gradients exist there as well.

A much larger intercellular voltage gradient has been reported between oocyte and nurse cell of the spirally cleaving polychaete, *Ophryotrocha labronica* [Emanuelsson and Arlock, 1985]. In this species, the oocyte is supported during vitellogenesis by a single nurse cell which is attached to the oocyte with a cytoplasmic bridge as in cecropia. An amazing 22–32 mV difference has been reported across the 3 μm wide cytoplasmic bridge with the oocyte more positive than the nurse cell. This system clearly merits further investigation into the role of this voltage gradient in the polarized transport.

2. Activating eggs. Four vertebrate eggs have been investigated during activation with either the extracellular vibrating probe or patch clamp techniques during fertilization: the eggs of the frogs *Xenopus laevis, Rana pipiens,* and

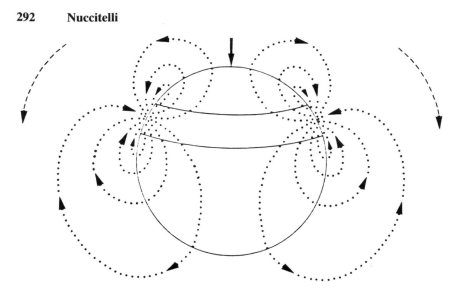

Fig. 8.   The ring-like current pattern inferred from the vibrating probe measurements. The arrow at the top indicates the site of activation. The dotted arrows to the left and right indicate the direction of the current wave which consists of a band of inward current with outward current on either side. (Courtesy of Kline and Nuccitelli [1985].)

Discoglossus pictus [Kline and Nuccitelli, 1985; Kline, 1986; Jaffe et al., 1985; Nuccitelli et al, 1988], and the egg of the medaka fish, *Oryzias latipes* [Nuccitelli, 1987]. In all but *D. pictus,* the activation current was found to enter the site of activation and spread across the egg in a ring-shaped wave (Fig. 8). However, the egg of *D. pictus* exhibits no such wave, despite the presence of a wave of increased intracellular calcium concentration. Current enters only at the animal pole in the dimple region and the Cl^- channels appear to be more highly localized than in any of these other three vertebrates. This egg appears to be the only one studied thus far that exhibits such a striking localization of the channels responsible for the fertilization potential. In the only other case in which the fertilization-gated channels appear to be localized, that of the invertebrate egg, *Urechis caupo* [Gould-Somero, 1981], poor spatial resolution makes a direct comparison with the egg of *D. pictus* difficult.

What is the physiological significance of this activation current? The two main consequences of the activation current are the accompanying changes in both the membrane potential (fertilization potential) and the intracellular concentrations of the ions carrying the current. In many species, the fertilization potential provides a fast block to polyspermy [Jaffe and Cross, 1986], however, the egg of *D. pictus* appears to be polyspermic [Talevi, 1989] and the egg of the medaka exhibits no electrical block to polyspermy [Nuccitelli, 1980]. Therefore, we have here an example of a transcellular ionic current that is the result of a localized permeability increase (probably mediated by an increase in intracellular Ca^{2+}) with unknown function.

D. Ionic Currents Around Multicellular Animal Tissues

This category contains the largest number of investigations of them all. Forty-four papers have appeared describing vibrating probe measurements around various multicellular animal tissues (Table IV). Because these currents are typically generated by an electrically coupled epithelial sheet of cells, the current density is generally larger than that generated by a single cell, making these measurements a bit easier to obtain. While most of the single cell currents are in the 1 μA/cm^2 range, 80% of the tissue currents are in the 10–100 μA/cm^2 range.

1. Ionic currents associated with epithelia undergoing morphogenesis. Most epithelia are composed of a sheet of cells with the same apical-basal polarity. Unidirectional ion transport can generate a voltage across the epithelial layer that will drive current out between all of the cells forming the layer, but this current will be larger where cells are less tightly apposed, leading to a lower resistance to current flow between them. Thus, rather large currents are measured leaving the primitive streak in chick [Jaffe and Stern, 1979] (Fig. 9) and mouse embryos [Winkel and Nuccitelli, 1989] (Fig. 2), where intercellular junctions are breaking down. These currents will generate an internal voltage gradient in the embryo that could be used to provide long-range spatial information to cells in need of guidance [Erickson and Nuccitelli, 1986], but thus far there is no direct evidence that the internal fields are being used in this way. However, there is some indirect evidence: many embryonic motile cells have been found to be exquisitely sensitive to weak voltage gradients, usually migrating toward the negative pole in imposed electric fields [Robinson, 1985; Nuccitelli, 1988b]. Since some differentiated motile cell types (such as BHK cells) do not exhibit this galvanotaxis, it is not a universal response of all motile cells so perhaps embryonic cells have developed the ability to sense these internal fields as one of the many cues that guide them to their destination. Also, neurons and muscle cells exhibit galvanotropism with a similar low threshold and may also utilize embryonic fields for orientation.

One of the most intriguing observations correlating ionic currents and epithelial morphogenesis is the predictive nature of the currents preceding limb bud formation in amphibians. In both axolotls [Borgens et al., 1983] and frogs [Robinson, 1983], outward current can be detected at the site of future limb bud formation long before any anatomical change. For more than a week prior to the emergence of the hind limb in axolotls, a steady ionic current is driven out of the ventrolateral flank and returns through the integument in adjacent regions of the body. A peak in the density of the outward current occurs over the exact area of hind limb formation 4–6 days prior to its appearance (Fig. 10). In *Xenopus* embryos, the region of largest outward current predicted the site of bud formation at least one day prior to the first thickening of ectoderm over the hind limb bud area. These outward currents indicate that the outer epithelium in the prospective limb bud region is becoming leaky long before bud

TABLE IV. Ionic Currents Around Multicellular Animal Tissues

Species	Tissue type	Current magnitude and pattern	Reference*
Noctiluca miliaris	Dinoflagellate sulcus	11 $\mu A/cm^2$ enters middle of sulcus during localized cytoplasmic movement associated with feeding	1
Pomacea	Eyestalk of mystery snail	5–60 $\mu A/cm^2$ enters eyestalk stump after amputation and declines steadily to 2 $\mu A/cm^2$ in 50 hours	2
Petromyzon marinus	Lamprey spinal cord	500 $\mu A/cm^2$ enters freshly cut surface of spinal cord and falls to 100 $\mu A/cm^2$ within an hour and 4 $\mu A/cm^2$ in 2 days	3
Oryzias latipes	Mid-epiboly fish embryos	No currents in intact embryos, but 5 $\mu A/cm^2$ current pulses enter wounds near pacemaker region within 90 s period	4
Sarotherodon mossambicus	Fish gill epithelium	400 $\mu A/cm^2$ enters epithelium in highly localized regions above chloride cells and is carried by Cl^- efflux	5–8
Bufo viridis *Rana pipiens*	Toad skin epithelium	5–100 $\mu A/cm^2$ leaves mitochondrial-rich cells carried by Cl^- influx	9–11
Necturus maculosus	Mud puppy gastric mucosa	Up to 500 $\mu A/cm^2$ enters oxyntic cells of gastric crypts carried by Cl^- efflux and 5 $\mu A/cm^2$ enters surface cells by Na^+ influx	12
Xenopus laevis	Neurula stage embryo	A few $\mu A/cm^2$ enters most of embryo's surface and 10 $\mu A/cm^2$ exits the blastopore carried largely by Na^+; wound-healing in these embryos depends on an endogenous Na^+ current	13–15
Xenopus laevis *Ambystoma mexicanus*	Developing hindlimb	7 $\mu A/cm^2$ leaves prospective hindlimb region of stage 43 *Xenopus* embryos and 0.5–3 $\mu A/cm^2$ exited the same region 1 week prior to placode formation in axolotl embryos	16,17
Notophthalmus viridescens	Regenerating newt limb	10–100 $\mu A/cm^2$ leaves stump of amputated limb during 5–10 days after amputation	18–23
Rana pipiens	Limb stump	20–40 $\mu A/cm^2$ leaves stump of amputated limb 1–10 days after amputation but currents are depressed in center	23,24

(continued)

TABLE IV. Ionic Currents Around Multicellular Animal Tissues *(continued)*

Species	Tissue type	Current magnitude and pattern	Reference*
Chick	Embryo; primitive streak stage	100 μA/cm^2 leaves whole streak and enters elsewhere through epiblast	25,26
Rat	Lumbrical muscle	3 μA/cm^2 enters end plate region of neonatal muscle up to 5 days after birth when the current reverses and 7 μA/cm^2 leaves end plate region	27–32
Rat	Lens	Up to 60 μA/cm^2 enters anterior pole and leaves equator of freshly removed lens	33–36
Rat	Cultured lung epithelia	Up to 30 μA/cm^2 exists localized areas of fluid accumulation	37
Mouse	7-day embryo	20–60 μA/cm^2 exits along the midline of the embryo with inward current along the extraembryonic ectoderm	38
Mouse	Metatarsal bone	10 μA/cm^2 enters articular surface of epiphyses with less dense current entering remaining epiphyseal regions. 100 μA/cm^2 enters fractured regions	39
Human	Regenerating fingertip	20 μA/cm^2 exits stump of finger tip amputation for an average of 8 days following amputation	40

*1: Nawata and Sibaoka [1987]. 2: Bever and Borgens [1988]. 3: Borgens et al. [1980]. 4: Fluck and Jaffe [1988]. 5: Foskett and Scheffey [1982]. 6: Foskett et al. [1983]. 7: Scheffey et al. [1983]. 8: Foskett and Machen [1985]. 9: Katz and Scheffey [1986]. 10: Scheffey and Katz [1986]. 11: Foskett and Ussing [1986]. 12: Demarest et al. [1986]. 13: Robinson and Stump [1984]. 13a: McCaig and Robinson [1982]. 14: Stump and Robinson [1986]. 15: Rajnicek et al. [1988]. 16: Robinson [1983]. 17: Borgens et al. [1983]. 18: Borgens [1977]. 19: Borgens et al. [1977b]. 20: Borgens et al. [1979a,d]. 21: Borgens et al. [1984]. 22: Eltinge et al. [1986]. 23: Borgens et al. [1989]. 24: Borgens et al. [1979b,c]. 25: Jaffe and Stern [1979]. 26: Kucera and de Ribaupierre [1989]. 27: Kinnamon et al. [1985]. 28: Betz et al. [1980]. 29: Betz and Caldwell [1984]. 30: Betz et al. [1984a,b]. 31: Caldwell and Betz [1984]. 32: Betz et al. [1986]. 33: Robinson and Patterson [1983]. 34: Parmalee et al. [1985]. 35: Parmalee [1986]. 36: Wind et al. [1988a,b]. 37: Sugahara et al. [1984]. 38: Winkel and Nuccitelli [1989]. 39: Borgens [1984]. 40: Illingworth and Barker [1980].

formation. Whether or not this current influences the development of this bud region is not known. However, Robinson has pointed out that supernumerary limbs can be induced in amphibians simply by making an incision in the skin which will allow current to leak out [Thornton, 1954]. Therefore, it is possible that the outward current might act to organize or stimulate limb development. This is a very promising and exciting area for further research.

2. Vertebrate regeneration. One of the most exciting areas in this field of

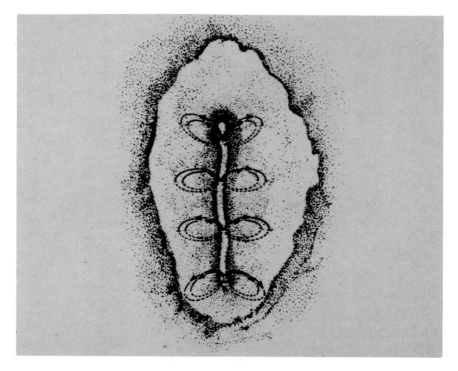

Fig. 9. Extracellular current pattern in stage 4 chick embryo (Courtesy of Jaffe [1981b].)

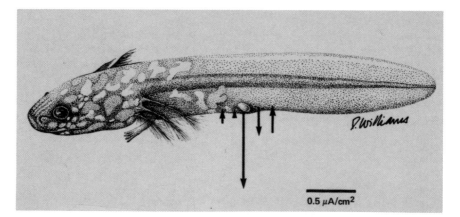

0.5 μA/cm²

Fig. 10. Pattern of current around a limb bud. Arrows indicate the direction and magnitude of current density. This picture was constructed from measurements made 2 days after the appearance of a limb bud was externally observed. This animal was approximately 1.2 cm nose to vent; the measurement positions depicted are about 0.8 mm apart. (Courtesy of Borgens et al. [1983].)

bioelectricity is the study of both natural and imposed currents during limb regeneration. The adult salamander is able to completely regenerate an amputated limb, while the closely related frog cannot. One of the differences between the two is the pattern of ionic currents that flows through the limb stump. In both species, current enters the skin at nearly all points of the limb surface prior to amputation. After amputation, there is an increase in the inward current density and a dramatic efflux of current from the cut surface of the stump [Borgens et al., 1977b, 1979c,d, 1984, 1989]. In the salamander, these stump currents have densities on the order of $10-100$ $\mu A/cm^2$ and persist with fluctuation until just about the early blastema formation about 2 weeks later. As regeneration continues, the outward current drops to less than 5 $\mu A/cm^2$. The current leaving frog stumps is of similar magnitude but the distribution of the current is not uniform. Instead, there is a large current leak between the skin and the tissues below, so that little current traverses the muscle and other subdermal tissues. This leak is absent in urodeles and the current densities are highest in the central and postaxial regions, rather than between the skin and the tissues below. Two lines of evidence support the hypothesis that when this current traverses subdermal tissues it helps to stimulate regeneration. The first is a series of experiments designed to reduce the magnitude of this current (found to be largely carried by Na^+) by either the topical application of the Na^+ influx blocker, amiloride, or removal of Na^+ from the external medium. Both of these treatments were found to interfere with normal regeneration [Borgens et al., 1979d]. The second line of evidence came from the implantation of a wick electrode in the stump region of a frog amputation connected to a battery implanted beneath the skin. This was designed to drive small currents through the subdermal tissues with the same polarity and magnitude as the currents in the salamander limb. The results of such implants are striking. Substantial regeneration occurs resulting in thorn-shaped epidermal papillae with nerve trunks within a cartilage core [Borgens et al., 1977a, 1979b]. These minute electrical currents clearly enhance regeneration and this may involve stimulating early nerve growth. There is a substantial literature indicating that nerve grows toward the negative pole in a small dc electric field [reviewed by Robinson, 1985], and now there are direct measurements of the electric fields within regenerating limbs, indicating that the distal stump is electronegative with respect to the base of the limb [McGinnis and Vanable, 1986]. Thus, galvanotropism should result in nerve growth into the limb. These experiments indicate that limb regeneration can be enhanced by electrical stimulation and leads to the exciting prospect that such currents might be used to enhance human regenerative abilities. Wound currents are widespread in mammals [Barker et al., 1982; Illingworth and Barker, 1980], and are similar in current density and direction. Therefore, there is some hope for practical benefits from these studies of bioelectricity and regeneration [Jaffe and Vanable, 1985].

3. Ionic currents associated with transporting epithelia. Many transporting epithelia, such as gill and skin, are composed of heterogeneous cell types. One use of the vibrating probe technique has been to identify which cells in a given epithelium are active in the transport of specific ions across the cell layer. The method was pioneered by Foskett and Scheffey [1982] when they showed that the current flow across the opercular membrane of fish gills was localized to the Cl$^-$ cells. Further work confirmed that the current was indeed carried by Cl$^-$ since removal of that anion from solution reduced the current by 98% [Scheffey et al., 1983]. Foskett and Machen [1985] then showed that epinephrine rapidly inhibits this Cl$^-$ current in a manner that is reversed by glucagon and phosphodiesterase inhibitors. Two other epithelia that have been studied with the vibrating probe technique include the mudpuppy gastric mucosa [Demarest et al., 1986] and frog skin epithelium [Foskett and Ussing, 1986; Scheffey and Katz, 1986; Katz and Scheffey, 1986]. In frog skin, Cl$^-$ enters the mitochondria-rich cells, generating a current density of up to 100 μA/cm^2. In the gastric mucosa, the Na$^+$ influx was shown to occur along the surface cells of the epithelium with the Cl$^-$ efflux localized to the oxyntic cells of the gastric crypts. This application of the vibrating probe has provided new spatial information about transepithelial fluxes that is not readily obtained using any other standard electrophysiological techniques.

4. Ionic currents in intact and damaged bone. One of the more exciting papers in this multicellular animal category appeared as a research article in *Science* in 1984. This was the first study of the endogenous current pattern around freshly dissected bone from the metatarsals of weanling mice. In undamaged bone, the current density is largest entering the articular surface of the epiphyses with less dense current entering the remaining epiphysial regions (Fig. 11). The terminal cartilaginous regions of bones showed current densities two- to six-fold larger than the diffuse current observed along the diaphysis. Fracture currents always entered the fracture and were initially in the 100 μA/cm^2 range, but quickly declined to about 5 μA/cm^2 which lasted several hours. These fracture currents are of the same polarity and magnitude as clinically applied currents that are successful in treating chronic nonunions in fractured bones, suggesting that the defect in these nonunions may reside in the

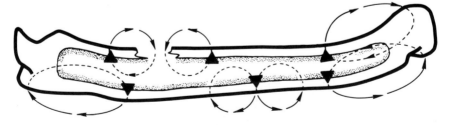

Fig. 11.   Current pattern around intact (**right**) and damaged (**left**) bone.

electrophysiology of repair. Borgens showed that these plateau currents were carried largely by Cl^- and were temperature-dependent indicating that they were probably driven by a cellular battery. This very interesting observation suggests that electrical phenomena may help to control natural tissue response to injury.

V. FUTURE APPLICATIONS AND IMPROVEMENTS

The vibrating probe reveals the total current flow outside of cells and tissues, but the ionic components of this flow must be extracted by indirect means such as by removing ionic species one at a time and determining how that affects the overall current level. A recent solution to this nonspecificity has been the development of ion-selective vibrating probes at the National Vibrating Probe Facility at Woods Hole. The idea is to move an ion-specific microelectrode back and forth between two points just outside of a cell very slowly so that any standing gradients in concentration resulting in current flow across the plasma membrane are minimally disturbed by the electrode movement. A calcium-specific electrode was the first to be developed [Jaffe and Levy, 1987; Kuhtreiber and Jaffe, 1990], but pH- and K^+-specific electrodes are not far behind because only the ion-specific resin in the electrode's tip need be changed to switch the selectivity. The vibrating Ca^{2+} electrode is now a working tool and is being applied to several systems in which Ca^{2+} fluxes are thought to play important roles in cellular function, including migrating *Dictyostelium* slugs and elongating pollen tubes.

A second future application that would be very useful is an aerial probe, one that can detect currents in systems which develop in air. Good examples of such systems are the shoots and leaves of most higher plants, most insect eggs and many fungi. In at least two cases, corn coleoptiles and *Drosophila* eggs, it has been shown that the developing system drives relatively steady currents through the outer conductive layer (and under the insulating one) which are big enough to maintain voltage gradients. Gravistimulated or photostimulated corn coleoptiles develop voltage drops of up to 80 mV across themselves [Grahm, 1964; Johnsson, 1965; Woodcock and Hertz, 1972] and *Drosophila* eggs maintain about 4 mV across themselves [Overall and Jaffe, 1985]. The main idea behind the aerial probe proposed by Lionel Jaffe (pers. commun.) is to vibrate a plate electrode perpendicular to the surface being studied. The electrode and the cell surface constitute a capacitor whose capacitance varies sinusoidally during vibration and any voltage difference between the surface and the electrode is then converted into an AC current through the electrode and the meter attached to it.

VI. CONCLUSIONS

Transcellular and transembryonic ionic currents are found in a wide variety of both plant and animal systems. They are associated with many cellular mechanisms, including nutrient uptake, signal transduction, and cell polarization. Studies of these currents over the past 15 years have generated over 200 papers and most of these investigations have raised more questions than they have answered. In some systems, such as fish gill epithelia, the role of these currents is very clear. In others, such as fungal hyphae, the current direction and magnitude are not directly correlated with growth and this raises questions regarding the significance of the current. The strongest cases for a direct role of ionic currents in development are still the polarizing fucoid egg where the current is carried in part by Ca^{2+} and generates an intracellular concentration gradient of this ion that orients the outgrowth, and the insect follicle in which an intracellular voltage gradient is responsible for the polarized transport from nurse cell to oocyte. However, in most of the systems studied, the experiments to determine if the observed ionic currents are required for a specific cellular function are yet to be done. Our experience with the fucoid egg and the fungal hypha of *Achlya* suggest that it is the change in the intracellular ion concentration resulting from the ionic current that is critical for morphogenesis [Harold, 1986b; Harold et al., 1986]. This was the motivation behind the development of the Ca^{2+}-specific vibrating probe, and has led many researchers including myself to begin using indicators of intracellular free $[Ca^{2+}]$ such as aequorin and fura-2 to visualize the distribution of free Ca^{2+} in cells.

This emphasis on the role of specific ion gradients does not mean that the vibrating probe is no longer useful. In most cases, it remains an effective strategy to survey the net current pattern first before investigating its components. While there are those cases in which it is the specific ion and not the current per se that is important for achieving the required effect, there are also examples where the net current is the important effector. Consider, for example, wound currents in bone or skin or currents through the primitive streak in chick and mouse embryos. These intense currents are almost certainly pumped by the epithelium and leak out through unselective spaces between streak cells. Here the voltage-sensitive vibrating probe should be more sensitive than ion-specific ones, and the overall current pattern and magnitude is of more interest than the ionic species carrying the current.

One non-biological application of the vibrating probe technique which should be mentioned here is the measurement of corrosion currents near corroding metallic surfaces [Issacs, 1986]. This could have practical applications in the development of improved plating techniques designed to avoid corrosion.

ACKNOWLEDGMENTS

I wish to thank Lionel Jaffe, Wiel Kuhtreiber, Keerti Rathore, and Ken Robinson for sharing unpublished work. The original work from my laboratory was supported by National Institutes of Health grants HD19966, HD21526 and HD22594.

VII. REFERENCES

Achenbach F, Weisenseel MH (1981) Ionic currents traverse the slime mould Physarum. Cell Biol Int Rep 5:375–379.

Armbruster B, Weisenseel MH (1983) Ionic currents traverse growing hyphae and sproangia of the mycelial water mold Achlya debaryana. Protoplasma 115:65–69.

Barker AT, Jaffe LF, Vanable JW, Jr (1982) The glabrous epidermis of cavies contains a powerful battery. Am J Physiol 242:R358–R366.

Beaver MM, Borgens RB (1988) Electrical responses to amputation of the eye in the mystery snail. J Exp Zool 245:43–52.

Behrens HM, Weisenseel MH, Sievers A (1982) Rapid changes in the pattern of electric current around the root tip of Lepidium sativum L. following gravistimulation. Plant Physiol 70:1079–1083.

Betz WJ, Caldwell JH (1984) Mapping electric currents around skeletal muscle with a vibrating probe. J Gen Physiol 83:143–156.

Betz WJ, Caldwell JH, Ribchester RR, Robinson KR, Stump RF (1980) Endogenous electric field around muscle fibers depends on the Na^+-K^+ pump. Nature 287:235–237.

Betz WJ, Caldwell JH, Kinnamon SC (1984a) Physiological basis of a steady endogenous current in rat lumbrical muscle. J Gen Physiol 83:175–192.

Betz WJ, Caldwell JH, Kinnamon SC (1984b) Increased sodium conductance in the synaptic region of rat skeletal muscle fibers. J Physiol (Lond) 352:189–202.

Betz WJ, Caldwell JH, Harris GL, Kinnamon SC (1986) A steady electric current at the rat neuromuscular synapse. Prog Clin Biol Res 210:205–212.

Bjorkman T (1989) The use of bioelectric currents to study gravity perception in roots. Biol Bull 176:49–55.

Bjorkman T, Leopold AC (1987a) An electric current associated with gravity sensing in maize roots. Plant Physiol 84:841–846.

Bjorkman T, Leopold AC (1987b) Effects of inhibitors of auxin transport and of calmodulin on a gravisensing-dependent current in maize roots. Plant Physiol 84:847–850.

Blatt MR, Weisenseel MH, Haupt W (1981) A light-dependent current associated with chloroplast aggregation in the alga, Vaucheria sessilis. Planta 152:513–526.

Blüh O, Scott BIH (1950) Vibrating probe electrometer for the measurement of bioelectric potentials. Rev Sci Inst 21:867–868.

Bohrmann J, Gutzeit H (1987) Evidence against electrophoresis as the principal mode of protein transport in vitellogenic ovarian follicles of Drosophila. Development 101:279–288.

Bohrmann J, Heinrich UR, Dorn A, Sander K, Gutzeit H (1984) Electrical phenomena and their possible significance in vittelogenic follicles of Drosophila melanogaster. J Embryol Exp Morphol 82:151.

Bohrmann J, Dorn A, Sander K, Gutzeit H (1986a) The extracellular electrical current pattern and its variability in vittelogenic Drosophila follicles. J Cell Sci 81:189–206.

Borhmann J, Huebner E, Sander K, Gutzeit H (1986b) Intracellular electrical potential measurements in Drosophila follicles. J Cell Sci 81:207–221.

Borgens RB (1977) Skin batteries and regeneration. Nat Hist 86:84–89.

Borgens RB (1982) What is the role of naturally produced electric current in vertebrate regeneration and healing? Int Rev Cytol 76:245–298.

Borgens RB (1984) Endogenous ion currents traverse intact and damaged bone. Science 225:478–482.

Borgens RB (1989a) Natural and applied currents in limb regeneration and development. In Borgens RB, Robinson KR, Vanable JW, Jr, McGinnis ME: Electric Fields in Vertebrate Repair: Natural and Applied Voltages in Vertebrate Regeneration and Healing. New York: Alan R. Liss, Inc., pp 27–75.

Borgens RB (1989b) Artificially controlling axonal regeneration and development by applied electrical fields. In Borgens RB, Robinson KR, Vanable JW, Jr, McGinnis ME: Electric Fields in Vertebrate Repair: Natural and Applied Voltages in Vertebrate Regeneration and Healing. New York: Alan R. Liss, Inc., pp 117–170.

Borgens RB, Vanable JW, Jr, Jaffe LF (1977a) Bioelectricity and regeneration. I. Initiation of frog limb regeneration by minute currents. J Exp Zool 200:403–417.

Borgens RB, Vanable JW, Jr, Jaffe LF (1977b) Bioelectricity and regeneration: II. Large currents leave the stumps of regenerating newt limbs. Proc Natl Acad Sci USA 74:4528–4532.

Borgens RB, Vanable JW, Jr, Jaffe LF (1979a) Bioelectricity and regeneration. Bioscience 29:468–494.

Borgens RB, Vanable JW, Jr, Jaffe LF (1979b) Small artificial currents enhance Xenopus limb regeneration. J Exp Zool 207:217–225.

Borgens RB, Vanable JW, Jr, Jaffe LF (1979c) Role of subdermal current shunts in the failure of frogs to regenerate. J Exp Zool 209:49–55.

Borgens RB, Vanable JW, Jr, Jaffe LF (1979d) Reduction of sodium dependent currents disturbs urodele limb regeneration. J Exp Zool 209:377–394.

Borgens RB, Jaffe LF, Cohen MJ (1980) Large and persistent electrical currents enter the transected lamprey spinal cord. Proc Natl Acad Sci USA 77:1209–1213.

Borgens RB, Rouleau MF, DeLanney LE (1983) A steady efflux of ionic current predicts hind limb development in the axolotl. J Exp Zool 228:491–503.

Borgens RB, McGinnis ME, Vanable JW, Jr, Miles ES (1984) Stump currents in regenerating salamanders and newts. J Exp Zool 231:249–256.

Borgens RB, McCaig CD (1989) Endogenous currents in nerve repair, regeneration and development. In Borgens RB, Robinson KR, Vanable JW, Jr, McGinnis ME: Electric Fields in Vertebrate Repair: Natural and Applied Voltages in Vertebrate Regeneration and Healing. New York: Alan R. Liss, Inc., pp 77–116.

Bowdan E, Kunkel JG (1990) Patterns of ionic currents around the developing oocyte of the german cockroach, Blatella germanica. Dev Biol 137:266–275.

Bowles EA, Allen NS (1984) Steady currents go through Acetabularia crenulata: A vibrating probe analysis. Bio Bull 167:501–502.

Bowles EA, Allen NS (1986) A vibrating probe analysis of light-dependent transcellular currents in Acetabularia. Prog Clin Biol Res 210:113–122.

Bowling DJF, Edwards MC, Gow NAR (1986) Electrical currents at the leaf surface of Commelina communis and their relationship to stomatal activity. J Exp Bot 37:876–882.

Brawley SH, Quatrano RS (1979) Sulfation of fucoidin in Fucus embryos. IV. Autoradiographic investigations of fucoidin sulfation and secretion during differentiation and the effect of cytochalasin treatment. Dev Biol 73:193–205.

Brawley SH, Robinson KR (1985) Cytochalasin treatment disrupts the endogenous currents associated with cell polarization in fucoid zygotes: Studies of the role of F-actin in embryo-genesis. J Cell Biol 100:1173–1184.

Brawley SH, Wetherell DF, Robinson KR (1984) Electrical polarity in embryos of wild carrot precedes cotyledon differentiation. Proc Natl Acad Sci USA 81:6064–6067.

Brownlee C (1989) Visualizing cytoplasmic calcium in polarizing zygotes and growing rhizoids of Fucus serratus. Biol Bull 176(S):14–17.

Brownlee C, Pulsford AL (1988) Visualization of the cytoplasmic Ca^{2+} gradient in Fucus ser-ratus rhizoids; correlation with cell ultrastructure and polarity. J Cell Sci 91:249–256.

Brownlee C, Wood JW (1986) A gradient of cytoplasmic free calcium in growing rhizoid cells of Fucus serratus. Nature 320:624–626.

Caldwell JH, Betz WJ (1984) Properties of an endogenous steady current in rat muscle. J Gen Physiol 83:157–174.

Campanella C, Talevi R, Kline D, Nuccitelli R (1988) The cortical reaction in the egg of *Discoglossus pictus:* A study of the changes in the endoplasmic reticulum at activation. Dev Biol 130:108–119.

Dazy A-C, Borghi H, Pichon Y (1981) Spontaneous bioelectrical activity in the unicellular alga Acetabularia-mediterranea. C R Heb Seances Acad Sci Ser D Sci Nat 291(7):637–640.

Dazy A-C, Borghi H, Garcia E, Puiseux-Dao S (1986) Transcellular current and morphogenesis in Acetabularia mediterranea grown in white, blue and red light. Prog Clin Biol Res 210:123–130.

De Loof A (1986) The electrical dimension of cells: The cell as a miniature electrophoresis chamber. Int Rev Cytol 104:251–352.

Demarest JR, Scheffey C, Machen TE (1986) Segregation of gastric sodium and chloride trans-port: A vibrating probe and microelectrode study. Am J Physiol 251:C643–C648.

Diehl-Jones W, Huebner E (1989) Pattern and composition of ionic currents around ovarioles of the hemipteran, Rhodnius prolixus (Stahl). Biol Bull 176:86–90.

Dittmann F, Ehni R, Engels W (1981) Bioelectric aspects of hemipteran telotrophic ovariole (Dysdercus intermedius). Wilh Rouxs Arch 190:221–225.

Dohman MR, Arnolds WJA, Speksnijder JE (1986) Ionic currents through the cleaving egg of Lymnaea stagnalis (mollusca, gastropoda, pulmonata). Prog Clin Biol Res 210:181–188.

Dorn A, Weisenseel MH (1982) Advances in vibrating probe techniques. Protoplasma 113:89–96.

Dorn A, Weisenseel MH (1984) Growth and the current pattern around internodal cells of Nitella flexilis L. J Exp Bot 35:373–383.

Eltinge EM, Cragoe EJ, Jr, Vanable JW, Jr (1986) Effects of amiloride analogues on adult Notophthalmus viridescens limb stump currents. J Comp Biochem Physiol 84A:39–44.

Emmanuelsson H, Arlock P (1985) Intercellular voltage gradient between oocyte and nurse cell in a polychaete. Exp Cell Res 161:558–561.

Erickson CA, Nuccitelli R (1986) Role of electric fields in fibroblast motility. Prog Clin Biol Res 210:303–309.

Ferrier JM, Lucas WJ (1986a) Ion transport and the vibrating probe. Biophys J 49:803–808.

Ferrier JM, Lucas WJ (1986b) Theory of ion transport and the vibrating probe. Prog Clin Biol Res 210:45–52.

Fisahn J, Lucas WJ (1990) Inversion of extracellular current and axial voltage profile in Chara and Nitella. J Mem Biol 113:23–30.

Fisahn J, McConnaughey T, Lucas WJ (1989) Oscillations in extracellular current, extend pH and membrane potential and conductance in the alkaline bands of Nitella and Chara. J Exp Bot (in press).

Fluck RA, Jaffe LF (1988) Electrical currents associated with rhythmic contractions of the blas-toderm of the medaka Oryzias latipes. Comp Biochem Physiol [A] 89:609–613.

Foskett JK, Machen TE (1985) Vibrating probe analysis of teleost opercular epithelium correlation between active transport and leak pathways of individual chloride cells. J Membr Biol 85:25–35.

Foskett JK, Scheffey C (1982) The chloride cell: Definitive identification as the salt secretory cell in teleosts. Science 215:164–166.

Foskett JK, Ussing HH (1986) Localization of chloride conductance to mitochondria rich cells in frog Rana pipiens skin epithelium. J Membr Biol 91:251–258.

Foskett JK, Bern HA, Machen TE, Conner M (1983) Chloride cells and the hormonal control of teleost fish osmoregulation. J Exp Biol 106:255–281.

Freeman JA, Manis PB, Snipes GJ, Mayes BN, Samson PC, Wikswo JP, Jr, Freeman DB (1985) Steady growth cone currents revealed by a novel circularly vibrating probe: A possible mechanism underlying neurite growth. J Neurosci Res 13:257–283.

Freeman JA, Manis PB, Samson PC, Wikswo JP, Jr (1986) Microprocessor controlled two- and three-dimensional vibrating probes with video graphics: Biological and electro-chemical applications. Prog Clin Biol Res 210:21–36.

Gorst J, Overall RL, Wernicke W (1987) Ionic currents traversing cell clusters from carrot suspension cultures reveal perpetuation of morphogenetic potential as distinct from induction of embryogenesis. Cell Differ 21:101–110.

Gould-Somero M (1981) Localized gating of egg Na$^+$ channels by sperm. Nature (London) 291:254–256.

Gow NAR (1984) Transhyphal electrical currents in fungi. J Gen Microbiol 130:3313–3318.

Gow NAR (1989) Relationship between growth and the electrical current of fungal hyphae. Biol Bull 176:31–35.

Gow NAR, McGillviray AM (1986) Ion currents, electrical fields and the polarized growth of fungal hyphae. Prog Clin Biol Res 210:81–88.

Gow NAR, Kropf DL, Harold FM (1984) Growing hyphae of Achlya bisexualis generate a longitudinal pH gradient in the surrounding medium. J Gen Microbiol 130:2967–2974.

Grahm L (1964) Measurements of geoelectric and auxin-induced potentials in coleoptiles with a refined vibrating electrode technique. Physiol Plant 17:231–261.

Gutzeit HO (1986) Transport of molecules and organelles in meroistic ovarioles of insects. Differentiation 31:155–165.

Harold FM (1986a) A Study of Bioenergetics. New York: W.H. Freeman.

Harold FM (1986b) Transcellular ion currents in tip-growing organisms: Where are they taking us? Prog Clin Biol Res 210:359–366.

Harold FM, Kropf DL, Caldwell JH (1985a) Why do fungi drive electric currents through themselves? Exp Mycol 9:183–186.

Harold FM, Schreurs WJA, Harold RL, Caldwell JH (1985b) Electrobiology of fungal hyphae. Microbiol Sci 2:363–366.

Harold FM, Schreurs WJA, Caldwell JH (1986) Transcellular ion currents in the water mold Achlya. Prog Clin Biol Res 210:89–96.

Horwitz BA, Weisenseel MH, Dorn A, Gressel J (1984) Electric currents around growing Trichoderma hyphae, before and after photoinduction of conidiation. Plant Physiol 74:912–916.

Huebner E, Sigurdson W (1986) Extracellular currents during insect oogenesis: Special emphasis on telotrophic ovarioles. Prog Clin Biol Res 210:155–164.

Hush JM, Overall RL (1989) Steady ionic currents around pea (Pisum sativum L.) root tips: The effects of tissue wounding. Biol Bull 176:56–64.

Isaacs HS (1986) Applications of current measurement over corroding metallic surfaces. Prog Clin Biol Res 210:37–44.

Illingworth CM, Barker AT (1980) Measurement of electrical currents emerging during the regeneration of amputated fingertips in children. Clin Phys Physiol Meas 1:87–89.

Iwabuchi A, Yano M, Shimizu H (1989) Development of extracellular electric pattern around Lepidium roots: Its possible role in root growth and gravitropism. Protoplasma 148:94–100.

Jaffe LA, Cross NL (1986) Electrical regulation of sperm-egg fusion. Annu Rev Physiol 48:191–200.
Jaffe LA, Kado RT, Muncy L (1985) Propagating potassium and chloride conductances during activation and fertilization of the egg of the frog, Rana pipiens. J Physiol (Lond) 368:227–242.
Jaffe LF (1958) Tropistic responses of zygotes of the fucaceae to polarized light. Exp Cell Res 15:282–299.
Jaffe LF (1966) Electrical currents through the developing Fucus egg. Proc Natl Acad Sci USA 56:1102–1109.
Jaffe LF (1979) Control of development by ionic currents. In Cone R, Dowling J (eds) Soc Gen Physiol Symp. New York: Raven Press, 33:199–231.
Jaffe LF (1981a) The role of ion currents in establishing developmental gradients. In Schweiger HG (ed): The International Cell Biology 1980–1981. Berlin: Springer-Verlag, pp 507–511.
Jaffe LF (1981b) The role of ionic currents in establishing developmental pattern. Philos Trans R Soc (Lond) 295:553–566.
Jaffe LF (1982) Developmental currents, voltages, and gradients. In Subtelny S, Green PB (eds): Developmental Order: Its Origin and Regulation. Symp Soc Dev Biol 40:183–218.
Jaffe LF (1986) Ion currents in development: An overview. Prog Clin Biol Res 210:351–358.
Jaffe LF, Levy S (1987) Calcium gradients measured with a vibrating calcium-selective electrode. Proc IEEE/EMBS Conference 9:779–781.
Jaffe LF, Nuccitelli R (1974) An ultrasensitive vibrating probe for measuring steady extracellular currents. J Cell Biol 63:614–628.
Jaffe LF, Nuccitelli R (1977) Electrical controls of development. Annu Rev Biophys Bioeng 6:445–476.
Jaffe LF, Stern CD (1979) Strong electrical currents leave the primitive streak of chick embryos. Science 206:569–571.
Jaffe LF, Vanable JW, Jr (1985) Electrical fields and wound healing. In Eaglstein WH (ed): Clinics and Dermatology. Philadelphia: Lippincott.
Jaffe LF, Walsby AE (1985) An investigation of extracellular electrical currents around cyanobacterial filaments. Biol Bull 168:476–481.
Jaffe LF, Woodruff RI (1979) Large electrical currents traverse developing Cecropia follicles. Proc Natl Acad Sci USA 76:1328–1332.
Jaffe LF, Robinson KR, Nuccitelli R (1974a) Transcellular currents and ion fluxes through developing fucoid eggs. In Zimmerman U, Dainty J (eds): Membrane Transport in Plants and Plant Organelles. Berlin: Springer-Verlag, pp 226–233.
Jaffe LF, Robinson KR, Nuccitelli R (1974b) Local cation entry and self-electrophoresis as an intracellular localization mechanism. Ann NY Acad Sci 238:372–389.
Jaffe LF, Robinson KR, Nuccitelli R (1975) Calcium currents and gradients as a localizing mechanism. In McMahon DM, Fox CF (eds): Proc 1975 ICN-UCLA Conf on Dev Biol. New York: Benjamin WA, 2:135–147.
Johnsson A (1965) Photoinduced lateral potentials in Zea mays. Physiol Plant 18:574–576.
Kataoka H, Weisenseel MH (1988) Blue light promotes ionic current influx at the growing apex of Vaucheria terrestris. Planta 173:490–499.
Katz U, Scheffey C (1986) The voltage-dependent chloride current conductance of toad Bufo viridis skin is localized to mitochondria rich cells. Biochim Biophys Acta 861:480–482.
Kinnamon SC, Betz WJ, Caldwell JH (1985) Development of a steady electric current in neonatal rat lumbrical muscle. Dev Biol 112:241–247.
Kline D (1986) A direct comparison of the extracellular currents observed in the activating frog egg with the vibrating probe and patch clamp techniques. Prog Clin Biol Res 210:189–196.
Kline D, Nuccitelli R (1985) The wave of activation current in the Xenopus egg. Dev Biol 111:471–487.

Kline D, Robinson KR, Nuccitelli R (1983) Ion currents and membrane domains in the cleaving Xenopus laevis egg. J Cell Biol 97:1753–1761.

Kropf DL (1986a) Electrophysiological properties of Achlya hyphae: Ionic currents studied by intracellular potential recording. J Cell Biol 102:1209–1216.

Kropf DL (1986b) Intracellular potential recording as a means to investigate the transhyphal current in Achlya. Prog Clin Biol Res 210:97–104.

Kropf DL, Lupa MDA, Caldwell JH, Harold FM (1983) Cell polarity: Endogenous ion currents precede and predict branching in the water mold Achyla. Science 220:1385–1387.

Kropf DL, Caldwell JH, Gow NAR, Harold FM (1984) Transcellular ion currents in the water mold Achlya bisexualis amino acid proton symport as a mechanism of current entry. J Cell Biol 99:486–496.

Kucera P, de Ribaupierre Y (1989) Extracellular electrical currents in the chick blastoderm. Biol Bull 176:118–122.

Kuhtreiber WM, Jaffe LF (1990) Detection of extracellular calcium gradients with a calcium-specific vibrating electrode. J Cell Biol (in press).

Kunkel JG (1986) Dorsoventral currents are associated with vitellogenesis in cockroach ovarioles. Prog Clin Biol Res 210:165–172.

Kunkel JG, Bowdan E (1989) Modeling currents about vitellogenic oocytes of the cockroach, Blatella germanica. Biol Bull 176:96–102.

Lucas WJ (1982) Mechanism of acquisition of exogenous HCO^{3-} by internodal cells of Chara corallina. Planta 156:181–192.

Lucas WJ, Kochian LV (1987) Ion transport processes in corn roots: An approach utilizing micro-electrode techniques. In Gensler WC (ed): Nato ASI (Advances Science Institutes) Series. Boston: Kluwer Academic, pp 402–426.

Lucas WJ, Nuccitelli R (1980) Bicarbonate and hydroxyl transport across the plasmalemma of Chara corallina spatial resolution obtained using extracellular vibrating probe. Planta (Berlin) 150:120–131.

Lucas WJ, Sanders D (1989) Ion transport in Chara. Methods Enzymol 174:443–478.

Lucas WJ, Keiffer DW, Saunders D (1983) Bicarbonate transport in Chara corallina: Evidence for cotransport of bicarbonate with protons. J Membr Biol 73:263–274.

Lucas WJ, Wilson C, Wright JP (1984) Perturbation of Chara corallina plasmalemma transport function by 2-4-2' 4' di chlorophenoxyphenoxyl propionic acid. Plant Physiol 74:61–66.

McCaig CD, Robinson KR (1982) The ontogony of the transepidermal potential difference in frog embryos. Dev Biol 90:335–339.

McGillviray AM, Gow NAR (1987) The transhyphal electrical current of Neurospora crassa is carried principally by protons. J Gen Microbiol 133:2875–2881.

McGinnis ME (1989) The nature and effects of electricity in bones. In Borgens RB, Robinson KR, Vanable JW, Jr, McGinnis ME: Electric Fields in Vertebrate Repair: Natural and Applied Voltages in Vertebrate Regeneration and Healing. New York: Alan R. Liss, Inc., pp 225–284.

McGinnis ME, Vanable JW, Jr (1986) Electrical fields in Notophthalmus viridescens limb stumps. Dev Biol 116:184–193.

Miller AL (1989) Ion currents and growth regulators in plant root development. Biol Bull 176:65–70.

Miller AL, Raven JA, Sprent JI, Weisenseel MH (1986) Endogenous ion currents traverse growing roots and root hairs of Trifolium repens. Plant Cell Environ 9:79–83.

Miller AL, Shand E, Gow NAR (1988) Ion currents associated with root tips, emerging laterals and induced wound sites in Nicotiana tabacum: Spatial relationship proposed between resulting electrical fields and phytophthoran zoospore infection. Plant Cell Environ 11:21–25.

Money NP, Brownlee C (1987) Structural and physiological changes during sporangial development in Achlya intricata beneke. Protoplasma 136:199–204.

Munz A, Dittmann F (1987) Voltage gradients and microtubules both involved in intercellular protein and mitochondria transport in the telotrophic ovariole of Dysdercus intermedius. Roux Arch Dev Biol 196:391–396.

Nawata T (1984) A simple method for making a vibrating probe system. Plant Cell Physiol 25:1089–1094.

Nawata T, Sibaoka T (1987) Local ion currents controlling the localized cytoplasmic movement associated with feeding initiation of Notiluca. Protoplasma 137:125–133.

Nawata T, Hishinuma T, Wada S (1989) Ionic currents during regeneration of thallus and rhizoid from cell segments isolated from the marine alga Bryopsis. Biol Bull 176:41–45.

Nuccitelli R (1978) Ooplasmic segregation and secretion in the Pelvetia egg is accompanied by a membrane-generated electrical current. Dev Biol 62:13–33.

Nuccitelli R (1980) The fertilization potential is not necessary for the block to polyspermy or the activation of development in the medaka egg. Dev Biol 76:499–504.

Nuccitelli R (1983a) Transcellular ion currents: Signals and effectors of cell polarity. In McIntosh JR (ed): Modern Cell Biology, Vol. 2. New York: Alan R Liss, Inc., pp 451–481.

Nuccitelli R (1983b) Steady transcellular ion currents. In Grinnell AD, Moody WJ, Jr (eds): The Physiology of Excitable Cells. New York: Alan R. Liss, Inc., pp 475–489.

Nuccitelli R (1984) The involvement of transcellular ion currents and electric fields in pattern formation. In Malasinski GM, Bryant SV (eds): Pattern Formation. New York: MacMillan Publishing, pp 23–46.

Nuccitelli R (1986) A two-dimensional vibrating probe with a computerized graphics display. Prog Clin Ciol Res 210:13–20.

Nuccitelli R (1987) The wave of activation current in the egg of the medaka fish. Dev Biol 122:522–534.

Nuccitelli R (1988a) Physiological electrical fields can influence cell motility, growth and polarity. Adv Cell Biol 2:213–233.

Nuccitelli R (1988b) Ionic currents in morphogenesis. Experientia 44:657–666.

Nuccitelli R, Jaffe LF (1974) Spontaneous current pulses through developing fucoid eggs. Proc Natl Acad Sci USA 71:4855–4859.

Nuccitelli R, Jaffe LF (1975) The pulse current pattern generated by developing fucoid eggs. J Cell Biol 64:636–643.

Nuccitelli R, Jaffe LF (1976a) The ionic components of the current pulses generated by developing fucoid eggs. Dev Biol 49:518–531.

Nuccitelli R, Jaffe LF (1976b) Current pulses involving chloride and potassium efflux relieve excess turgor pressure in Pelvetia embryos. Planta 131:315–320.

Nuccitelli R, Wiley LM (1985) Polarity of isolated blastomeres from mouse morulae: Detection of transcellular ion currents. Dev Biol 109:452–463.

Nuccitelli R, Poo M-M, Jaffe LF (1977) Relations between ameboid movement and membrane-controlled electrical currents. J Gen Physiol 69:743–763.

Nuccitelli R, Kline D, Busa WB, Talevi R, Campanella C (1988) A highly localized activation current yet wide-spread intracellular calcium increase in the egg of the frog, Discogloussus pictus. Dev Biol 130:120–132.

Oiki S, Ohno-Shosaku T, Okada Y (1989) Electric currents associated with directed migration of fibroblasts. Biol Bull 176:123–125.

Overall RL, Jaffe LF (1985) Patterns of ionic current through Drosophila follicles and eggs. Dev Biol 108:102–119.

Overall RL, Wernicke W (1986) Steady ionic currents around haploid embryos formed from tobacco pollen in culture. Prog Clin Biol Res 210:139–146.

Parmalee JT (1986) Measurement of steady currents around the frog lens. Exp Eye Res 42:433–442.

Parmalee JT, Robinson KR, Patterson JW (1985) Effects of calcium on the steady outward currents at the equator of the rat lens. Invest Ophthalmol Vis Sci 26:1343–1348.

Payne R, Fein A (1986) Localization of the photocurrent of limulus ventral photoreceptors using a vibrating probe. Biophys J 50:193–196.

Peng HB, Zhu D-L (1989) Ionic control of postsynaptic differentiation in muscle. Biol Bull 176:126–129.

Quatrano RS (1973) Separation of processes associated with differentiation of two-celled Fucus embryos. Dev Biol 30:209–213.

Racusen RH, Cooke TJ, Ketchum KA (1986) Ionic basis of tip growth in the fern gametophyte. Prog Clin Biol Res 210:131–138.

Racusen RH, Ketchum KA, Cooke TJ (1988) Modifications of extracellular electric and ionic gradients preceding the transition from tip growth to isodiametric expansion in the apical cell of the fern gametophyte. Plant Physiol 87:69–77.

Rajnicek AM, Stump RF, Robinson KR (1988) An endogenous sodium current may mediate would healing in *Xenopus* neurulae. Dev Biol 128:290–299.

Rathore KS, Robinson KR (1989) Ionic currents around developing embryos of higher plants in culture. Biol Bull 176:46–48.

Rathore KS, Hodges TK, Robinson KR (1988) Ionic basis of currents in somatic embryos of Daucus carota. Planta 175:280–289.

Rathore KS, Hotary KB, Robinson KR (1990) A two-dimensional vibrating probe study of currents around lateral roots of *Raphanus sativus* developing in culture. Plant Physiol 92:543–546.

Raven JA (1979) The possible role of membrane electrophoresis in the polar transport of IAA and other solutes in plant tissues. New Phytol 82:285–291.

Robinson KR (1979) Electrical currents through full-grown and maturing Xenopus oocytes. Proc Natl Acad Sci USA 76:837–841.

Robinson KR (1983) Endogenous electrical current leaves the limb and prelimb region of the Xenopus embryo. Dev Biol 97:203–211.

Robinson KR (1985) The responses of cells to electrical fields: A review. J Cell Biol 101:2023–2027.

Robinson KR (1989) Endogenous and applied electrial currents: Their measurement and application. In Borgens RB, Robinson KR, Vanable JW, Jr, McGinnis ME: Electric Fields in Vertebrate Repair: Natural and Applied Voltages in Vertebrate Regeneration and Healing. New York: Alan R. Liss, Inc., pp 1–25.

Robinson KR, Cone R (1980) Polarization of fucoid eggs by a calcium ionophore gradient. Science 207:77–78.

Robinson KR, Jaffe LF (1975) Polarizing fucoid eggs drive a calcium current through themselves. Science 187:70–72.

Robinson KR, Patterson JW (1983) Localization of steady currents in the lens. Curr Eye Res 2:843–848.

Robinson KR, Stump RF (1984) Self-generated electrical currents through Xenopus neurulae. J Physiol (Lond) 352:339–352.

Saunders MJ (1986a) Correlation of electrical current influx with nuclear position and division in Funaria calonema tip cells. Protoplasma 132:32–37.

Saunders MJ (1986b) Cytokinin activation and redistribution of plasma-membrane ion channels in Funaria. Planta 167:402–409.

Scheffey C (1986a) Pitfalls of the vibrating probe technique, and what to do about them. Prog Clin Biol Res 210:XXV.

Scheffey C (1986b) Tutorial: Electric fields and the vibrating probe, for the uninitiated. Prog Clin Biol Res 210:XXV.

Scheffey C (1988) Two approaches to construction of vibrating probes for electrical current measurement in solution. Rev Sci Instr 59:787–792.

Scheffey C, Katz U (1986) Current flow measurements from the apical side of toad skin. A vibrating probe analysis. Prog Clin Biol Res 210:213–222.

Scheffey C, Fosektt JK, Machen TE (1983) Localization of ionic pathways in the teleost Sarotherodon mossambicus opercular membrane by extracellular recording with a vibrating probe. J Membr Biol 75:193–204.

Schreurs WJA, Harold FM (1988) Transcellular proton current in Achlya bisexualis hyphae: Relationship to polarized growth. Proc Natl Acad Sci USA 85:1534–1538.

Slayman CL, Slayman CW (1962) Measurement of membrane potentials in Neurospora. Science (Wash DC) 136:876–877.

Speksnijder JE, Weisenseel MH, Chen TH, Jaffe LF (1989) Calcium buffer injections arrest fucoid egg development by suppressing calcium gradients. Biol Bull 176:9–13.

Stump RF, Robinson KR (1986) Ionic currents in Xenopus embryos during neurulation and wound healing. Prog Clin Biol Res 210:223–230.

Stump RF, Robinson KR, Harold RL, Harold FM (1980) Endogenous electrical currents in the water mold Blastocladiella emersonii during growth and sporulation. Proc Natl Acad Sci USA 77:6673–6677.

Sugahara K, Caldwell JH, Mason RJ (1984) Electrical currents flow out of domes formed by cultured epithelial cells. J Cell Biol 99:1541–1546.

Sun Y-A, Wyman RJ (1989) The Drosophila egg chamber: External ionic currents and the hypothesis of electrophoretic transport. Biol Bull 176:79–85.

Takeuchi Y, Schmid J, Caldwell JH, Harold FM (1988) Transcellular ion currents and extension of Neurospora crassa hyphae. J Membr Biol 101:33–41.

Talevi R (1989) Polyspermic eggs in the anuran Discoglossus pictus develop normally. Development 105:343–350.

Telfer WH, Woodruff RI, Huebner E (1981) Electrical polarity and cellular differentiation in meroistic ovaries. Am Zool 21:675–686.

Thiel R, Schreurs WJA, Harold FM (1988) Transcellular ion currents during sporangium development in the water mould Achlya bisexualis. J Gen Microbiol 134:1089–1097.

Thornton CS (1954) The relation of epidermal innervation to limb regeneration in Amblystoma larvae. J Exp Zool 127:577–601.

Toko K, Iiyama S, Tanaka C, Hayashi K, Yamafuji K (1987) Relation of growth process to spatial patterns of electric potential and enzyme activity in bean roots. Biophys Chem 27:39–58.

Troxell CL (1989) Transcellular ionic currents during primary cell wall morphogenesis in Micrasterias and Closterium. Biol Bull 176:36–40.

Troxell CL, Scheffey C, Pickett-Heaps JD (1986) Ionic currents during wall morphogenesis in Micrasterias and Closterium. Prog Clin Biol Res 210:105–112.

Vanable JW, Jr (1989) Integumentary potentials and wound healing. In Borgen RB, Robinson KR, Vanable JW, Jr, McGinnis ME: Electric Fields in Vertebrate Repair: Natural and Applied Voltages in Vertebrate Regeneration and Healing. New York: Alan R. Liss, Inc., pp 171–224.

Van Brunt J, Caldwell JH, Harold FM (1982) Circulation of potassium across the plasma membrane of Blastocladiella emersonii potassium channel. J Bacteriol 150:1449–1461.

Verachtert B, De Loof A (1986) Electrical fields around the polytrophic ovarian follicles of Sarcophaga bullata and the panoistic follicles of Locusta migratoria. Prog Clin Biol Res 210:173–180.

Verachtert B, De Loof A (1988) Experimental reversal of the electric field around vitellogenic follicles of Sarcophaga bullata. Comp Biochem Physiol 90A:253–256.

Verachtert B, De Loof A (1989) Intra- and extracellular electrical fields of vitellogenic polytrophic insect follicles. Biol Bull 176:91–95.

Waaland SD, Lucas WJ (1984) An investigation of the role of transcellular ion currents in morphogenesis of Griffithsia pacifica Kylin. Protoplasma 123:184–191.

Weisenseel MH, Jaffe LF (1976) The major growth current though lily pollen tubes enters as K^+ and leaves as H^+. Planta 133:1–7.

Weisenseel MH, Kicherer RM (1981) Ionic currents as control mechanism cytomorphogenesis. In Kiermayer O (ed): Cytomorphogenesis in Plants. New York: Springer-Verlag, pp 379–399.

Weisenseel MH, Nuccitelli R, Jaffe LF (1975) Large electrical currents traverse growing pollen tubes. J Cell Biol 66:556–567.

Weisenseel MH, Dorn A, Jaffe LF (1979) Natural H^+ currents traverse growing roots and root hairs of barley (Hordeum vulgare L). Plant Physiol 64:512–518.

Wiley LM, Nuccitelli R (1986) Detection of transcellular currents and effect of an imposed electric field on mouse blastomeres. Prog Clin Biol Res 210:197–204.

Wind BE, Walsh S, Patterson JW (1988a) Equatorial potassium currents in lenses. Exp Eye Res 46:117–130.

Wind BE, Walsh S, Patterson JW (1988b) Effect of quabain on lens equatorial currents. Invest Ophthalmol Vis Sci 29:1753–1755.

Winkel GK, Nuccitelli R (1989) Large ionic currents leave the primitive streak of the 7.5-day mouse embryo. Bio Bull 176:110–117.

Woodcock AER, Hertz CH (1972) The geoelectric effect in plant shoots. J Exp Bot 23:963–957.

Woodruff RI (1989) Charge-dependent molecular movement though intercellular bridges in Drosophila follicles. Biol Bull 176:71–78.

Woodruff RI, Anderson KL (1984) Nutritive cord connection and dye-coupling of the follicular epithelium to the growing oocytes in the telotrophic ovarioles in Oncopeltus fasciatus, the milkweed bug. Wilh Rouxs Arch 193:158–163.

Woodruff RI, Telfer WH (1973) Polarized intercellular bridges in ovarian follicles of the cecropia moth. J Cell Biol 58:172–188.

Woodruff RI, Telfer WH (1980) Electrophoresis of proteins in intercellular bridges. Nature 286:84–86.

Woodruff RI, Huebner E, Telfer WH (1986a) Electrical properties of insect ovarian follicles: Some challenges of a multi-cellular system. Prog Clin Biol Res 210:147–154.

Woodruff RI, Huebner E, Telfer WH (1986b) Ion currents in Hyalophora cecropia ovaries: The role of the epithelium and the intercellular spaces of the trophic cap. Dev Biol 117:405–416.

Woodruff RI, Kulp JH, LaGaccia ED (1988) Electrically mediated protein movement in Drosophila follicles. Rouxs Arch Dev Biol 197:231–238.

Youatt J, Gow NAR, Gooday GW (1988) Bioelectric and biosynthetic aspects of cell polarity in Allomyces marcrogynus. Protoplasma 146:118–126.

Zivkovic D, Dohmen MR (1989) Ionic currents in Lymnaea stagnalis eggs during maturation divisions and first mitotic cell cycle. Biol Bull 176:103–109.

Noninvasive Techniques in Cell Biology: 311–326
© 1990 Wiley-Liss, Inc.

12. Optical Scattering Studies of Muscle Contraction

Gerald H. Pollack

Division of Bioengineering, University of Washington, Seattle, Washington 98195

I. INTRODUCTION

From the time the practical optical microscope was developed in the seventeenth century, muscle has been a favorite specimen for observation. Its striated pattern was distinct, crisp, analyzable, and most of all, it provided a clue as to the inner workings of nature's motor.

With the advent of modern optical sensors to augment the microscope, it would be logical to presume that the inner workings might finally have been explored in enough depth to settle on the mechanism. This is not so. While numerous optical (and other) studies have been carried out on muscles, the pattern of shortening of the sarcomeres is still debated as is the underlying driving mechanism [Pollack, 1983]. One theory holds dominance [A.F. Huxley, 1957; H.E. Huxley, 1969]. Many other theories have been advanced [e.g., Iwazumi, 1970; Tirosh et al., 1979; Harrington, 1971; Pollack, 1990] in an attempt to explain a broader range of observations.

My goal in this chapter is to describe briefly two areas of optical study of muscle, and the controversy that has surrounded their application. The first is optical imaging of the striations; the second is the study of the dynamics of shortening by optical diffraction. At stake in both cases are fundamental issues vis-à-vis contraction.

II. ANATOMY OF THE SARCOMERE

The structure of the sarcomere, the fundamental unit of contraction, is diagrammed in Figure 1. Thick and thin filaments are familiar elements. They are constructed principally of the proteins myosin and actin, respectively, though accessory proteins are present in each. Newly discovered are the connecting filaments [Maruyama, 1976; Wang, 1984], which interconnect either end of the thick filament with the neighboring Z-line.

The structures of Figure 1 are detected most readily with the electron microscope. Many EM studies have been directed at elucidating the morphological changes that accompany contraction. However, time resolution is a problem. Because specimens need to be fixed, a time series is not possible; specimens can be examined only at a single instant—one that is not particularly well defined because of the span of time required for fixation. Quick-freeze methods have begun to ameliorate the ambiguity, but still, EM methods are only of limited value in this dynamically varying system.

More appropriate—albeit with attendant resolution limitations—is the optical microscope, where dynamic events can be followed. Figure 1 shows that

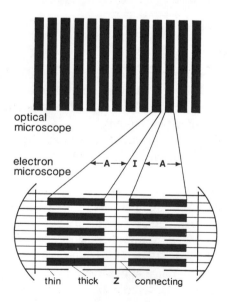

Fig. 1. A- and I-bands, as seen in the optical microscope, shown diagrammatically. Protein filaments corresponding to these bands are diagrammed below.

the thick filaments correspond to A-bands, intervening zones to I-bands. The label A derives from the fact that the designated band is anisotropic: because filament rodlets are well registered, this band is strongly birefringent. The I-band, being less dense, and possibly less regular, is less anisotropic—thus, the designation I, for isotropic. A- and I-bands can be tracked during contraction. From the pattern of changes, the molecular events that may be occurring during contraction can be inferred.

With regard to birefringent properties of A- and I-bands, it should be recognized that in addition to the rodlets, there is a transverse network of inter-filaments links. Rung-like links are found between thick filaments in the A-band [Pollack, 1983; Magid et al., 1984; Suzuki and Pollack, 1986; Baatsen et al., 1988], and finer links are found throughout the I-band [Trombitas et al., 1988]. These transverse links dominate the EM pattern. As yet, they have not figured into any analysis of the sarcomere's pattern of birefringence.

III. OPTICAL IMAGING OF STRIATIONS

From early on, the question of compelling interest to muscle scientists has been: what changes of the striation pattern accompany sarcomere shortening? Using relatively primitive methods, scientists of the nineteenth century could follow the time course of sarcomere shortening, but it was not until the first half of the twentieth century that a coherent picture began to emerge [cf. A.F. Huxley, 1980, for review].

The emergent picture was that active sarcomere shortening was brought about by shortening of both A- and I-bands. This pattern was confirmed by detailed studies in a number of independent laboratories. Because A- and I-band shortening were found consistently among *all* scientists of the first half of the twentieth century [cf. A.F. Huxley, 1980], the field grew to accept this feature as a working paradigm. Efforts were redirected toward uncovering the mechanism underlying these changes of banding pattern.

But in 1954, the paradigm abruptly shifted. Working independently, H.E. Huxley and Hanson [1954] and A.F. Huxley and Niedergerke [1954] studied the changes in banding pattern using more up-to-date methods, and concluded that active shortening was associated with A-band changes alone—not A- and I-band changes, as had previously been found. A brief review of these classical experiments, published in more detail several years later [Huxley and Hanson, 1955; A.F. Huxley and Niedergerke, 1958] proves illuminating.

In the work by H.E. Huxley and Hanson, isolated myofibril bundles of rabbit psoas (tenderloin) muscle were examined in the phase microscope. Fibrils were initially in the rigor state. When exposed to ATP, they contracted. The extent of contraction was found to increase with the concentration of ATP. Thus, the extent of contraction could be titrated, and because the fibril remained quasi-static at its terminal length, it was relatively easy to record the striation pattern.

The results showed that most of the shortening induced in this manner took place in the I-band. Although some A-band shortening was noted, the authors lent little significance to this finding because of its limited magnitude and because of emerging electron-micrographic studies. The EM studies were beginning to show that the sarcomere was built of separate filament sets (Fig. 1), and that changes of sarcomere length were brought about by sliding of thin filaments over thick—at least during manual stretch or release in the relaxed state.

The experiments of A.F. Huxley and Niedergerke were more extensive. Intrigued by optical microscopes, A.F. Huxley constructed his own interference microscope which at the time constituted a genuine advance in the field. Huxley and Niedergerke used the interference microscope to examine the striation pattern in single frog skeletal muscle fibers during various contractile maneuvers. Unlike the experiments of H.E. Huxley, these contractions were rapid, and it was necessary to use a strobe to capture the striation pattern on film.

In the main, A.F. Huxley's results were similar to those of H.E. Huxley. During forcible stretch of unactivated sarcomeres, most of the change took place in the I-band; the A-band remained approximately constant. Nor did the A-band width vary substantially during isometric contraction, or during local activation. In all of these experimental maneuvers, the minor A-band width variation that was observed could be easily relegated to secondary influences. All of the band changes, it appeared, were consistent with emergent EM data.

On the other hand, in experiments in which the sarcomere underwent active shortening, the results were qualitatively different. In this maneuver, the sarcomere was first stretched to some initial length between 2.2 and 3.2 μm. As it shortened under constant load, the banding pattern was recorded, then analyzed. Once the sarcomere shortened on average to 2.4 μm or below, A-band shortening was detected. As contraction continued, A- and I-bands both shortened.

It is regrettable that the information provided in this classical paper is inadequate to allow full interpretation. Although the initial sarcomere length ranged between 2.2 and 3.2 μm, the mean was not given. If, for example, the mean fell at 2.7 μm—midway between the two extremes—then significant A-band shortening occurred after 0.3 μm of sarcomere shortening (2.7 minus 2.4). Given the era's attendant resolution limitations, this finding is exactly what would be expected if A-band shortening had begun from the onset of contraction.

The observation of A-band shortening during isotonic contraction was not paid much heed, despite consistency with *all* results published during the preceding half-century. Huxley and Niedergerke felt that the noted A-band shortening was likely to be an artifact associated with the microscope's limited resolution. Because the I-band was bisected by an optically dense Z-line, Huxley and Niedergerke argued that the microscope was unable to resolve I-band shortening beyond a certain degree, and therefore, that any changes in band width had to be detected in the A-band—artifactually. The microscope, in other words, could not ever see the I-band draw to zero width.

Although the classical experiments cited above are broadly cited as having proved that A-bands remain constant during active sarcomere shortening, the results evidently show otherwise. A-band shortening was observed, but dismissed. On the other hand, the compelling feature of both these experiments, by A.F. Huxley and H.E. Huxley, was that the A-band width did remain virtually constant in the relaxed state. Imposed stretch or retraction had little effect on the relaxed A-band. This latter finding was beautifully consistent with EM observations which showed filament sliding and constant thick filament length under similar conditions.

What seems to have been inappropriately extrapolated from these observations is the conclusion that A-bands remain constant under all conditions—passive as well as active. Although filament sliding as a universal mechanism is an attractively simple paradigm, evidence for universality is lacking: during active sarcomere shortening, A-band shortening is consistently noted.

The issue of what molecular events underlie these changes in banding pattern has also been addressed by electron microscopy. EM studies are perhaps of only limited interest in a paper devoted to optical methods, so I will not dwell on them. A comprehensive review of EM evidence from the mid-fifties on shows that thick filament shortening is observed in the majority of reports [Pollack, 1983]. This, in contrast to the widely held view that A-bands/thick filaments do not shorten during contraction.

Few modern optical microscopic studies have been brought to bear on the question of the A- and I-band dynamics, probably because the problem appeared to have been solved. The exception is a very recent study [Periasamy et al., 1990] in which the striations of frog skeletal were examined with polarization microscopy. Advances in electro-optical sensors allowed the striation pattern to be studied with improved spatial and temporal resolution. Image-enhancement methods and de-skewing algorithms made it possible to analyze the pattern with increased reliability. The results were comparable to those of Huxley and Niedergerke: active sarcomere shortening was accompanied by A-band shortening as well as I-band shortening, and concomitant EM studies showed that the A-band shortening was associated with thick filament shortening.

To recapitulate, optical studies of the striation pattern have been carried out by numerous investigators. Results obtained with diverse imaging methods are more-or-less consistent. In relaxed specimens, A-band width remains approximately constant, irrespective of the degree of stretch or retraction. In actively contracting specimens, the A-band shortens. Both features are largely confirmed by EM studies. Yet, inexplicably, the significance of the latter findings seems to have escaped the attention of the field's notables. Filament sliding as a universal mechanism remains the accepted paradigm.

IV. OPTICAL DIFFRACTION

A second and complementary method of investigating the dynamics of contraction is to examine the optical diffraction pattern. Diffraction patterns can be obtained from striated muscle because the striations are regular. In fact, the striation pattern closely resembles a diffraction grating—a planar surface with parallel, closely spaced slits. When a beam of collimated light, such as the one obtained from a laser, is passed through a muscle (or a grating), a diffraction pattern is obtained (see Fig. 2).

The principal value of using the diffraction pattern is that mean striation spacing can be inferred. All that is required is the Bragg equation. In early diffraction applications, only static values of sarcomere length could be obtained. With modern methodology, it has been possible to trace the dynamics of sarcomere shortening in real time [Iwazumi and Pollack, 1979].

Such diffraction-based studies have also proved controversial—seemingly provoking more questions than answers. Studies carried out during the late 1970s challenged two of the underlying bases of the accepted cross-bridge theory, and thereby sparked a controversy. I touch briefly on each of them.

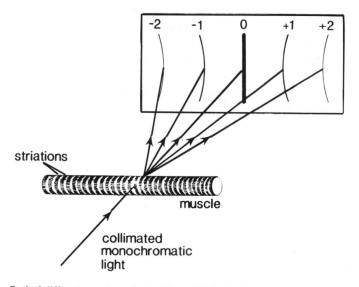

Fig. 2.   Optical diffraction pattern obtained from striated muscle, shown diagrammatically. As the striations shorten, the bands move apart.

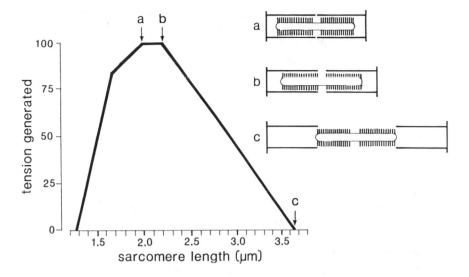

Fig. 3. Length-tension curve, after Gordon et al. [1966]. Diagrams at right denote sarcomeric configurations at key points along the curve.

V. LENGTH-TENSION RELATIONS

The classical length-tension relation, obtained by Gordon et al. [1966] is shown in Figure 3. Because the shape of the "descending limb" fits the expectation of the sliding filament/cross-bridge theory (Fig. 1), this results has long been taken as strong evidence favoring that theory. The result was obtained by placing markers along the surface of the fiber, and using servo-control to keep the spacing between any two markers constant—truly isometric contraction.

On the other hand, the result harbored some anomalaies. Tension did not attain steady state. Instead of a "flat" tension trace, tension kept increasing with time, a kind of creep of tension. The absence of a plateau prompted the investigators to plot a transient value of tension. The tension level prior to creep was selected—the creep being hypothesized to arise from an increase of striation inhomogeneity, and therefore to an experimental artifact.

Whether the increase of inhomogeneity really took place was the question addressed in the diffraction study [ter Keurs et al., 1978]. The issue was a critical one. If the inhomogeneity were confirmed, then Huxley's approach would be validated; if not, the plotting of pre-steady-state tension would have no basis—it would be more appropriate to wait until creep terminated, and plot the ensuing steady-tension level.

The diffraction study by ter Keurs and colleagues confirmed that inhomogeneity did not develop, at least in the extremely uniform fibers selected for that study. Scans along the fiber during the plateau of isometric contraction showed that the sarcomere-length distribution remained tight. Absence of local inhomogeneity was confirmed by examining the profile of the first order: absence of significant broadening implied that inhomogeneity did not develop. The authors concluded that while inhomogeneity could certainly develop in some fibers (particularly near the ends), inhomogeneity did not necessarily develop in all fibers. Since, in their fibers, inhomogeneity was not present but creep of tension was, the creep did not necessarily arise out of inhomogeneity.

As a consequence of these findings, the authors deemed it more reasonable to plot steady-state tension instead of the early phase of tension rise that was plotted in Huxley's experiments. Their result is shown in Figure 4. Over approximately half the relevant range, steady-state tension is independent of the degree of thick and thin filament overlap.

This result inevitably triggered controversy. The experiment was criticized on the basis that the diffraction method was flawed. While diffaction could reasonably detect mean striation spacing, the method was considered potentially unable to detect local heterogeneity [Julian and Morgan, 1979]. The diffraction signal arises from striations that are regular. If the illuminated volume contains a region that is regular and another that is less regular, the diffraction signal will be dominated by the former. If the irregular region shortened, but the regular region did not, no apparent length change would be detected,

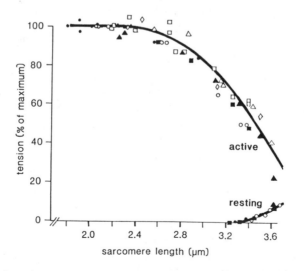

Fig. 4. Length-tension curve, after ter Keurs et al. [1978]. Tension varies little over the range 2.0 to 2.8 μm.

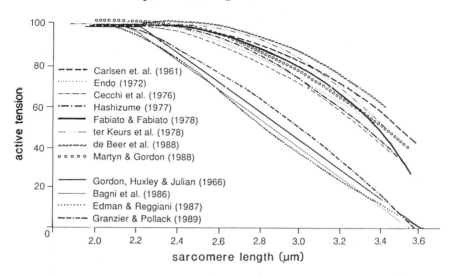

Fig. 5. Summary of length-tension curves measured with servo-control of sarcomere length (**bottom** group) and without servo-control (**top** group). In the upper group, some sarcomere shortening was permitted. Slight shortening resulted in higher tension.

whereas, in reality, some length change took place. Uncertainty of this sort led to a questioning of the new length-tension results.

The controversy inspired a number of re-investigations of the length-tension curve. The results of most of the newer and older studies are shown in Figure 5. They reveal much the same disparity: in servo-control experiments, the results are consistently like those of the classical experiments, whereas in non-servo-controlled experiments in which homogeneity was carefully checked, length-tension curves are higher and flatter. Which result is correct?

A new observation promises, finally, to settle the issue [Horowitz et al., 1989a,b]. It appears that a minute amount of sarcomere shortening is required for generation of maximum tension. Sarcomeres that remain absolutely iso-metric—as in the servo-control experiments—generate sub-maximal tension. Sarcomeres that shorten slightly—by 50 nm or less—and then remain isometric, generate considerably more tension. That such minuscule shortening can induce so much more tension is unexpected.

On the other hand, that finding may settle the length-tension controversy. In servo-control experiments, sarcomeres remain absolutely isometric, so generated tension is sub-maximal (Fig. 5, lower curves). In non-servo-control experiments (Fig. 5, upper curves), though gross inhomogeneity was averted, sarcomeres inevitably shortened or lengthened by a small amount—0.1 μm is not atypical. Such shortening is apparently sufficient to inspire development of much higher isometric tension. Thus, the higher, flatter length-tension curve.

The impact of the results of Horowitz et al. [1989] is too preliminary to judge. If the results stand the test of time—if the flatter length-tension curve is the valid one—the cross-bridge theory would lose one of its primary pillars of support.

VI. STEPWISE SHORTENING

A second application of optical diffraction is in the measurement of the time course of sarcomere shortening. With the advent of high-bandwidth optical sensors, it has become possible to project the first order diffraction peak onto an optical position sensor, and follow the time course of sarcomere shortening with high temporal and spatial resolution. Again, new information has been obtained, and again the information triggered a controversy.

The first high-resolution study of sarcomere dynamics by optical diffraction was carried out on isolated rat cardiac muscle [Pollack et al., 1977]. In such cardiac preparations, end regions are destroyed by clamping, and serve as series compliance—stretched by shortening sarcomeres in the specimen's mid-region. Mid-region sarcomeres were illuminated by a compressed laser beam, and the first-order peak of the diffraction pattern was projected onto a photodiode array. The centroid of the first order was computed on line, with 256 microsecond time resolution and spatial resolution of a few nanometers per sarcomere. Thus, sarcomere shortening could be accurately followed.

The results were unexpected—if not astonishing (Fig. 6). Sarcomere shortening appeared to be stepwise: steps of shortening were interspersed with pause-

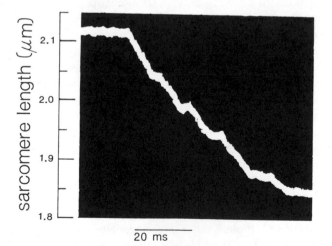

Fig. 6. Time course of sarcomere shortening measured by optical diffraction. Note pauses interspersed between steps of shortening.

like periods, during which there was little or no shortening. The shortening waveform resembled a staircase.

Why is the finding astonishing? According to the cross-bridge theory, the action of bridges is stochastic [A.F. Huxley, 1957]. Each cross-bridge operates independently of the others—especially from sarcomere to sarcomere. With stochastic action of many elements, the shortening waveform should be smooth. The steps, on the other hand, imply that the action is not at all stochastic. The numerous myofibrillar sarcomeres contained within the optical field must step and pause in synchrony. Unless all sampled sarcomeres pause simultaneously, the measured pause would not be "flat."

To check against artifact, a series of control experiments was carried out. These controls included use of a second optical sensor based on a different principle, applied simultaneously. The Schottky barrier photodiode gave stepwise patterns indistinguishable from those obtained with the photodiode array.

Nevertheless, artifact was suspected for a number of possible reasons. Rüdel and Zite-Ferenczy [1979], for example, argued that muscle is closer to a volume grating than a planar grating. Because of the fiber's thickness along the optical axis, the effects of the three-dimensional nature of the grating had to be considered.

More specifically, these investigators were concerned with the possible effect of "Bragg reflections." Bragg reflections can arise when the striations are skewed. When the angle between the optical axis and the plane containing the tilted striation is equal to the critical Bragg angle, a strong reflection results. Rüdel and Zite-Ferenczy argued that such reflections dominate the diffraction pattern. If the Bragg condition were satisfied several times during the course of shortening, they argued, the resulting waveform might be stepwise—like that of Figure 6.

The Bragg-reflection argument was given additional force by experiments carried out using white-light diffraction [Goldman, 1987]. Because each constituent wavelength in this scheme was incident at a slightly different angle, the effects of Bragg reflections were expected to be minimized, and indeed, the shortening pattern measured under these Bragg-reflection-free conditions was often free of steps.

A totally different kind of potential artifact was considered to arise from the effects of specimen translation [Altringham et al., 1984]. During contraction, sarcomeres inevitably translate across the optical axis. As one sarcomere passes out of the field and another enters, the scattered light signal may suffer a fluctuation. Such fluctuations, superimposed on an otherwise smooth shortening waveform, could confer a step-like feature on the waveform—or so it was argued.

These two classes of potential artifact are discussed in detail in a piece by Sir Andrew Huxley [1986], written in rebuttal of a review article on stepwise shortening [Pollack, 1986]—the two published back-to-back.

Is stepwise shortening an artifact of inappropriate use of optical diffraction?

As with most controversial results, acceptance rests on validation by alternative methods, and by confirmation in other laboratories. In the case of stepwise shortening, the result has been confirmed with several alternative methods (see below)—including simultaneously applied pairs of methods—and with still other methods applied in several independent laboratories.

The first alternative method was high-speed cinemicroscopy. The striations of contracting-frog-skeletal muscle were photographed at 4,000 frames per second. Frame-by-frame analysis revealed stepwise shortening [Delay et al., 1981]. Shortening waveforms were similar (though not identical) to those obtained by diffraction during the same contraction. The high-speed cine method was also applied in another laboratory, and the steps were confirmed [Toride and Sugi, 1989]. In this latter instance, translation-based artifacts were circumvented: the same striations could be visualized throughout the course of contraction, and the steps remained in evidence.

A second alternative method used an electronic sensor that captured the striation pattern in real time [Myers et al., 1982]. Mean striation spacing within a designated spatial window was computed on the fly using a phase-locked loop circuit. The advantage of this method is not only high temporal and spatial resolution but the ability to sample a far smaller field than with diffraction. The results were again stepwise [Jacobson et al., 1983]. In fact, evidence was found that implied quantal step size.

Although this latter method is theoretically sensitive to translation artifacts [Altrignham et al., 1984; but see Pollack, 1984, followed by Huxley, 1984], the method was recently applied in a situation where such artifacts were precluded. When single isolated myofibrils were examined near their fixed end, where translation was demonstrably negligible, stepwise shortening was noted [Bartoo et al., 1988]. Thus, the steps do not arise out of translation artifact. They do arise at the level of the myofibril.

A third alternative method complements all striation-based approaches by measuring the interval between surface markers positioned along the length of the fiber. Following the pioneering work of Gordon et al. [1966], Granzier et al. [1987] placed surface markers along single isolated skeletal muscle fibers and followed the time course of spacing between hairs as the fiber actively shortened. Steps and pauses were clearly noted in these segments. Optical diffraction, applied simultaneously, revealed a largely similar stepping pattern. Because the two methods are so different, and yet give smaller results, the steps were deemed unlikely to arise from hidden methodological artifact.

Additional segment measurements carried out in independent laboratories with different types of markers have also revealed stepwise shortening [Housmans, 1984; Tameyasu et al., 1985]. Although measurements were not as extensive as those cited above, the character of the records is qualitatively similar.

Finally, a fourth alternative method relied on measurements of shortening of the fiber itself. When isometrically contracting single fibers are suddenly released to a load below isometric, fiber shortening often shows a series of step-like fluctuations. Generally, these fluctuations die out [Civan and Podolsky, 1966], but in occasional instances they continue undamped throughout the period of shortening [Granzier et al., 1990]. Such fluctuations correlate well with steps measured along the fiber by optical diffraction.

Evidently, diffraction-based measurements of stepwise shortening are confirmed by a phalanx of alternative methods. Although each method could have its own source of hidden, step-generating artifact (the obvious sources of artifact were checked), the fact that steps were observed when pairs of methods were used simultaneously implies that hidden gremlins are extremely unlikely.

Why then are steps often obliterated when white light diffraction is used [Goldman, 1987]? How is this anomaly explained if diffraction-based steps are genuine?

One possible route to reconciliation rests on the domain theory [Sundell et al., 1986]. Suppose the region illuminated by the laser contained several domains, each containing striations inclined at a different angle to the fiber axis. Depending on the angle of incidence of the illuminating beam, some domains will contribute more than others to the diffraction pattern. When contributions from a series of incident angles are summed, the collective behavior of all domains will be represented. If the steps in each domain are not perfectly synchronized with one another, the summed waveform will be smoother than the one obtained using a single angle of incidence.

This hypothesis may explain why steps are often obliterated when non-collimated white light is used [Goldman, 1987]. The explanation presupposes less than perfect synchrony—a supposition that follows naturally from the observation (with all methods) that not all shortening waveforms are stepwise. It also explains why tracking of one first order can sometimes yield a shortening waveform different from that obtained by tracking the other first order [Pollack et al., 1984].

Stepwise shortening is evidently a basic feature of striated muscle—and it is not restricted to muscle alone. Other motile systems operate in stepwise fashion. The bending of cilia has been shown to occur in discrete steps [Baba, 1979], and kinesin-studded microspheres moves past tubulin in steps [Gelles et al., 1988]. In the latter case at least, step size correlates with molecular dimensions. Thus, step-like action may be a central feature of all biological motile systems.

VII. SUMMARY AND CONCLUSIONS

I have dealt with only two of the many types of optical scattering methods that have been applied to muscle contraction: striation imaging and optical diffraction. In each case, controversy has arisen, largely on the basis of methodological issues.

In the area of striation imaging, the work A.F. and H.E. Huxley in the mid-1950s appeared to have overturned the conclusions of the first half of the twentieth century. But closer examination reveals little or no difference. A-band shortening during active sarcomere shortening was noted consistently, and has been reinforced by complementary electron micrographic evidence of thick filament shortening. These observations contradict the prevailing view, and X-ray diffraction considerations do not mitigate the contradiction [cf. Pollack, 1990]. It appears that thick filaments/A-bands shorten during contraction.

In the realm of optical diffraction, the results of high-resolution applications were first challenged when they failed to support the prevailing view. Length-tension relations were inexplicably flat. Shortening was stepwise. Hypotheses were advanced to imply that diffraction-based results could not be unambiguously interpreted, and that the diffraction-based results were therefore invalid. While these hypotheses certainly have intrinsic merit, they fail to compromise the two results cited above, for similar results have been obtained in different laboratories through use of multiple independent methods. The length-tension curve (obtained when sarcomeres generate maximal tension) is relatively flat, and the shortening waveform is stepwise.

These conclusions are fundamental. they have major impact on theories of how muscles contract [Pollack, 1990].

VIII. REFERENCES

Altringham JD, Bottinelli R, Lacktis JW (1984) Is stepwise sarcomere shortening an artifact? Nature 307:653–655.

Baatsen PHWW, Trombitas K, Pollack GH (1988) Thick filaments of striated muscle are laterally interconnected. J Ultrastruct Mol Struc Res 98:267–280.

Baba S (1979) Regular steps in bending cilia during the effective stroke. Nature 282:717–720.

Bartoo ML, Tameyasu T, Burns DH, Pollack GH (1988) Stepwise shortening in single myofibrils. Biophys J 53:370a.

Civan MM, Podolsky RJ (1966) Contraction kinetics of striated muscle fibres following quick changes of load. J Physiol 184:511–534.

Delay MJ, Ishide N, Jacobson RC, Tirosh R, Pollack GH (1979) Stepwise sarcomere shortening: Analysis by high-speed cinemicrography. Science 213:1523–1525.

Gelles J, Schnapp BJ, Sheetz MP (1988) Tracking kinesin-driven movements with nanometre-scale precision. Nature 331:450–453.

Goldman Y (1987) Measurement of sarcomere shortening in skinned fibers from frog muscle by white light diffraction. Biophys J 52:57–68.

Gordon AM, Huxley AF, Julian FJ (1966) The variation in isometric tension with sarcomere length in vertebrate muscle fibres. J Physiol (London) 184:170–192.

Granzier HLM, Mattiazzi A, Pollack GH (1990) Sarcomere dynamics during isotonic velocity transients in single frog muscle fibers. Am J Physiol (in press).

Harrington WF (1971) A mechanochemical mechanism for muscle contraction. Proc Natl Acad Sci USA 68 (3): 685–689.

Horowitz A, Caljouw C, Pollack GH (1989) Length-tension relation for sightly shortened sarcomeres (1-3%) is higher than for isometric sarcomeres. Proc IUPS (Helsinki), in press.

Horowitz A, Caljouw CJ, Pollack GH (1989) Force-length relations of "almost isometric" sarcomeres. Biophys J 55:409a.

Housmans P (1984) Discussion in Pollack GH, Sugi H (eds): Contractile Mechanisms in Muscle. New York: Plenum Press, pp 782–784.

Huxley AF (1957) Muscle structure and theories of contraction. Prog Biophys Biophys Chem 7:255–318.

Huxley AF (1980) Reflections on Muscle. Liverpool: Liverpool University Press.

Huxley AF (1983) Reflections on Muscle. Sherrington Lectures XIV. Liverpool: Liverpool University Press.

Huxley AF (1984) Comments on "Is stepwise sarcomere shortening an artifact?"—a response. Nature 309:713–714.

Huxley AF (1986) Comments on Quantal mechanisms in cardiac contraction. Circ Res 59:9–14.

Huxley AF, Niedergerke R (1954) Structural changes in muscle during contraction. Interference microscopy of living muscle fibres. Nature 173:971–973.

Huxley AF, Niedergerke R (1958) Measurement of the striations of isolated muscle fibres with the interference microscope. J Physiol 144:403–425.

Huxley HE (1969) The mechanism of muscular contraction. Science 164:1356–1366.

Huxley HE, Hanson J (1954) Changes in the cross-striations of muscle during contraction and stretch and their structural interpretation. Nature 173:973–976.

Huxley HE, Hanson J (1955) The structural basis of contraction in striated muscle. Symp Soc Exp Biol 9:228–264.

Iwazumi T (1970) A new theory of muscle contraction. Ph.D. dissertation, University of Pennsylvania.

Iwazumi T, Pollack GH (1979) On-line measurement of sarcomere length from diffraction patterns in cardiac and skeletal muscle. IEEE Trans Biomed Eng 26(2):86–93.

Jacobson RC, Tirosh R, Delay MF, Pollack GH (1983) Quantized nature of sarcomere shortening steps. J Mus Res Cell Motil 4:529–542.

Julian FJ, Morgan DL (1979) The effect on tension of non-uniform distribution of length changes applied to frog muscle fibers. J Physiol 293:379–392.

Magid A, Ting-Beall HP, Carvell M, Kontis T, Lucaveche C (1984) Connecting filaments, core filaments, and side-struts: A proposal to add three new load-bearing structures to the sliding filament model. In Pollack GH, Sugi H (eds): Contractile Mechanisms in Muscle. New York: Plenum Press, pp 307–332.

Maruyama K (1976) Connectin, an elastic protein from myofibrils. J Biochem 80:405–407.

Myers J, Tirosh R, Jacobson RC, Pollack GH (1982) Phase-locked loop measurement of sarcomere length with high time resolution. IEEE Trans Biomed Eng BME-29(6):463–466.

Periasamy A, Burns DH, Holdren DN, Pollack GH, Trombitas K (1990) A-band shortening in single fibers of frog skeletal muscle. Biophys J 57:815–828.

Pollack GH (1983) The cross-bridge theory. Physiol Rev 63(3):1049–1113.

Pollack GH (1984) Response to "Is stepwise sarcomere shortening an artifact?" Nature 309(5970):712–714.

Pollack GH (1986) Quantal mechanisms in cardiac contraction. Circ Res 59:1–8.

Pollack GH (1990) Muscles and Molecules: Uncovering the Principles of Biological Motion. Seattle: Ebner and Sons (in press).

Pollack GH, Iwazumi T, Ter Keurs HEDJ, Shibata EF (1977) Sarcomere shortening in striated muscle occurs in stepwise fashion. Nature 268:757–759.

Pollack GH, Tirosh R, Brozovich FV, Lacktis JW, Jacobson RC, Tameyasu T (1984) Stepwise shortening: Evidence and implications. In Pollack GH, Sugi H (eds): Contractile Mechanisms in Muscle. New York: Plenum Press, pp 765–786.

Rüdel R, Zite-Ferenczy F (1979) Do laser diffraction studies on striated muscle indicate step-wise sarcomere shortening? Nature 278:573–575.

Sundell CL, Goldman YE, Peachey LD (1986) Fine structure in near-field and far-field laser diffraction patterns from skeletal muscle fibers. Biophys J 49:521–530.

Suzuki S, Pollack GH (1986) Bridgelike interconnections between thick filaments in stretched skeletal muscle fibers observed by the freeze-fracture method. J Cell Biol 102:1093–1098.

Tameyasu T, Toyoki T, Sugi H (1985) Non-steady motion in unloaded contractions of single frog cardiac cells. Biophys J 48:461–465.

Ter Keurs HEDJ, Iwazumi T, Pollack GH (1978) The sarcomere length-tension relationi skeletal muscle. J Gen Physiol 72:565–592.

Tirosh R, LIron N, Oplatka A (1979) A hydrodynamic a mechanism for muscular contraction. In Sugi H, Pollack GH (eds): Cross-Bridge Mechanism in Muscle Contraction. Tokyo: University of Tokyo Press.

Toride M, Sugi H (1989) Stepwise sarcomere shortening in locally activated frog skeletal muscle fibers. Proc Japan Acad 65. Ser B, No. 3, 49–52.

Trombitas K, Baatsen PHWW, Pollack GH (1988) I-bands of striated muscle contain lateral struts. J Ultrastruct Mol Struct Res 100:13–30.

Wang K (1984) Cytoskeletal matrix in striated muscle: the role of titin, nebulin and intermediate filaments. In Pollack GH, Sugi H (eds): Contractile Mechanisms in Muscle. New York: Plenum Press.

Noninvasive Techniques in Cell Biology: 327–352
© 1990 Wiley-Liss, Inc.

13. Macrospectrofluorometric Methods for Studying Cellular Activation and Function

Bruce Seligmann

Department of Inflammation and Osteoarthritis Research, Ciba-Geigy
Pharmaceuticals Division, Summit, New Jersey 07901

I. INTRODUCTION

Light spectroscopy permits investigations of cellular metabolic, activation, and functional states through the use of indirect probes or direct measurements of spectroscopic properties of cellular components. This review covers the application of spectrophotometric methods to the study of cellular function. Because this topic encompasses numerous cell types, the reviewed material will concentrate on recently reported methodology and techniques which are currently in wide use or in which there is increasing interest. This review is divided into two broad classifications of related approaches: fluorescence and chemiluminescence. In a review of this type it is difficult to avoid simply cataloguing methods, and it is not feasible to present and assess the actual data which has been obtained with these methods in all the cell types studied. Therefore the review is organized according to specific types of studies, methodology for each is discussed, and applications appropriate for a variety of cell types and preparations are discussed. Microscopic and cytometric methods are discussed elsewhere and are not included in this chapter.

II. FLUORESCENCE

A. Membrane Potential Studies

Spectroscopic measurement of membrane potential has been based primarily on the use of two types of lipophilic probes, the positively charged cyanine dyes such as 3,3′-dipentyloxacarbocyanine [diO-C_5(3)] and the negatively charged oxonol dyes such as bis-[1,3-diisopropyl-2-barbiturate-(5)]-trimethin oxonol [di-isoC$_3$-BA(3)]. The distribution of both the cyanine and oxonol dyes depends upon the plasma membrane potential.

With regard to cyanine dyes there are at least two mechanisms which can be taken advantage of to measure the amount of probe inside cells, and hence membrane potential [Seligmann et al., 1980; Seligmann and Gallin, 1983; Waggoner, 1976]. The initial mechanism described in the literature was a phenomenon of self-aggregation that occured when the concentration of probe reached a critical level and that resulted in quenching of fluorescence. In this type of experimental protocol a high extracellular concentration of probe is used. As dye accumulates in the cells, the intracellular concentration rapidly rises above the critical concentration, causing aggregation and a loss of total fluorescence. This loss of fluorescence is proportional to the membrane potential gradient across the plasma membrane. The second mechanism of dye action occurring at concentrations below that at which aggregation occurs is the enhancement of the fluorescence of dye molecules in the cytosol of cells, relative to that in the extracellular aqueous buffer. By this method the fluorescence increases proportionately as the cellular membrane potential becomes more negative and dye is accumulated. Special care must be taken to assure that there is no cross-over of mechanisms. If the study is designed to use the cyanine dyes in the fluorescence enhancing mode, the intracellular concentration of probe should not reach levels where it begins to form aggregates and become quenched upon depolarization of the membrane potential. If this occurs the results will be uninterpretable. To determine that the cyanine dyes are not responding by a mixed fluorescence enhancement/quenching mechanism, the cellular response should be measured over a range of dye concentrations, from the minimum giving detectable membrane potential changes to the maximum causing fluorescence quenching. A concentration of dye should be selected that guarantees only a single mode of action [Seligmann and Gallin, 1982]. Most investigators currently use the fluorescence enhancement mechanism to keep intracellular dye concentrations below toxic levels. Oxonol dye fluorescence operates by the same solvent enhancement mechanism as described for cyanine dyes.

The measurement of membrane potential with either class of fluorescent dye is straightforward using a single excitation wavelength and a single emission wavelength spectroscopy. However, there are many technical considerations and limitations in the use of indirect probes to measure membrane potential. The probes tend to be toxic, inhibiting numerous functions in a variety of cell types [Seligmann and Gallin, 1983; Wilson et al., 1985; Miller and Koshland, 1978; Montecucco et al., 1979; Simons, 1979; Rink et al., 1980]. Usually toxicity can be avoided by using sufficiently low concentrations, but then the quenching method of measuring cyanine dye accumulation cannot be used. It is important that investigators determine for their specific conditions whether the dyes have any inhibitory effects. To date there has been no indication that the measurement of membrane potential is affected by these toxic effects of (high) dye concentraiton. The choice between using either a cyanine or oxonol dye may be based on differential toxicity [Wilson et al., 1985; Wilson and Chused, 1985; Rink et al., 1980].

Besides having an awareness of toxicity, it is important to calibrate the dependence of dye fluorescence on membrane potential on a routine basis. Quite significant changes in fluorescence can occur, resembling stimulated membrane potential changes, which are not membrane potential dependent [Seligmann et al., 1988]. Affecting this is the dye/cell ratio. With both the cyanine and oxonol dyes there is a dye/cell concentration ratio which is optimal and at which artifacts are minimized. To account for day-to-day variation it is best to calibrate the membrane fluorescence response in each experiment. With the cyanine dye this is readily accomplished by preparing one cuvette of cells in high potassium (120 mM) containing buffer and another in physiologic (5 mM) potassium and then recording fluorescence before and after addition of valinomycin and/or a representative stimulus. The results should appear as depeicted in Figure 1A. These results were obtained using neutrophils at a concentration of 0.5 million cells per ml and the cyanine dye diO-C_5(3) at a concentration of 50 nM. We have observed that it is possible to have a dye/cell ratio for which the resting fluorescence of the cell appears to be quite sensitive to membrane potential, and yet the response of cells driven by a depolarizing stimulus contains a significant artifact which is not membrane potential sensitive (Fig. 1B). These results were obtained using the same conditions as the experiment shown in Fig. 1A with only one change, increasing the cell concentration to 2 million cells per ml [Seligmann et al., 1988]. The origin of this membrane potential insensitive component is not clear, but by varying the dye/cell ratio it is possible to arrive at conditions where 75% to 90% of the recorded fluorescence is membrane potential sensitive.

A.

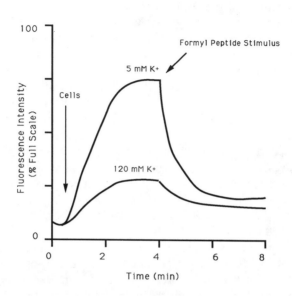

B.

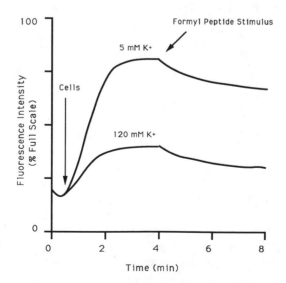

The classical method of calibrating cyanine dye fluorescence, whether used in a fluorescence quenching or enhancing mode, is the "null point" method [Seligmann et al., 1980; Laris et al., 1976; Philo and Eddy, 1978]. In this method the extracellular potassium concentration is varied from 5 to 120 mM in discrete steps. The fluorescence at each level of potassium is recorded before and after addition of the potassium ionophore valinomycin. This ionophore increases potassium permeability to the point where other endogenous ion permeabilities become insignificant, converting the cell membrane potential to a purely potassium regulated state described by the Nernst equation:

$$\text{Membrane potential} = (RT/F) \ln([K^+_{ext}]/[K^+_{int}])$$

It is necessary to make an estimate of the intracellular potassium concentration. With this value the relationship between membrane potential (calculated from the Nernst equation) to dye fluorescence in the presence of valinomycin can be plotted. This fluorescence calibration curve can then be used to determine the resting membrane potential and changes in membrane potential induced by the stimulation of cells.

Calibration of oxonol dye fluorescence is complicated because there is an interaction between the oxonol dyes and valinomycin. If the resting membrane potential of the cells being investigated is high, then calibration can be achieved by addition of gramicidin or calcium ionophore to drive the membrane potential to the sodium or calcium equilibrium potential. This must be accompanied by manipulation of the extracellular sodium or calcium concentration to demonstrate that in the presence of these ionophores the fluorescence changes are actually sensitive to membrane potential [Seligmann and Gallin, 1983; Simons, 1979]. Care should be exercised if any of these ionophores stimulate an active response from the cells being investigated. However, in practice it is much more difficult to calibrate oxonol dy fluorescence than cyanine dye fluorescence.

Besides the problem of compensating for membrane potential insensitive fluorescence, there are inherent problems with regard to buffers when using both the cyanine dyes and oxonol dyes in the fluorescence-enhancing mode: namely, addition to the buffer of certain agents and proteins, such as albumin, substantially increases the fluorescence of the dye in the extracellular media. Since there is no physical discrimination of extracellular from intracellular

---

Fig. 1. Membrane potential sensitive and insensitive cyanine dye fluorescence. Cells were added to a cuvette containing di-O-C_5(3) (50 nM) and either a 5 mM or 120 mM potassium buffer. After equilibration the cells were stimulated with the formyl peptide N-formyl-methionylleucyl-phenylalanine (1 μM). **A:** The membrane potential dependent response obtained when the cell concentration used was 0.5 million cells per ml. **B:** The membrane potential insensitive response obtained when the cell concentration was 2 million cells per ml.

dye fluorescence in the cuvette format of spectroscopy, raising the fluorescence of dye in the extracellular media without a comparable increase in the fluorescence of dye within the cells reduces the detectable membrane potential dependent fluorescence and may mask changes in intracellular dye concentration. Selection of dyes which have a maximal fluorescence enhancement (such as di-isoC$_3$-BA(3) as an oxonol dye) or modification of the buffer are the only ways to avoid this interference. We have found that stimuli which normally must be dissolved in albumin to prevent aggregation or loss due to sticking tubes, can instead be dissolved in tween 20 (0.1% and lower), which does not affect dye fluorescence.

Charged dyes bind to intracellular sites, and may be sequestered. The postively charged cyanine dyes are particularly plagued by accumulation in mitochondria which have a membrane potential more negative than the plasma membrane potential [Laris et al., 1975; Johnson et al., 1981]. Thus, in cells with numerous mitochondria, measurements with cyanine dyes reflect a component due entirely to the mitochondrial membrane potential. In these instances the mitochondria must either be poisoned with a protonophore or other mitochondrial poison such as a combination of oligomycin, antimycin, and 2,4-dinitrophenol, or the measurements must be made using a negatively charged oxonol dye which will not be accumulated by mitochondria [Wilson and Chused, 1985; Philo and Eddy, 1978].

The manner in which cells are handled will affect their ionic state and hence membrane potential. Sufficient equilibration time at 37°C must be allowed not only to permit equilibration of the probe but also to allow equilibration of ions. We, and others (unpublished), have observed that even when the dye fluorescence reaches a steady state, ionic mechanisms may remain out of balance, leading to bogus results upon stimulation. An easy check is to verify that the response obtained after the minimum incubation used is identical to the response obtained at various times during prolonged incubation (i.e., after 20, 30, 40 min). It is generally a good practice to check the response of cells periodically during the assay with any of the methods reviewed in this chapter, if only to confirm that the responsiveness of the cells is not changing over the time that the experiment is being conducted.

In summary, there are two oppositely charged probes which can be used to measure membrane potential. Either probe may cause unique problems which preclude their use, with toxicity and mitochondrial interferences being the greatest concerns. However, in situations where either probe can be used cyanine dyes are the better choice because they give greater changes in fluorescence, are easier to calibrate, and are less subject to cuvette equilibration artifacts.

B. Ion Studies

The spectrophotometric measurement of intracellular ion concentrations depends upon the use of indirect probes which chelate the ion of interest and exhibit a change in absorbance or fluorescence following the binding of the ion. The first generally used probes, aequorin and arsenazo for measuring calcium concentration based on absorbance, were of limited value because they had to be injected or loaded into permeabilized cells. Measurement of calcium binding sites has been achieved with very limited success using terbium, which fluoresces as a result of energy transfer when the protein to which it is bound absorbs light at 280 nm excitation. Chlortetracycline has also been used to measure bound calcium, and while it is cell permeable there are no mechanisms to permit its accumulation and retention by cells. Thus there is a continuous leakage of chlortetracycline out of the cell during measurement, and there is no way to quantitate the fluorescence signal with respect to calcium concentration.

Ion studies have experienced a tremendous growth since the introduction of "sequestered" cell-permeant fluorescent probes to study the free intracellular calcium concentration. The first of the fluorescent probes was Quin-2 [Tsien et al., 1982], which upon chelation of calcium exhibits enhanced fluorescence. The basic chemistry upon which this initial probe was based has been applied subsequently to an extended family of probes. Chelation of positively charged ions is achieved by the carboxylate anion groups of the molecule through carboxylic acid residues. Incorporation of probe into cells is accomplished simply by incubating the esterified form of the molecule with cells. Once the esterified molecule enters the cell it is converted to the calcium-binding free acid following complete cleavage of the esters from the molecule by endogenous cellular esterases. The free acid is not only capable of binding calcium, but is impermeant and trapped inside the cells. Probes are commercially available to measure calcium, magnesium, sodium, potassium, and chloride based on a variety of wavelength combinations. Besides simple enhancement of fluorescence upon binding of select ions, several probes exhibit a significant shift in either the excitation or emission maximum upon ion binding. Thus three general types of measurement protocols are in use. The simplest measures probe fluorescence at a single excitation wavelength and single emission wavelength. Fluorescence intensity is at a minimum in the absence of calcium and reaches a maximum in the presence of calcium (Quin-2-type probes). The other two methods are based on probes which exhibit either a shift in their excitation maximum (dual excitation, Fura-2-type probes) or emission maximum (dual emission, Indo-1-type probes) upon chelation of calcium [Grynkiewicz et al., 1985].

The requirement to measure fluorescence at multiple excitation and emission wavelengths has been met by the development of a new generation of fluorescence instruments. There are give general configurations. When necessary to excite at a single wavelength and minotor two wavelengths, a "T" format or split beam emission format can be used which permits continuous monitoring of two emission wavelengths. Selection of wavelength can be accomplished either with monochromators or filters. Alternatively, an entire range of emission wavelengths can be continuously monitored using a diode array spectro-photometer, though these instruments currently lack the sensitivty to handle turbid cell suspensions, due to light scatter. When the choice of probes requires multiple excitation wavelengths, two basic methods can be used. The method giving the greatest time resolution is to ''chop'' the emission from two different monochromators or filter systems with a mechanical device, sequentially pass-ing each wavelength to the sample compartment at alternating rates as high as 5,000 Hz. Alternatively, a single monochromator can be programmed to slew sequentially between selected wavelengths at rates up to several hundred nm/s. The advantage of this latter method is that more than two wavelengths can be used, but the disadvantage is that current configurations limit acquisition cycle time to approximately 2 Hz (for Fura-2). Hybrid systems are available which permit the combination of both chopping and high speed monochromator slewing.

Irrespective of instrumentation, the use of each probe has specific limita-tions and features with which the investigator must be familiar. Any indirect method is very susceptible to artifact. Therefore, the limitations and precau-tions which must be used are discussed in so far as is currently understood by the scientific community. The majority of this discussion centers around the permeant calcium-sensitive probes, but the discussion is applicable to all of these probes irrespective of the ion being measured. To convert fluorescemce measurements quantitatively into ion concentration on fluorescence must be calibrated. With single-wavelength probes such as Quin-2, Fluo-3, and Rhod-2, this calibration must be performed with each cuvette of cells since any varia-tion in the intracellular concentration of the probe changes and the correlation between fluorescence and calcium concentration. The intracellular probe con-centration is different for each batch of loaded cells and can change during the assay while the cells are stored before use. The calibration is performed by making the intracellular trapped probe accessible to external ionic man-ipulation of calcium conentration, either by lysing the cells to release the probe, or by use of a calcium ionophore to permit the free equilibration of calcium across the plasma membrane. The latter has particular problems should the ionophore or calcium entry activate the cells, or should the cell possess highly active mechanisms (Ca ATPase's) to sequester calcium or pump it back out of the cell. In this case it is necessary to poison the Ca ATPase and/or use supramaximal concentrations of ionophore. For these reasons the method of choice for the types of spectroscopic studies discussed in this section is the lysis of cells and release of intracellular probe into the extracellular medium. Once the intracellular probe is made available to the manipulation of extracellular calcium, measurements of the fluorescence in the absence of calcium and under conditions of calcium saturation are made and the following equation can then be employed to calculate the concentration of calcium at any measured fluorescence:

$$[Ca] = K_d (F-F_{min})/(F_{max} - f)$$

The K_d is the affinity of the probe for calcium, F_{min} is the fluorescence in the absence of calcium, F_{max} is the fluorescence in the presence of a saturating amount of calcium, and F is the experimental fluorescence value.

A similar approach is used when dual-wavelength probes are employed. However, in this case advantage is taken of the fact that measurements are made at two wavelengths which are differentially affected by calcium. Typical of these probes is Fura-2, for which the calcium dependence of the excitation spectrum is shown in Figure 2. Three things are to be noted. Fluorescence increases as a function of calcium concentration at wavelengths below approximately 365 nm (noting that unless the spectra are corrected each instrument will produce somewhat different wavelength numbers). The fluorescence at 365 nm is not sensitive to calcium concentration, and is the isosbestic point for this probe. Fluorescence decreases as a function of increasing calcium concentration at wavelengths above 365 nm. Using a ratio measurement of the fluorescence at 350 nm divided by fluorescence at around 380 nm, a similar method of calibration can be used as described above which permits correction for differences in probe concentration:

$$[Ca] = K_d (S)(R - R_{min})/(R_{max} - R)$$

In this equation R is the experimental 350/380 ratio value, R_{min} is the ratio in the absence of calcium, R_{max} is the raio in the presence of a saturating amount of calcium, and S is the fluorescence ration at 380 nm derived from the fluorescence in the absence of calcium divided by the fluorescence in the

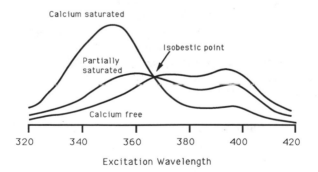

Fig. 2. Calcium-depe..dent Fura-2 excitation spectra. The excitation spectra of Fura-2-free acid was obtained in the presece of 2 mM EGTA (calcium-free tracing), following addition of a subsaturating amount of calcium (partially saturated tracing), or following addition of a saturating amount of calcium (calcium-saturated tracing). The isosbestic point, independent of calcium concentration, is indicated.

presence of calcium. The actual 350/380 ration "R" will change as a function of probe concentration. Provided the cell number is kept constant, the correction factors R_{min}, R_{max}, and S will compensate for changes in Fura-2 concentration from one batch of cells to another. If ratio values are to be compared between different preparations of cells, or if accurate calibration is not possible (as in tissues) when the isobestic point must be used as one of the wavelengths in the ratio (i.e., use the 350/365 ratio).

The quality of reagent is crucial. Incompletely esterified material will not enter the cell properly. Incompletely de-esterified material may not bind calcium at all or do so with altered affinity, and may redistribute into subcellular compartments. Thus, to assure consistent results it is important to verify the equivalency of reagents when changing batches or manufacturers. This type of information can be obtained from nuclear magnetic resonance spectra. However, a side-by-side comparision of new probe with old in the same batch of cells can give the necessary functional information as well. Not only should the loading be compared by determining F_{min}, F_{max}, and S values, but the response to stimulus should be determined and a comparative calcium saturation curve should be determined using the material freed from cells following lysis and titrating with calcium in the presence of a fixed amount (2 nM) of EGTA.

There are a number of additional concerns which can have a major impact on the results obtained using these approaches. First, the probes all act by chelating a portion of the free intracellular calcium, and therefore add buffering capacity to the cell. It is thus desirable to use the minimum amount of probe which gives reliable measurement of fluoresence since not only may the resting level of calcium be raised artificially, but calcium transients will be buffered, altering their kinetics and magnitude. In addition, the cleaved esters of acetoxymethyl ester derivatized probes are converted intracellularly to formaldehyde, which may be toxic. For example, for every molecule of Fura-2 trapped inside the cell five molecules of formaldehyde are generated. A control for these effects should be performed by varying the amount of probe loaded into the cells over a reasonable range while measuring representative cellular function(s) and noting any changes in the calcium responses. Instrument sensitivity, background fluorescence and cell autofluorescence ultimately determine the minimum probe concentration to be used.

Besides calcium buffering, the redistribution of de-esterified probe is a concern. The data indicate that these probes can leak out of the cell into the calcium rich extracellular medium, as well as possibly become sequestered in intracellular compartments. The possibility that either of these processes are occurring must be examined with each cell preparation and stimulus condition, since the latter may alter the intracellular and membrane dynamics. The presence of probe in the extracellular media can be determined by examining whether addition of a calcium chelator such as EGTA has an immediate effect on recorded fluorescence. Alternatively, since EGTA may have some immediate effects on intracellular calcium levels, manganese has been used to measure the contribution of extracellular probe to total fluorescence. Manganese binds to the calcium probes without causing a change in fluorescence. While there is risk that manganese may block calcium transport mechanisms and/or reach the cell interior, conditions have been published using manganese to eliminate the contribution of extracellular Fura-2 calcium binding from the measurements. It must be noted that quantitative calculations remain compromised unless account is taken of the contribution to the (380 nm) fluorescence of this calcium desensitized pool to total intracellular and extracellular (380 nm) fluorescence McDonough and Button, 1989].

The sequestration into subcellular compartments of anionic probes such as Fura-2 has been demonstrated [McDonough and Button, 1989; Arkhammar et al., 1989; Di Virgilio et al., 1988]. This is due apparently to anion transporters, and is dependent upon the cell type and the specific probe being used. However, it appears that the process can be blocked using an anion transport inhibitor, probenecid (1 mM), though the effect of probenecid on each specific cell type and response must be assessed before including this agent in the assay medium.

Both Fura-2 and Indo-1 (and presumably any esterified probe) may be incompletely hydrolyzed within the cells. When this occurs the probe cannot chelate calcium, yet is trapped within the cell and contributes to fluorescence as a calcium insensitive component. For Indo-1 this interference is in the region of the 480 nm wavelength typically used in the measurement [Luckhoff, 1986]. For Fura-2, the interference affects the entire spectrum [Scanlon et al., 1987; Oakes et al., 1980; Highsmith et al., 1985]. Thus the cell preparations should be examined for calcium-insensitive fluorescence. This is done by comparing the calcium-saturated minus calcium-depleted difference spectra of authentic free acid to the spectra of cell-derived free acid. If there is a significant calcium insensitvie component the difference spectra will not be the same. In addition the isobestic point is likely to be shifted since this is a unique colligative property of each molecular species, and in the case of a calcium insensitive component, there are two fluorescent species present. If there is a significant calcium-sensitive component to the fluoresence, then this must be subtracted before calculating any ratio values.

Other factors which are important to consider is whether autofluorescence and stimulus-induced changes in autofluorescence or light scattering affect the fluorescence measurements. The effect of autofluorescence and light scatter can be devastating, particularly if the autofluorescence artifact is seen with intact cells and not with the lysed cell preparation used for calibration. A uniform change in autofluorescence affecting all wavelengths equally can be compensated for by employing the isobestic point as one of the wavelengths used in the calculation of the ratio, subtracting any change in fluorescence at the isobestic point from the fluorescence measurements before calculation of the ratio. However, if autofluorescence intensity is not the same at all wavelengths, or if stimulation causes extensive changes in endogenous fluorescence or absorbance (flavin, etc.), then wavelength-specific interferences may have a significant impact on measurements of probe fluorescence. This is particularly true at low loading concentrations of probe. While it may still be possible to correct for these interferences using a multi-point correction routine, it may also become necessary to raise probe concentration even though this may increase the intracellular buffering of the ion being measured. Regardless, as in the correction for calcium-insensitive fluorescence, autofluorescence artifacts must be subtracted before calculating any ratio values.

The last major variable in the use of cell-permeant probes is the method used to "load" the probes into the cells. Variations abound, and each one is likely to have its appropriate place. However, it is also likely that many of the discrepant results which have been reported stem from differences in methods. These include how the cells are loaded with probe and stored before use or failure to control for sequestration. In addition, different intracellular concentrations of probe may cause differences in ion buffering. DMSO is a good solvent for these probes, and aqueous solutions made from DMSO stocks can be used for loading cells. Incorporation of Pluronic F-127 in the incubation media may facilitate loading [Drummond et al., 1987; Owen, 1988; Lee et al., 1987]. However, the major variations occur in the relationship of probe concentration to cell concentration during loading, temperature during loading, whether calcium is present during loading, whether sufficient time is allowed for complete hydrolysis following loading, and most important, whether sufficient time is permitted following storage (on ice or at room temperature) and before assay (at 37°C) to permit the cells to restore their intracellular ionic milieu. Loading cells (at 37°C) in the absence of extracellular calcium or storage at 4°C in the absence of calcium can result in depletion of intracellular calcium stores. Prolonged incubation at 37°C may be necessary for the cells to recover from this state, and in fact irreversible or poorly reversible effects on function may result.

Variability in the probe/cell ratio used for loading affects the final intracellular concentration of probe. Loading can be limited either by diffusion into the cells or by the rate of hydrolysis. If the rate of diffusion is limiting, then the length of incubation time and the extracellular probe concentration will determine the extent of loading. If instead the rate of hydrolysis is rate-limiting, then the situation is complicated and length of incubation becomes more important than probe concentration. The investigator should be aware of the factors controlling the loading of probe into the cells being studied. This can be accomplished by simply varying length of incubation and extracellular probe concentration in individual experiments, and by determining the spectrum and calcium sensitivity of intracellular material (released by lysis) at various times to determine that all the material loaded into the cells is de-esterified and sensitive to calcium.

In summary, the use of fluorescent probes to measure intracellular calcium is now in widespread use. Because the probes perturb the cells the investigator must be certain that the best conditions have been chosen to minimize these perturbations. In addition to calcium there are probes to measure intracellular hydrogen ions (pH), magnesium, sodium, potassium, and chloride [Negulescu et al., 1988; Moore et al., 1988; Murphy et al., 1989; Raju et al., 1989; Harootunian et al., 1988]. The same concerns pertain to the use of these probes. Since measurement of pH constitutes a large body of literature, it is dealt with separately below. Probes that measure other cations are all relatively new (SPFI, sodium; PBFI, potassium, mag-Fura 2, magnesium). In addition, mag-Indo-1 is a dual emission probe available for measuring magnesium. Chloride can be measured using the single-wavelength probe SPQ [Illsley and Verkman, 1987; Chen and Verkman, 1988; Chen et al., 1988].

C. pH Studies

The measurement of intracellular pH with fluorescent probes is a common method. Two classes of measurement exist—those for which a single excitation wavelength is used, and those which can be used in the ratio mode by measuring the fluorescence at two wavelengths. Taking advantage of the pH sensitivity of fluorescein, a family of fluorescein ester derivatives were developed which are trapped in cells as the free fluorescein acid [Grinstein et al., 1984, 1985]. The development of these probes preceeded and laid the foundation for the development of the calcium sensitive probes discussed above. Biscarboxyethyl-5(6)-carboxyfluorescein acetoxymethyl ester (BCECF) is one such derivative which has been used extensively. Calibration of pH-dependent fluorescence can be achieved by lysing the cells at the end of each experiment with Triton X-100 and titrating the released fluorescence with Tris buffer. However, intracellular solvent effects can cause a shift in fluorescence, leading to artifacts. This can be evaluated by scanning the fluorescence spectrum of intact loaded cells versus the lysate, and also by calibrating the fluorescence of intact cells without lysis. The hydrogen/potassium exchanging ionophore nigericin can be used for this purpose. In the presence of nigericin under conditions where the intracellular potassium concentration is equal to the extracellular potassium concentration, the intracellular proton concentration will be equal to the extracellular proton concentration. A correction factor can be calculated at the start of each experiment and used to permit the routine use of the lysate calibration method [Grinstein et al., 1984; Thomas et al., 1979]. There is one caution. Nigericin will bind to certain proteins, so inclusion of albuminin the media will prevent nigericin permeation into the cell membrane and will in fact extract it from the membranes.

As with the ion probes discussed above, the effect of BCECF on cell function must be determined. Formaldehyde is generated intracellularly upon hydrolysis of the probe [Mills et al., 1985].

Because these probes are all trapped within the cells and there is no extracellular probe present during the measurements, it is possible to assess the pH (and intracellular calcium concentration) of adherent cells in the spectrofluorometer format [Moolenaar et al., 1983; Mayer et al., 1989; Malgaroli et al., 1987]. Typically this has been accomplished by growing or adhering cells to coverslips which are then placed at an angle in a cuvette. Media can be exchanged, permitting a complex series of experiments to be conducted on the same cell sample. Depending upon the instrument and the degree to which scattered light is excluded from the emission measurement, the angle at which the specimen is mounted can be critical. A 45° angle scatters the maximum amount of light into the emission slit. In contrast, when the cover clip is mounted at a 30° angle, the amount of light scattered into the emission slit is minimized. An optional accessory we have found useful is a "Front Face Accessory," designed to record fluorescence spectra from solid samples capable of holding a cuvette. Cells can either be grown on one face of the cuvette, or on coverslips placed against the front face of the cuvette with the cell layer facing inward. In addition, we have used this accessory to record signals from live tissues. Responses of adherent cells grown on microcarrier beads and assayed in suspension have also been recorded [Hesketh et al., 1985].

A specialized application of pH studies is the measurement of phagosomal pH using fluorescein labeled, opsonized particles [Mayer et al., 1989]. Nonadherent particles are washed off the cells before initiating the study, leaving only the adherent and internalized particles. Since lysis could not be relied upon to release the particles from the phagolysozomes to permit individual calibration of each cuvette, the authors chose instead to construct a single calibration curve and normalize fluorescence between this curve and the cell samples by measuring fluorescence emission at two excitation wavelengths (450 nm and 495 nm). Calculating the ratio of these two wavelengths produced a value independent of particle concentration, and permitted the construction of a pH calibration curve. For this method to work either an isobestic point must be used, or the dependence (slope function) of fluorescence on pH must be different at each wavelength. An alternative approach is to use particles labeled with both the pH-sensitive fluorescein indicator and a pH-insensitive indicator such as rhodamine. The concentration of fluorescein can then be normalized based on rhodamine fluorescence.

In summary, the measurement of intracellular pH is straightforward. However, the availability of new pH indicators should greatly improve and simplify these studies. True dual emission probes are now available, SNARF 1 and derivatives SNARF 2, SNARF 6, and SNARF x [Whitaker et al., 1988]. Dual excitation wavelength probes are also available, SNAFL 1 and SNAFL 2.

D. Membrane Fluidity Studies

Membrane fluidity can be measured with fluorescent probes responding by two basic mechanisms. First, the freedom of rotation of molecules within the membrane fluid phase is a measure of fluidity. Rotational freedom of a fluorescent molecule can be determined by exiting the specimen with polarized light and recording the fractional depolarization of the emitted fluorescence [Adler and Tritton, 1988]. The greater the freedom of orientation, the greater the extent to which the polarized excitation light is depolarized [Van der Meer et al., 1986; Shinitzky and Barenholz, 1978; Beccerica et al., 1988; Adler and Tritton, 1988]. Diphenylhexatriene (DPH, ex 348 nm, em 426nm) has been used extensively for measuring membrane fluidity by this method. Alternatively, families of impermeant fluorescent molecules, which have hydrocarbon chain tails of various lengths and hydrophilic "head" portions, can be used to measure membrane fluidity based on the ability of these probes to enter but not to cross the plasma membrane. Long chain cyanine dyes have been used to measure membrane fluidity based on this principle. Whether the changes in fluorescence represent actual changes in membrane fluidity, local membrane reorganization, or changes in intracellular membrane components labeled by cell permeant probes (such as DPH), is unclear from the fluorescence changes alone. Increases in membrane surface area, which occur during secretion and blastogenesis, will increase the fluorescence when cell impermeant dyes are used [Bronner et al., 1986]. Measurements of lipis composition are necessary to confirm the biochemical basis for the monitored changes in fluorescence, but these probes represent a convenient way to obtain an indication that membrane remodeling is occurring. Once it is established what membrane reorganizational events are reflected by the fluorescence changes, these probes are a convenient method of following these changes.

Though not precisely a measure of fluidity, the movement of particles along the surface of the plasma membrane and the formation and internalization of patches of particles can be followed by fluorescence quenching and energy transfer. In this method a mixture of two fluorochromes, or a fluorochrome and chromophore, is inserted into the membrane at a concentration such that the probes are sufficiently separated in the plane of the membrane so that there is no optical interaction between them. However, the stimulated formation of patches containing these probes can result in an optical interaction between them. Depending on the specific fluorochromes/chromophore used, either fluorescence is quenched or fluorescence is generated which is proportional to the number of molecules in the patch. For example, if a chromophore is selected which absorbs at the emisison wavelength of a counter fluorochrome, then excitation of the fluorochrome will be quenched upon patching. Similarly, if two fluorochromes are used, one of which (designated "b") is excited at the emission wavelength of the other (designated "a"), then excitation of "a" will result in emission of "b" and reciprocal loss of emission of "a" upon patching.

E. Receptor Dynamics

Binding of ligands to cells and consequent signal transduction is an exciting area in cellular physiology. Measurement of ligand binding dynamics is a noninvasive technique because the ligand binding event is not perturbed by the measurement method. This an exciting and powerful application of fluorescence spectroscopy which is too seldom taken advantage of. Several basic types of studies have been pursued with macrospectrofluorometric methods. In all it is necessary that the ligand be coupled to a fluorochrome or synthesized as a fluorochrome. This is one of the major limitations of the method, since not all ligands lend themselves to this modification. Furthermore, each ligand is unique, and the methods must be adapted to each case.

Quenching of ligand fluorescence upon binding to the receptor is the simplest event which can be used to measure ligand-receptor interactions [Heidmann and Changeux, 1979b; Conti-Tronconi et al., 1982; Sklar et al., 1989]. Energy transfer between the receptor and ligand is a second method by which ligand-receptor interactions can be monitored [Heidmann and Changeux, 1979a]. A third, and perhaps more universally applicable, approach is based on use of an antibody to the fluorochrome which quenches the fluorochrome fluorescence upon binding. With this tool a binding assay can be developed so long as the antibody does not bind (or binds poorly) to the fluorochrome when the ligand is bound to its receptor. If this is the case, then the antibody will bind exclusively to unbound ligand and only quench the fluorescence of the free ligand. This principle has been exploited in studies of the binding of the formyl peptide ligand to phagocytic cells [Sklar et al, 1984; Sklar, 1987]. Addition of antibody to the reaction mixture results in the quenching of the fluorescence of unbound ligand, with no effect on the fluorescence of the ligand molecules bound to the cell surface. The amount bound at any time can thus be determined at the point antibody is added. Because antibody also prevents the binding of free ligand to its receptor, this protocol is identical to infinitely diluting the free ligand concentration, permitting the instantaneous and, subsequently, continuous measurement of ligand dissociation rates. For the results to be quantitative, it is essential to know the relationship between fluorescence intensity and ligand concentration. This relationship can be determined spectroscopically using the extinction coefficient of the fluorochrome to determine the concentration of fluorescent ligand, or through the application of mass spectroscopy. Reliance on fluorescent emission of the ligand/fluorochrome complex compared to standard solutions of fluorochrome are not reliable since the quantum yield of the fluorochrome may be altered when incorporated into the ligand molecule.

The antibody is not likely to quench fluorescence completely. Therefore, the extent to which the fluorochrome fluorescence is quenched must be determined in solutions containing the free ligand and varied amounts of added antibody. A correction factor can then be calculated to permit the accurate determination of amount of bound ligand remaining after addition of antibody.

A fourth approach to measure ligand binding results in a discrete change in polarization due to restricted rotation. The total change in polarization is directly proportional to the amount of ligand bound to the receptor. However, if additional steps in the ligand/receptor interaction involve additional changes in ligand mobility, further changes in polarization can occur. This method measures instantaneous binding events and is a very powerful technique. Exact calibration of number of binding events to degree of polarization is difficult, so that exact numbers of bound ligand are not calculated, but rather percent occupancy is used as the quantitative measure. Quenching methods combined with polarization, however, affords the ability to calculate the actual number of bound ligand molecules which give rise to a discrete amount of fluorescence depolarization. With polarization methods the instantaneous rate of binding can be determined at any point, including both the association rate and (in the presence of quenching antibody or infinite dilution) the dissociation rate [Sklar, 1987; Sklar et al., 1989; Cheung et al., 1984].

Fluorescence methods can also be used to obtain information about what residues interact between the ligand and its receptor, and what residues are confined within the receptor binding domain. These studies require the incorporation of fluorochromes at specific sites on the ligand molecule and the determination of changes in either fluorescence quenching, degree of fluorescence depolarization, or differential accessibility to anti-fluorochrome antibody of the receptor bound ligand. This last property has been used to define the receptor pocket for formyl peptides [Sklar et al., 1990].

F. Oxidative Metabolism Studies

Products of oxidative metabolism can be measured by fluorescence methods. In addition, the oxidative state of cofactors (i.e., flavin) can also be determined, but this will not be discussed here. Hydrogen peroxide can be measured through the peroxidase-catalyzed transfer of electrons from hydrogen peroxide to a fluorescent acceptor. Three methods based on this have been developed using different fluorescence substrates as electron acceptors: scopoletin (ex 350 nm, em 460 nm), homovanillic acid (ex 315 nm, em 420 nm), and para-hydroxyphenylacetic acid (ex 340 nm, em 400 nm) [Root et al., 1975; Metcalf et al., 1986; Hyslop and Sklar, 1984; Roos et al., 1983]. Some preparations of peroxidase contain glucose oxidase. Hence, when added to media containing glucose, superoxide/hydrogen peroxide is produced from this source, causing a significant artifact evidenced by a baseline change in fluorescence. This is solved by using a different source of peroxidase. This artifact is also easily diagnosed by simply omitting glucose from the assay buffer. If this manipulation eliminates the baseline fluorescence change, then the presence of glucose oxidase activity is implicated. Besides the measurement of hydrogen peroxide, total oxidant production can be measured by incorporating superoxide dismutase in the assay medium, converting superoxide to hydrogen peroxide.

While these methods are sufficiently robust and straightforward that the assays can be performed with any filter or monochromator based fluorometer/spectrofluorometer, recent advances in fluorescence instrumentation offer an alternative format, that of 96-well microtiter plate readers [Nathan, 1989]. These instruments lack the sensitivity of top-of-the-line fluorometers, and currently do not permit the continuous measurement of oxidant production, but they permit the simultaneous reading of the equivalent of 96 cuvettes, at a relatively rapid periodic time interval. While previously this method suffered from the limitation that polystyrene plates had to be used which themselves stimulate phagocytic cells, this problem can be overcome by approximately coating these plates or by using polypropylene plates (Corning, Inc.). There is one technical caution. The addition of stimuli to the microtiter plate wells must be made while shaking the plates to assure rapid and thorough mixing, avoiding the exposure of cells to local high concentrations of stimulus at the end of the pipette tip. Inexpensive incubated microplate shakers are available (Denley, Inc.). Furthermore, to reduce surface tension effects and bubble formation, and to make the shaking more effective the addition of Tween 20 (0.001%) or pre-treatment of wells with 0.1% Tween 20 is advisable. This amount of Tween does not have adverse effects on phagocytic cells, but certainly should be tested for effect on other cells.

G. Enzyme Measurements and the Secretory Process

It is not possible to review each specific enzyme which can be monitored using fluorescent methods, but it is useful to mention the use of fluorescent substrates to monitor the secretion of enzymes from cells. The basis for these types of measurements is the cleavage of a fluorescent substrate to a non-fluorescent product, non-fluorescent substrate to a fluorescent product or simply the exocytosis of a fluorochrome incorporated into the granule compartment. The methodology is nearly identical regardless of the enzyme. One example is the assay developed to measure the secretion of elastase from neutrophils [Sklar et al., 1982].

III. CHEMILUMINESCENCE

Applications of chemiluminescence are increasing. Its utilitization in monitoring the oxidative metabolism of cells was spurred by the introduction of specific chemiluminometers which afforded temperature control, stirring, and in some cases injection ports for stimuli and reagents. The instrumental limitation of chemiluminescence is that the photomultiplier tube must be close to the sample to avoid loss of the emitted photons. Thus, initial chemiluminescence studies were performed using scintillation counters or modified fluorometers. Dedicated instruments permit simultaneous kinetic assays and simplify data reduction. Besides monitoring endogenous chemiluminescence, agents such as luminol, lucigenin, or pholasin can be employed to enhance emission [Campbell, 1988]. Luminol chemiluminescence results from the hydrogen-peroxide-dependent, peroxidase-catalyzed formation of aminopthalate plus light from the luminol substrate. While peroxidase can be added exogenously, in the case of phagocytic cells myeloperoxidase serves this function. Both the intracellular and extracellular pools of myeloperoxidase are involved in the chemiluminescence response. Thus secretion affects the response as well as the generation of reactive oxygen metabolites [Dahlgren and Stendahl, 1983; De Chatelet et al., 1982; Briheim et al., 1984; Cheung et al., 1983; Dahlgren, 1987a,b; Johansson and Dahlgren, 1989]. Luuminol can also interact directly with superoxide, singlet oxygen, and hypochlorous acid. Luminol-enhanced luminescence is inhibited by catalase, D-mannitol, sodium benzoate, and the peroxidase poisons azide and cyanide. There is a negligible effect of superoxide dismutase (converting superoxide to hydrogen peroxide) on luuminol enhanced chemiluminescence. Pholasin, the protein-bound luciferin from bivalve mollusc *Pholas dactylus,* reacts in the same manner as luminol but has 50–100-fold greater luminescent yield.

Lucigenin reacts directly with superoxide and the hydroxyl radical to form a lucigenin radical, which in turn reacts with oxygen to form the dioxetan which decays to N-methyl acridone accompanied by the release of photons [Campbell, 1988]. Lucigenin chemiluminescence is thus inhibited by superoxide dismutase [Gyllenhammer, 1987; Van Kessel et al., 1989].

Since luminescence intensity is a function of the rate at which photons are emitted, the intensity is a direct measure of the rate of reaction. A positive slope to the chemiluminescence tracing indicates that the rate of reaction is increasing, while a negative slope indicates that the rate of reaction is decreasing. Thus analysis of chemiluminescence data is different than for absorbance data. Integration of the area under the curve, however, yields the amount of chemiluminescence over the particular time interval chosen.

Great care must be taken when conducting experiments because many agents can quench chemiluminescence; hence careful controls must always be performed before deciding that a treatment of cells either enhances or inhibits the oxidant production leading to chemiluminescence. For instance, since luminol- and pholasin-enhanced chemiluminescence is also dependent upon myeloperoxidase, the kinetics of reaction are influenced by the secretion kinetics of this peroxidase. There is a distinct difference between the kinetics of luminol- and pholasin-enhanced chemiluminescence from phagocytic cells which is not explained by any known differences in their reaction mechanisms. Instead these differences may be due to differences in availability of luminol and pholasin to the cell membrane and cytosol [Campbell, 1988].

Instruments and reagents are being introduced which may further revolutionize the place of chemiluminescence in the study of cellular functions. The instruments include microtiter plate readers which permit the sequential stimulation of each well and recording of response, as well as instruments which aspirate the sample and then record chemiluminescence in a rapid mixing flow chamber upon simultaneous addition of reagent. A variation of this latter instrument is one which incorporates an electrical discharge plate and performs electrochemical chemiluminescence measurements. The reagents being developed permit the coupling to ligands/proteins which when exposed to appropriate substrates or electrical ionization produce chemiluminescence. Little has been published to date using these novel instruments and reagents but applications include measurement of cell surface antigense and receptors, and measurement of responses from adherent cells grown in a 96-well microtiter plate format.

IV. COMBINED ASSAYS

All the assays discussed above have been treated as isolated methods. Even though assays described in this review are rapid, sequential assays require that cells be incubated for differing periods of time during which the cell responses may change. This problem can be reduced by taking advantage of an intrinsic aspect of photometric methods; namely, they permit inherent optical selectivity. Thus if there are no optical interferences more than one assay can be performed simultaneously with the same sample [Omann and Sklar, 1989; Seligmann et al., 1987]. The key is careful evaluation to confirm that there are no optical interferences. Optical interferences originate from two main sources, absorbance of excitation light by more thn one reagent, and absorbance of emitted light by a reagent included in the reaction mixture. The rules applied to standard lamp spectrometers are different than those which can be applied to laser based systems with regard to excitation. Laser-based systems inherently eliminate the concern that probes will interfere with one another as a result of absorbance at the same wavelength, or emission by one probe at a wavelength where another probe absorbs. Laser emission is sufficient to saturate the electron energy levels of every probe in solution, so that each will be maximally excited and not capable of absorbing any emitted photons. Thus a dye that absorbs broadly from 460 to 520 nm can be saturated by laser excitation at 480 nm such that it will not absorb emission (from another probe) at 510 nm. However, with a standard lamp instrument where excitation intensity is limiting, if two probes absorb at the same wavelength, then the amount of photons absorbed by each will be reduced. While this does not preclude simultaneous assays, it certainly makes interpretation more difficult and necessitates particularly detailed experiments using each probe singly and demonstration that their use simultaneously, while affecting the emission intensity, can be compensated for through appropriate quantitative analysis. Of greater concern is the situation which may develop if one probe absorbs at a wavelength where another emits. When this occurs, it essentially precludes simultaneous assays using standard (not laser-based) spectrophotometry. Ideal conditions for simultaneous assays are those where neither the excitation nor emission wavelengths overlap, or where other techniques such as light scatter and transmission are combined with fluorescence. Simultaneous assays have been successfully applied to measure calcium rise, secretion, oxidant production, light scatter, and membrane potential [Omann and Sklar, 1989; Seligmann et al., 1987]. While simultaneous assays have definite utility, they are inherently more complicated than single assays, and may not gain the researcher any advantage. For instance, if the dose-response curve for each assay is sufficiently different, it may be more efficient to perform the assays individually.

V. CONCLUSIONS

The material covered above indicates the diverse spectrum of cellular functions which can be conveniently monitored by quantitative optical methods.

There is considerable room for new developments and wider applications of the methods which provide unique types of information such as the real-time measurement of ligand binding and the measurement of the free intra-cellular concentration of ions in small cells. This is an area driven by the development of new probes and instrumentation, both of which are occurring at a fast pace.

The methods reviewed are powerful yet the results are subject to misinter-pretation if the proper controls are not performed and the precautions indicated in this review are not kept in mind. While these methods are not overtly inva-sive, nonetheless, the use of an indirect probe does perturb the cellular milieu, and this must be carefully examined before publishing results. The methods which have been reviewed have opened areas of research to investigation not just to a limited number of scientists, but to any investigator with the appro-priate instrumentation. It is simple to perform most of the assays based on the methods described in the literature without any specialized knowledge or training. Besides serving to shape experimental design, the points made in this review should be kept in mind when critically reading the literature. Before trusting the interpretation of data as presented in a manuscript, it is necessary that there be an indication that the proper controls were performed and pre-cautions taken to avoid misinterpretation of the data.

VI. REFERENCES

Adler M, Tritton TR (1988) Fluorescence depolarization measurements on oriented membranes. Biophys J 53:989–1005.

Arkhammer P, Nilsson T, Berggren P-O (1989) Glucose-stimulated efflux of Fura-2 in pancre-atic B-cells is prevented by probenecid. Biochem Biophys Res Commun 159:223–228.

Beccerica E, Piergiacomi G, Curatola G, Ferretti G (1988) Changes of lymphocyte membrane fluidity in rheumatoid arthritis: A fluorescence polarization study. Ann Rheum Dis 47:472–477.

Briheim G, Stendahl O, Dahlgren C (1984) Intra- and extracellular events in luminol-dependent chemiluminescence of polymorphonuclear leukocytes. Infect Immun 45:1–5.

Bronner C, Landry Y, Fonteneau P, Kuhry J-G (1986) A fluorescent hydrophobic probe used for monitoring the kinetics of exocytosis phenomena. Biochemistry 25:2149–2154.

Campbell AK (1988) Chemiluminescence: Principles and Applications in Biology and Medi-cine. New York: VCH Publishers, Inc. pp 302–305, 334–373.

Chen PY, Illsley NP, Verkman AS (1988) Renal brush border chloride transport mechanisms characterized using a fluorescent indicator. Am J Physiol 254:F114–F120.

Chen PY, Verkman AS (1988) Sodium-dependent chloride transport in basolateral membrane vesicles isolated from rabbit proximal tubule. Biochemistry 27:655–660.

Cheung AT, Johnson DA, Taylor PA (1984) Kinetics of interaction of N epsilon-fluorescein isothiocyanate-lysine-23-cobra alpha-toxin with the acetylcholine receptor. Biophys J 45:447–454.

Cheung K, Archibald AC, Robinson MF (1983) The origin of chemiluminescence produced by neutrophils stimulated by opsonized zymosan. J Immunol 130:2324–2329.

Conti-Tronconi BM, Dunn SMJ, Raftery MA (1982) Independent sites of low and high affinity for agonists on Torpedo californica acetylcholine receptor. Biochem Biophys Res Commun 107:123–129.

Dahlgren C, Stendahl O (1983) Role of myeloperoxidase in luminol-dependent chemiluminescence of polymorphonuclear leukocytes. Infect Immun 39:736–741.

Dahlgren C (1987a) Polymorphonuclear leukocyte chemiluminescence induced by formylmethionyl-leucyl-phenylalanine and phorbol myristate acetate: Effects of catalase and superoxide dismutase. Agents Actions 21:102–112.

Dahlgren C (1987b) Difference in extracellular radical release after chemotactic factor and calcium ionophore activation of the oxygen radical-generating system in human neutrophils. Biochim Biophys Acta 930:33–38.

De Chatelet LR, Long GD, Shirley PS, Bass DA, Thomas MJ, Henderson FW, Cohen MS (1982) Mechanism of the luminol-dependent chemiluminescence of human neutrophils. J Immunol 129:1589–1593.

Di Virgilio F, Steinberg TH, Swanson JA, Silverstein SC (1988) Fura-2 secretion and sequestration in macrophages: A blocker of organic anion transport reveals that these processes occur via a membrane transport system for organic anions. J Immunol 140:915–920.

Drummond IAS, Lee AS, Resendez E, Jr, Steinhardt RA (1987) Depletion of intracellular calcium stores by calcium ionophore A23187 induces the genes for glucose-regulated j proteins in hamster fibroblasts. J Biol Chem 262:12801–12805.

Grinstein S, Cohen S, Rothstein A (1984) Cytoplasmic pH regulation in thymic lymphocytes by an amiloride-sensitive Na^+/H^+ antiport. J Gen Physiol 83:341–369.

Grinstein S, Rothstein A, Cohen S (1985) Mechanism of osmotic activation of Na^+/H^+ exchange in rat thymic lymphocytes. J Gen Physiol 85:765–787.

Grynkiewicz G, Poenie M, Tsien RY (1985) A new generation of Ca^{2+} indicators with greatly improved fluorescence properties. J Biol Chem 260:3440–3450.

Gyllenhammer H (1987) Lucigenin chemiluminescence in the assessment of neutrophil superoxide production. J Immunol Methods 97:209–213.

Harootunian A, Eckert B, Minta A, Tsien RY (1988) Ratio imaging using the newly developed fluorescent sodium indicators in rat embryo fibroblasts. FASEB J 2:2508 (abstr).

Heidmann T, Changeux JP (1979a) Fast kinetic studies on the interaction of a fluorescent agonist with the membrane-bound acetylcholine receptor from Torpedo marmorata. Eur J Biochem 94:255–279.

Heidmann T, Changeux JP (1979b) Fast kinetic studies on the allosteric interactions between acetylcholine receptor and local anesthetic binding sites. Eur J Biochem 94:281–296.

Hesketh TR, Moore JP, Morris JDH, Taylor MV, Rogers J, Smith GA, Metcalfe JC (1985) A common sequence of calcium and pH signals in the mitogenic stimulation of eukaryotic cells. Nature 313:481–484.

Highsmith S, Bloebaum P, Snowdowne KW (1985) Sarcoplasmic reticulum interacts with the Ca^{2+} indicator precursor Fura-2-AM. Biochem Biophys Res Commun 138:1153–1162.

Hyslop PA, Sklar LA (1984) A quantitative fluorometric assay for the determination of oxidant production by polymorphonuclear leukocytes: Its use in the simultaneous fluorometric assay of cellular activation processes. Anal Biochem 141:280–286.

Illsley NP, Verkman AS (1987) Membrane chloride transport measured using a chloride-sensitive fluorescent probe. Biochemistry 26:1215–1219.

Johansson A, Dahlgren C (1989) Characterization of the luminol-amplified light-generating reaction induced in human monocytes. J Leukocyte Biol 45:444–451.

Johnson LV, Walsh ML, Bockus BJ, Chen LB (1981) Monitoring of relative mitochondrial membrane potential in living cells by fluorescence microscopy. J Cell Biol 88:526–535.

Laris PC, Bahr DP, Chaffee RRJ (1975) Membrane potentials in mitochondrial preparations as measured by means of a cyanine dye. Biochim Biophys Acta 376:425–425.

Laris PC, Pershadsingh HA, Johnstone RM (1976) Monitoring membrane potential in Ehrlich ascites tumor cells by means of a fluorescent dye. Biochim Biophys Acta 436:475–488.

Lee H-C, Smith N, Mohabir R, Clusin WT (1987) Cytosolic calcium transients from the beating mammalian heart. Proc Natl Acad Sci USA 84:7793–7797.

Luckhoff A (1986) Measuring cytosolic free calcium concentration in endothelian cells with Indo-1: The pitfall of using the ratio of two fluorescence intensities recorded at different wavelengths. Call Calcium 7:233–248.

Malgaroli A, Milani D, Meldolesi J, Pozzan T (1987) Fura-2 measurement of cytosolic free Ca^{2+} in monolayers and suspensions of various types of animal cells. J Cell Biol 105:2145–2155.

Mayer SJ, Keen PM, Craven N, Bourne FJ (1989) Regulation of phagolysosome pH in bovine and human neutrophils: The role of NADPH oxidase activity and an Na^+/N^+ antiporter. J Leukocyte Biol 45:239–248.

McDonough PM, Button DC (1989) Measurement of cytoplasmic calcium concentration in cell suspensions: Correction for extracellular Fura-2 through use of Mn^{2+} and probenecid. Call Calcium 10:171–180.

Metcalf JA, Gallin JI, Nauseef WM, Root RK (1986) Laboratory Manual of Neutrophil Function. New York: Raven Press Books, Ltd.

Miller JB, Koshland DE (1978) Effects of cyanine dye membrane probes on cellular properties. Nature 272:83–84.

Mills GB, Cragoe EJ, Jr, Gelfand EW, Grinstein S (1985) Interleukin 2 induces a rapid increase in intracellular pH through activation of a Na^+/H^+ antiport. J Biol Chem 260:12500–12507.

Montecucco C, Pozzan T, Rink TJ (1979) Dicarbocyanine fluorescent probes of membrane potential block lymphocyte capping, deplete cellular ATP, and inhibit respiration of isolated mitochondrial. Biochim Biophys Acta 552:552–557.

Moolenaar WH, Tsien RY, Van der Saag PT, De Laat SW (1983) Na^+/H^+ exchange and cytoplasmic pH in the action of growth factors in human fibroblasts. Nature 304:645–648.

Moore EDW, Tsien RY, Minta A, Fay FS (1988) Measurement of intracellular sodium with SBFP, a newly developed sodium sensitive fluorescent dye. FASEB J 2:2660 (abstr).

Murphy E, Freudenrich CC, Levy LA, London RE, Lieberman M (1989) Monitoring of cytosolic free magnesium in cultured chicken heart cells by use of the fluorescent indicator FURAPTRA. Proc Natl Acad Sci USA 86:2981–2984.

Nathan CF (1989) Respiratory burst in adherent human neutrophils: triggering by colony-stimulating factors CSF-GM and CSF-G. Blood 73:301–306.

Negulescu PA, Harootunian A, Minta A, Tsien RY, Machen TE (1988) Intracellular sodium regulation in rabbit gastric glands determined using a fluorescent sodium indicator. J Gen Physiol 92: 26a (abstr 53).

Oakes SG, Martin WJ, II, Lisek CA, Powis G (1988) Incomplete hydrolysis of the calcium indicator precursor Fura-2 pentaacetoxymethyl ester (Fura-2 AM) by cells. Anal Biochem 169:159–166.

Omann GM, Sklar LA (1989) Spectrofluorometric analysis of cell responses: Activation of neutrophils by chemoattractants and hexachlorocyclohexanes. In Goldberg MC (ed): Fluorescence Spectroscopy, Biological, Hydrological, and Environmental Applications. ACS Publications, in press.

Owen CS (1988) Quantitation of lymphocytes intracellular free calcium signals using Indo-1. Cell Calcium 9:141–147.

Philo RD, Eddy AA (1978) The membrane potential of mouse ascites-tumor cells studied with the fluorescent probe 3,3′-dipropyloxacarbocyanine: Amplitude of the depolarization caused by amino acids. Biochem J 174:801–810.

Raju B, Murphy E, Levy LA, Hall RD, London RE (1989) A fluorescent indicator for measuring cytosolic free magnesium. Am J Physiol 256:C540–C548.

Rink TJ, Montecucco C, Hesketh TR, Tsien RY (1980) Lymphocyte membrane potential assessed with fluorescent probes. Biochim Biophys Acta 595:15–30.

Roos D, Voetman AA, Meerhof LJ (1983) Functional activity of enucleated human polymorphonuclear leukocytes. J Cell Biol 97:368–377.

Root RK, Metcalf JA, Oshino N, Chance B (1975) H_2O_2 release from human granulocytes during phagocytosis. J Clin Invest 55:945–955.

Scanlon M, Williams DA, Fay FS (1987) A Ca^{2+}-insensitive form of Fura-2 associated with polymorphonuclear leukocytes. J Biol Chem 262:6308–6312.

Seligmann BE, Gallin EK, Martin DL, Shain W, Gallin JI (1980) Interaction of chemotactic factors with human polymorphonuclear leukocytes: Studies using a membrane potential-sensitive cyanine dye. J Membr Biol 52:257;–272.

Seligmann B, Gallin JI (1982) Neutrophil activation studies using two indirect probes of membrane potential which respond by different fluorescence mechanisms. In Rossi F, Patriarca P (eds) Biochemistry and Function of Phagocytes. New York: Plenum Press, Inc., pp 335 394.

Seligmann BE, Gallin JI (1983) Comparison of indirect probes of membrane potential utilized in studies of human neutrophils. J Cell Physiol 115:105–115.

Seligmann B, Patel K, Haston WO, Rediske JI (1987) Fluorometer based multi-parameter analysis of phagocytic cell activation. Agents Actions 21:375–378.

Seligmann B, Haston WO, Wasvary JS, Rediske JR (1988) Measurement of membrane potential responses elicited from blood cells: Effect of the dye/cell ratio and the presence of an intracellular calcium probe. In Cell Physiology of Blood. New York: The Rockefeller University Press, Inc.

Shinitzky M, Barenholz Y (1978) Fluidity parameters of lipid regions determined by fluorescence polarization. Biochim Biophys Acta 515:376–394.

Simons TJB (1979) Action of a carbocyanine dye on calcium-dependent potassium transport in human red cell ghosts. J Physiol (Lond) 288:481–507.

Sklar LA, McNeil VM, Jesaitis AJ, Painter RG, Cochrane CG (1982) A continuous, spectroscopic analysis of the kinetics of elastase secretion by neutrophils. The dependence of secretion upon receptor occupancy. J Biol Chem 257:5471–5475.

Sklar LA, Finney DA, Oades ZG, Jesaitis AJ, Painter RG, Cochrane CG (1984) The dynamics of ligand-receptor interactions: Real-time analyses of association, dissociation, and internalization of an N-formyl peptide and its receptors on the human neutrophil. J Biol Chem 259:5661–5669.

Sklar LA (1987) Real time, spectroscopic analysis of ligand-receptor dynamics. Annu Rev Biophys Biophys Chem 16:479–506.

Sklar LA, Mueller H, Swann WN, Omann GM, Bokoch GM (1989) Ligand-receptor-G protein dynamics and its relationship to neutrophil respons. Adv Leukocyte Biol 14: in press.

Sklar LA, Fay SP, Seligmann BE, Freer RJ, Muthukumaraswamy N, Mueller H (1990) Fluorescence analysis of the size of a binding packet of a peptide receptor at a natural abundance. Biochemistry 29:313–316.

Thomas JA, Buschbaum RN, Zimniak A, Racker E (1979) Intracellular pH measurements in Ehrlich ascites tumor cells utilizing spectroscopic probes generated in situ. Biochemistry 18:2210–2218.

Tsien RY, Pozzan T, Rink TJ (1982) Calcium homeostasis in intact lymphocytes: Cytoplasmic free calcium monitored with a new, intracellularly trapped fluorescent indicator. J Cell Biol 94:325–334.

Van Kessel KPM, Van Strijp JAG, Miltenburg LM, Van Kats-Renaud HJ, Fluit AC, Verhoef J (1989) Antibody-coated target cell membrane-induced chemiluminescence by human poly-morphonuclear leukocytes. J Immunol Methods 118:279–285.

Van der Meer BW, Van Hoeven RP, Van Blitterswijk WJ (1986) Steady-state fluorescence polar-ization data in membranes: Resolution into physical parameters by an extended perrin equation for restricted rotation of fluorophores. Biochim Biophys Acta 854:38–44.

Waggoner A (1976) Optical probes of membrane potential. J Membr Biol 27:317–334.

Whitaker JE, Haugland RP, Prendergast FG (1988) Seminaptho-fluoresceins and -rhodamines: Dual fluorescence pH indicators. Biophys J 53:197a.

Wilson HA, Seligmann B, Chused TM (1985) Voltage-sensitive syanine dye fluorescence sig-nals in lymphocytes: Plasma membrane and mitochondrial components. J Cell Physiol 125:61–71.

Wilson HA, Chused TM (1985) Lymphocyte membrane potential and Ca^{2+}-sensitive potas-sium channels described by oxonol dye fluorescence measurements. J Cell Physiol 125:72–81.

Noninvasive Techniques in Cell Biology: 353–374
© 1990 Wiley-Liss, Inc.

14. Use of Flow Cytometry to Study Ion Transport in Lymphocytes and Granulocytes

Louis B. Justement, John C. Cambier, Ger T. Rijkers, and Claus Fittschen

Division of Basic Sciences, Department of Pediatrics, National Jewish Center for Immunology and Respiratory Medicine, Denver, Colorado 80206 (L.B.J., J.C.C., C.F.); Department of Immunology, University Hospital for Children and Youth, "Het Wilhelmina Kinderziekenhuis," 3501 Utrecht, The Netherlands (G.T.R.)

I. INTRODUCTION

Maintenance of intracellular ion homeostasis is crucial to the viability of all living systems. The alkali metal cations Na^+ and K^+ and the alkaline earth metal cations Mg^{2+} and Ca^{2+} are ubiquitous in living cells; their importance is dependent upon or related to their unequal distribution across the plasma membrane. While K^+ is the predominant cytoplasmic monovalent cation, Na^+, on the other hand, is found predominantly in the extracellular fluid.

An appropriate ratio of Na_o^+/K_i^+ is important for the maintenance of transmembrane potential, and is regulated by Na^+/K^+ and Na^+/H^+ antiporters. The energy derived from the inward-directed Na^+ gradient can also be used by cells for transmembrane transport of other ions and small molecules. A similar differential distribution is observed for Mg^{2+} and Ca^{2+}, respectively: Mg^{2+} is more abundant intracellularly than is Ca^{2+}, with the reverse being true for the extracellular fluid. The concentration of these divalent cations is regulated by mechanisms that include exchange between intracellular storage organelles and the cytoplasm, as well as between the cytoplasm and extracellular fluid. Calcium influx is mediated predominantly by ion channels, some of which may also facilitate the movement of Mg^{2+} across the plasma membrane. The biological role of Mg^{2+} and Ca^{2+} is potentially much more varied than that of the monovalent cations due to the fact that they can bind to macromolecules, polyelectrolytes and biological surfaces. However, an extensive discussion of these interactions is beyond the scope of this chapter.

The prime interest of our laboratory is the study of changes in intracellular processes associated with, and important for, cellular activation following ligand-receptor interaction. In many cases, cellular activation involves changes in the distribution and/or concentration of the cations mentioned above. For example, in the lymphocyte, ligand binding (i.e., specific antigen or polyclonal mitogen) has been shown to cause an alteration in the surface charge of the membrane, changes in transmembrane potential, increased ion flux across the membrane, and increased intracellular free Ca^{2+} [Chandy et al., 1985a]. Similar processes occur in granulocytes as well. Over 50 hormones, growth factors, and other ligands have been described to date which cause transient increases in free intracellular Ca^{2+} ($[Ca^{2+}]_i$) following binding to their respective receptors. For example, binding of serotonin in blowfly salivary gland, f-methionyl-leucyl-phenylalanine (FMLP) to neutrophils, acetylcholine to neuronal tissue, and antigen to the antigen receptors on T and B lymphocytes all lead to an increase in $[Ca^{2+}]_i$. In several of these systems, it has been shown that ligation of receptor induces the phospholipase C-catalyzed breakdown of phosphatidylinositol 4,5-bisphosphate to yield inositol 1,4,5-trisphosphate and diacylglycerol, both of which serve as second messengers [Berridge, 1987]. Inositol-1,4,5-trisphosphate diffuses to the endoplasmic reticulum where it mediates the release of sequestered Ca^{2+} into the cytosol [Berridge, 1987]. Recently, a cytoplasmic kinase has been identified which phosphorylates inositol 1,4,5-trisphosphate on the 3 position, to yield inositol 1,3,4,5-tetrakisphosphate [Irvine et al., 1986]. This kinase is activated in T lymphocytes following ligation of the cell's antigen receptor, leading to an increase in inositol 1,3,4,5-tetrakisphosphate levels [Imboden and Pattison, 1987]. In sea urchin eggs, and presumably lymphocytes, inositol 1,3,4,5-tetrakisphosphate in turn mediates a sustained increase in $[Ca^{2+}]_i$ by stimulating influx of extracellular Ca^{2+} [Irvine and Moore, 1986]. Mobilized Ca^{2+}, apart from activating calmodulin-dependent protein kinase(s), also appears to act in concert with diacylglycerol to induce the high-affinity association of protein kinase C (PKC) with the plasma membrane [Nishizuka, 1984]. Now active by virtue of its association with Ca^{2+}, diacylglycerol, and phosphatidylserine, PKC has been shown to alter the activity of various ion transport mechanisms, presumably via phosphorylation [Molenaar, 1986].

Another event which has been observed following ligand-receptor interaction, and one which may be important for regulation of subsequent cellular responses, is an alteration in the cell's cytoplasmic pH (pH$_i$). Cytoplasmic pH is regulated by an interplay of ion antiporters localized in the plasma membrane which, in resting cells, maintain the pH near 7.2 [Roos and Boron, 1981]. The function of these antiporters is best illustrated experimentally by the cell's response to manipulation of its pH$_i$. Cytoplasmic acidification for example, has been shown to activate the Na$^+$/H$^+$ antiporter and, in some cases the Na$^+$-coupled HCO$^-_3$/Cl$^-$ exchanger which exchanges extracellular Na$^+$ and HCO$^-_3$ for intracellular Cl$^-$ and, perhaps, H$^+$. Both mechanisms function to reestablish the basal pH$_i$ level. Conversely, cytoplasmic alkalinization is counteracted by Na$^+$-independent HCO$^-_3$/Cl$^-$ exchange [Aicken, 1986; Boron, 1986; Molenaar, 1986; Grinstein et al., 1989]. In addition to its participation in the homeostasis of pH$_i$, the Na$^+$/H$^+$ antiporter contributes to the regulation of cell volume [Grinstein et al., 1989] and to signal transduction. Thus, Na$^+$/H$^+$ antiport and the associated increase in pH$_i$ play an important role in growth factor-induced mitogenesis in a variety of cell types, including lymphocytes [Gerson et al., 1982; Grinstein et al., 1989] and, perhaps, in the activation of blood neutrophils by chemoattractants [Simchowitz, 1985; Molski et al., 1980].

In addition to carrier-mediated and active transport devices, ion transport is also mediated by ion channels. Ion channels exist in either an open or closed state, thus regulating the transport of ions across the plasma membrane. The process of regulating a channel's state, i.e., open or closed, is called "gating." Channel gating can be controlled by altrations in membrane potential, and thus the channels are voltage-gated, or via binding of intra- or extracellular ligands, hence the term "ligand-gated channels" [Chandy et al., 1985b; Gallin, 1986]. Since ligand binding to plasma membrane receptors can cause changes in membrane potential, and/or stimulate the generation of second messengers, both of which may open ion channels resulting in transport of ions, it is important to examine the role of these channels in cellular activation. For discussion of the various ion channels found on lymphoid and myeloid cells, and the function of these channels, see the reviews by Chandy et al., [1985a,b] and Gallin et al., [1986].

Many of the processes described above have been delineated using invasive techniques such as microinjection and patch clamping. The development of a series of ion-sensitive fluorescent dyes has greatly facilitated the study of these processes using noninvasive techniques. These fluorescent dyes are available in the form of membrane-permeant esters and as such can be loaded into cells without affecting cell viability or function. A great number of techniques can be used to monitor intracellular ion concentrations with these ion specific chelators, ranging from fluorimetric analysis of cell suspensions to microscopic imaging of single cells. In this chapter, we describe the use of flow cytometry for the analysis of intracellular ion concentrations. Measurement of $[Ca^{2+}]_i$, and pH_i in cells of the lymphoid and myeloid lineage are given as examples of sensitive, versatile and reproducible techniques that allow kinetic analysis of intracellular ion concentrations at the single cell level. Furthermore, we have chosen to focus our discussion on two recent additions to the existing list of intracellular ion indicators, fluo-3 AM, which is used to monitor $[Ca^{2+}]_i$ and seminaphthorhodafluor acetoxymethyl ester (SNARF-1 AM), which detects changes in pH_i.

II. MEASUREMENT OF INTRACELLULAR pH
A. Methodology

1. Reagents and neutrophil preparation. Due to the sensitivity of neutrophils to modulation by bacterial lipopolysaccharides (LPS), all plasticware and reagents used in our studies were routinely tested to assure minimal LPS contamination using the Limulus Amoebocyte Lysate Kit (Associates of Cape Cod, Woods Hole, MA) which can detect LPS at ≥ 0.01 ng/ml. No LPS was detected on sterile plastics or in reagents at the concentrations used. The salts for preparation of Krebs-Ringer phosphate buffer (pH 7.35) were obtained from Mallinkrodt (Paris, KY) and had undetectable levels of LPS. The buffer was supplemented with LPS-free dextrose (purchased as 5% dextrose in 0.2% sodium chloride, Abbott Laboratories, North Chicago, IL) at a final concentration of 0.2%. Since the predominant salt in this buffer is NaCl (135 mM) it is subsequently referred to as Na-buffer. The Krebs-Ringer phosphate buffer described above was modified by substituting its NaCl component and sodium phosphate salts with equimolar concentrations of KCl, K_2HPO_4, and KH_2PO_4, respectively (K-buffer). The resultant K-buffer was supplemented with 0.2% dextrose (Sigma Chemical Co., St. Louis, MO). Both buffer solutions were also supplemented with 0.25% human serum albumin (Biocell Laboratories, Carson, CA) which was selected from several batches because of its low LPS content (0.3 ng LPS/ml when the albumin was dissolved at 0.25% [w/v]).

The chemoattractant FMLP (Vega Biochemicals, Tucson, AZ) was dissolved in DMSO (Fisher Scientific Co., Fairlawn, NJ) at 10^{-3} M and kept frozen at $-18°C$. Before use, this stock was diluted in Na-buffer containing neutrophils to give a final concentration of 10^{-7}M. Nigericin (Sigma) stock solutions were prepared at a concentration of 2×10^{-2}M in absolute ethanol, and stored at 4°C. A working solution was obtained by diluting 5 μl of the nigericin stock in 10 ml K-buffer (final concentration 10^{-5}M).

Blood was obtained from normal donors after informed consent. Neutrophils were isolated by an LPS-free, plasma-Percoll method described previously [Haslett et al., 1985]. The cell preparation thus obtained consisted of no less than 95% neutrophils which were 99% viable based on trypan blue exclusion. Neutrophils isolated by this procedure exhibited less spontaneous shape change, lower baseline secretion of superoxide anion and granule enzymes, and a greater chemotactic response than did cells prepared by Ficoll-Hypaque methods [Haslett et al., 1985].

2. Loading procedure. Carboxy SNARF-1 AM (Molecular Probes, Eugene, OR) stock solutions were prepared by dissolving 50 μg dye in 50 μl DMSO and were stored at −18°C. Working solutions were prepared fresh daily by transfer of appropriate aliquots (i.e., 12 μl carboxy-SNARF-1 AM for a 2 μM solution) into 10 ml of albumin-containing Na-buffer (pH 7.3). Unless otherwise indicated, neutrophils (5×10^6/ml) were incubated for 10 min at 37°C in this solution, centrifuged at 800g, 22°C, and washed twice with Na-buffer.

3. Flow cytometry. Cytoplasmic pH was estimated in neutrophils loaded with SNARF-1 AM using an Epics 751 cytofluorograph (Coulter, Hialea, FL) with a Biosense Tip to increase its sensitivity and a Coherent argon laser (Innova 90-5, Innova, Palo Alto, CA) with an excitation wavelength of 480 nm, set at 250 mW. The system was interfaced with a Cytomation Cicero computer system (Cytomation Inc., Englewood, CO). Samples were prewarmed to 37°C for 5 min and maintained at this temperature during analysis in a Coulter Viable Sample Handler connected to a 37°C waterbath. Green fluorescence was collected using a 575 nm band-pass filter (PMT-1) and separated from red wavelengths (detected with a 610 nm long-pass filter ([PMT-2]) by a 600 nm dichroic short-pass filter. The ratio of linear green/red emission was calculated by the Cicero computer system. The main cell population of interest (neutrophils) was defined on the scattergram (forward scatter vs. red fluorescence) to exclude contaminant cell types and damaged cells. The selected population was gated to the histogram depicting cell number vs. fluorescence ratio, or to the cytogram of fluorescence ratio vs. time, respectively.

Standard curves depicting fluorescence ratio vs. pH were established for each experiment. To this end, neutrophils, 3×10^6/ml, were suspended in K-buffer at pH 6.7, 7.0, 7.3, 7.6, and 7.9, respectively, and treated with 10^{-5} M nigericin to equilibrate the pH$_i$ with that of the K-buffer. For measurement of changes in pH$_i$ following addition of chemoattractant, neutrophils were washed and resuspended in Na-buffer and stimulated with FMLP. In these experiments, determination of the resting or basal pH$_i$ was carried out prior to analysis of stimulus-induced cytoplasmic alkalinization. Data were collected during sequential 1 min intervals, and the mean fluorescence ratio for each interval transformed into "true" pH_i values using the standard curve.

4. Microscopic fluorescence imaging. Neutrophils loaded with SNARF-1 AM under optimal conditions were examined using the N2 filter in a Leitz Diaplan UV microscope (Leitz GmbH, Wetzlar, West Germany) at a magnification of $\times 500$.

B. Results

Methodologies for the measurement of pH_i have changed dramatically over the years, progressing from direct pH_i estimates in bulk cell homogenates to analysis of single impaled cells. In contrast to these invasive techniques, indirect measurements using membrane-permeant compounds such as weak acids and weak bases were developed [reviewed by Roos and Boron, 1981] which had the advantage of being noninvasive but required correction for water trapped in the extracellular space [Boron and Roos, 1976; Simchowitz, 1985], since these probes equilibrated between the intra- and extracellular fluids. Within the past decade several new fluorescent esterified dyes have been developed which initially are membrane permeant, but are cleaved intracellularly by cytoplasmic esterases to yield an impermeant form of the fluorophore which accumulates within the cell [Thomas et al., 1979]. Since residual extracellular dye is removed by washing, estimation of trapped extracellular water is unnecessary. Several of the early probes, however, still leaked from the cells at a problematic rate [Kolber et al., 1988] or required an unphysiologically low pH during the loading procedure to facilitate dye uptake [Thomas et al., 1979]. the chromophore biscarboxyethyl-carboxyfluorescein (BCECF) has proven to be well retained in a variety of cell types during relatively short incubation periods [Rink et al., 1982; Kolber et al., 1988]. However, incubation of BCECF-loaded cells for more than 60 min was associated with substantial leakage [Kolber et al., 1988]. This presents a problem when single-wavelength fluorescence intensity measurements are made, because decreased fluorescence due to leakage can not be differentiated from changes in pH_i. In contrast, as is the case with SNARF-1, estimates obtained by measurement of fluorescence ratios derived from distinct emission wavelengths are relatively unaffected by minor variations in fluorophore concentration. Here we describe the use of a new fluorophore, SNARF-1 AM, which we found to be retained in neutrophils without significant loss during storage for 6 h at 22°C or for 2 h at 37°C. SNARF-1 AM has been reported to have a pKa of about 7.3–7.4 at 37°C [Haugland, 1989] and to measure changes in pH_i within a range from pH 6.3 to 8.6.

Human blood neutrophils, known to undergo marked cytoplasmic alkalinization upon stimulation [Simchowitz, 1985; Grinstein and Furuya, 1986], were chosen for studies designed to examine the use of SNARF-1 to measure

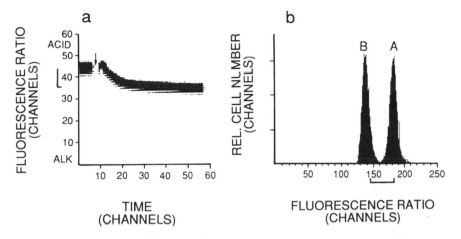

Fig. 1. Effect of FMLP on the pH_i of neutrophils. The fluorescence ratio of neutrophils loaded with SNARF-1 AM (1 μM) was measured following stimulation with FMLP at a concentration of 1×10^{-7} M. **a:** Cytogram—plotting fluorescence ratio (green/red) against time. Arrow indicates addition of FMLP. Note that the ordinate provides 64 channels for representation of the fluorescence ratio. The bracket indicates a shift of 9 channels in the fluorescence ratio after stimulation with FMLP. On the abscissa, ten channels equal 2.66 min. **b:** Histogram—plotting cell number against fluorescence ratio (green/red). Note that the abscissa provides 256 channels for representation of the fluorescence ratio. Resting cells (A) have a mean fluorescence ratio of 182. After stimulation with FMLP (B) the mean fluorescence ratio shifts to 146.

pH_i. We first considered software-related factors that affect the sensitivity of measurements. Changes in the fluorescence ratio of SNARF-1 may be recorded as either a cytogram (fluorescence ratio of green/red vs. time, Fig. 1a), or a histogram (cell number vs. fluorescence ratio, Fig. 1b) at selected timepoints. As the cytogram described above utilizes an ordinate with only 56 channels for expression of the fluorescence ratio, most of our calculations of pH_i were made using data in the histogram format which exhibits 256 channels on the abscissa and, therefore, has four times the resolution of the cytogram.

After establishing the optimal loading conditions for SNARF-1 AM, we analyzed neutrophils stimulated with the chemoattractant FMLP. These cells exhibited a brief cytoplasmic acidification phase followed by an increase in pH_i typically from 0.4 to 0.6 pH units (Fig. 2), that reached a peak between 10 and 15 min. Within the next 40–45 min the pH_i returned to a level just above that of the original basal pH_i. This biphasic response was observed for the total neutrophil population and is in agreement with the change in pH_i reported by several laboratories using fluorimetric analysis [Grinstein and Furuya, 1986; Weissman et al., 1987; Nasmith and Grinstein, 1988] or the DMO method [Molski et al., 1980; Simchowitz, 1985]. Interestingly, the initial acidification phase could be spuriously mimicked and/or exaggerated by

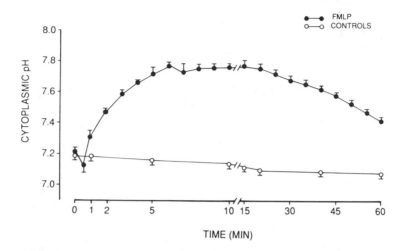

Fig. 2. Measurement of actual pH$_i$ in neutrophils loaded with SNARF-1 AM and stimulated with chemoattractant. At 0 min neutrophils were exposed to either 10^{-7} M FMLP (•——•) or buffer alone (O——O). Raw data were obtained in the form of fluorescence ratio values (i.e., mean channel number out of a 250 channel linear scale) corresponding to the mean fluorescence ratio for a population of cells at any given time point. The mean fluorescence ratio was converted to reflect actual pH$_i$ values using a standard curve depicting fluorescence ratio vs. pH. Each point represents the mean ± SE of three experiments.

boosting sample flow after addition of stimulus. This "apparent" acidification phase lasted approximately 2 minutes, and was not only observed in neutrophils, but also with polystyrene beads that had been coupled to SNARF-1 (Flow Cytometric Standards Corp., Research Triangle Park, NC), indicating that the "acidification" phase was not necessarily a cell physiologic response to stimulus but that it may be an artifact.

Analysis of changes in intracellular ion concentrations in cell populations using fluorimetric techniques is subject to the drawback that it is not possible to distinguish subpopulations which exhibit differential responses to ligand stimulation. Flow cytometry, on the other hand, overcomes this problem by allowing one to analyze large numbers of cells at the single cell level. This facilitates the identification of one or more subpopulations which differ in their physiological responsiveness. Occasionally, in our studies an unresponsive neutrophil subpopulation was observed (Fig. 3a,b). The identification of such unresponsive subpopulations, which have previously been reported to compose up to 40% of total neutrophil preparations [Simchowitz, 1985], is of considerable importance since calculation of the pH$_i$ for the total population leads to an underestimate of the change in pH$_i$ for the responsive cells. Moreover, if it is assumed that the unresponsive population is not artifactually created

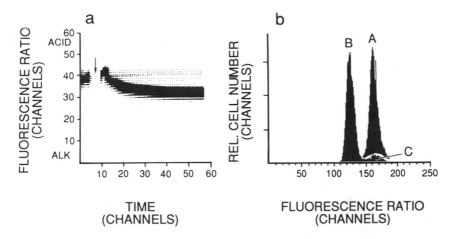

Fig. 3. Identification of distinct neutrophil subpopulations based on their response to FMLP.
a: Cytogram—before addition of FMLP, neutrophils exhibit a homogeneous fluorescence ratio,
and thus appear as one population. After stimulation (arrow) with FMLP (10^{-7}M), the main
cell population exhibits a decrease in its fluorescence ratio (i.e., becomes alkaline), whereas the
fluorescence ratio of a small subpopulation remains unchanged. **b:** Histogram—resting cells
(A) have a mean fluorescence ratio of 168. After exposure to FMLP a majority of the neutrophil
population (B) shifts (mean fluorescence ratio = 135) leaving a small subpopulation at the
point of origin (C).

due to the method used for cell preparation, then the presence of such a subpopulation could be of diagnostic significance, particularly when cells are
examined from patients who exhibit abnormal immune responses. This example illustrates a great advantage of flow cytometry over fluorimetry, i.e., the
capability to identify subpopulations and to analyze them separately.

As mentioned above, experiments were carried out to determine optimal
loading conditions for SNARF-1 AM, including variations in concentration
of dye, buffer pH, loading time, and cell concentration. Results are depicted
as standard curves in order to facilitate examination of the SNARF-1 fluorescence ratio over the full range of cytoplasmic pH values tested.

Standard curves, established using neutrophils loaded with SNARF-1 AM at
concentrations from 0.25 µM to 8.0 µM, were found to be linear over the pH
range examined regardless of the dye concentration used (Fig. 4). The decreased
fluorescence intensity associated with lower SNARF-1 AM concentrations could
be compensated for by parallel increases in gain settings of both photomultiplier
tubes (PMT 1 & 2). Additionally, at SNARF-1 AM concentrations of 0.25 and
0.5 µM, a small peak of dimly fluorescent material was frequently observed
in the lowest fluorescence ratio channels (after approximately 60–120 min),
which may have been due to dead or lysed cells. This second population was
found to interfere with accurate calculations of the mean fluorescence ratio.

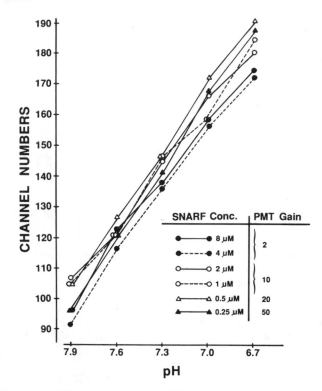

Fig. 4. Effect of increasing SNARF-1 AM concentration on measurement of pH_i. Neutrohpils, 5×10^6/ml, were incubated with different SNARF-1 AM concentrations (0.25–8 μM) for 10 min at 37°C. Standard curves, obtained by equilibration of pH_i and pH_o using K-buffers with varying pH values and 10^{-5}M nigericin, were linear regardless of the SNARF-1 AM concentration used for loading.

Although such interference could be prevented by excluding the lowest five channels from the calculation, we prefer to avoid these complications by loading at higher (1–2 μM) SNARF-1 AM concentrations. Loading cells at higher concentrations of SNARF-1 AM was found to enhance the distinction between live and dead cells or debris, thus eliminating the above problem.

Neutrophils were optimally loaded after 10 min incubation with SNARF-1 AM (2 μM). Longer periods of incubation did not improve the sensitivity or resolution of the pH_i measurement (Fig. 5). Efficiency of loading with 2 μM SNARF-1 AM was also independent of the cell densities studied (range 10^6–15×10^6 neutrophils/ml), indicating that the cells did not compete for dye under these conditions. Neutrophils loaded with SNARF-1 AM under optimal conditions characteristically exhibited uniform fluorescence throughout

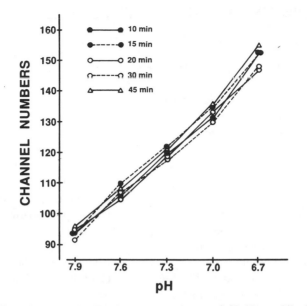

Fig. 5. Effect of increased loading time on measurement of pH$_i$. Neutrophils, 5×10^6/ml, were loaded for 10–45 min with SNARF-1 AM at a concentration of 2 μM. Standard curves depicting the fluorescence ratio versus pH were similar regardless of the length of the loading period.

the cytoplasm when examined microscopically. Compartmentalization of dye, suggestive of fluorophore exclusion from or preferential accumulation by cytoplasmic granules, was not observed.

The supplier of SNARF-1 AM recommends loading cells in buffer with a pH of 7.0 followed by an equal "neutralizing" volume of buffer at pH 7.4. Since we were concerned that such exposure to low pH conditions might alter subsequent physiologic responses of a highly reactive cell type such as the neutrophil, we compared the effect of loading cells at pH 7.0 with loading at pH 7.3 (other conditions were constant: SNARF-1 AM concentration 2.0 μM; incubation time 10 min; 5×10^6 cells/ml). We found no difference between the cells loaded at pH 7.0 versus those loaded at pH 7.3 with respect to pH$_i$ measurement (data not shown), indicating that the active dye penetrates the plasma membrane and is adequately converted into its nonpermeant form at pH 7.3.

In summary, SNARF-1 AM proved to be a useful dye for estimation of cytoplasmic pH in neutrophils. SNARF-1 AM was taken up rapidly and provided excellent results when used at a concentration of 1–2 μM for loading. Flow cytometric measurements of pH$_i$ were sensitive and not technically difficult and, in our hands, proved especially useful for the identification of neutrophil subpopulations which differed in their responses to the stimulant FMLP.

III. MEASUREMENT OF INTRACELLULAR Ca^{2+}

A. Methodology

1. Mice. C57BL/6 $\times$ DBA$_2$ F$_1$ mice (BDF$_1$) were obtained from The Jackson Laboratories (Bar Harbor, ME) or were produced at the National Jewish Center for Immunology and Respiratory Medicine. All experiments were carried out with BDF$_1$ mice between the ages of 6 and 8 weeks.

2. B Cell preparation and loading procedure. B cells were isolated as previously described [Justement et al., 1989] from spleens which had been removed aseptically and disrupted in balanced salts solution (BSS) containing 5 nM dextrose, 0.2 nM KH_2PO_4, 1.8 nM Na_2HPO_4, 1.2 nM $CaCl_2$, 5.3 nM KCl, 137 nM NaCl, 1.4 nM $MgSO_4$, and 0.5% phenol red, pH 7.2. Red blood cells were removed by treatment with Gey's solution for 5 min at 4°C. T cells were depleted from the lymphocyte population by incubating cells in a cocktail containing the monoclonal antibodies H013.4.9 [Marshak-Rothstein et al., 1979] and T24/40 [Dennert et al., 1980] which are directed against Thy-1, plus rabbit complement (GIBCO Laboratories, Grand Island, NY), 10 μg/ml DNAse (Sigma), and 5 μM $MgCl_2$ at 37°C for 40 min. The T-depleted cells were washed three times in BSS, followed by isolation of resting B cells by centrifugation through Percoll [Ratcliffe and Julius, 1982]. B cells with a density greater than 1.079 g/ml were used for the studies described here.

Resting B cells were loaded with either fluo-3 AM or indo-1 AM as follows: Cells were washed two times and suspended in buffer A [Iscove's modified Dulbecco's medium, [MDM (Sigma), containing 10 mM Hepes, pH 7.0] at a concentration of 5 $\times$ 10^6/ml. Fluo-3 AM or Indo-1 AM (Molecular Probes) were dissolved in DMSO (J.T. Baker, Chemical Co., Phillipsburg, NJ) at a concentration of 1 mM and were added directly to cells suspended in buffer A at a final concentration of 0.1 to 5 μM and 5 μM, respectively. Cells were then incubated at 37°C for 30 min unless otherwise indicated. Following this incubation period, an equal volume of buffer B [IMDM containing 10 mM Hepes, pH 7.4, and 5% fetal bovine serum (FBS, HyClone, Logan, UT] was added and the cells were incubated for an additional 10 min at 37°C. Excess fluo-3 AM and indo-1 AM were removed by washing cells two times in buffer C (IMDM containing 10 mM Hepes, pH 7.2, 5% FBS and 10 μg/ml DNAse). Cells were then resuspended at a final concentration of 1 $\times$ 10^6/ml in buffer C and were maintained at room temperature until analysis. Measurement of $[Ca^{2+}]_i$ was carried out at 25°C for the experiments described here.

3. Flow cytometric analysis. Flow cytometric analysis of $[Ca^{2+}]_i$ with the Ca^{2+}-sensitive dye indo-1 AM was carried out using an Ortho System 50H with a 2150 computer (Ortho Diagnostic Systems, Westwood, MA) and a 5 W argon laser (Coherent, Palo Alto, CA) set for 364 nm excitation at 300 mW. Fluorescence emissions were separated by a 465 nm long-pass steering dichroic element (Ortho) into two component emissions, violet and blue, by passage through 390 nm or 490 nm-centered 10 nm band-pass filters, respectively (Oriel Corp., Stratford, CT). The ratio of emissions for 390/490 was determined using the 2150 computer and a cytogram of fluorescence ratio vs. time constructed using data collected during consecutive 9 s increments ($\sim$7,200 cells analyzed per increment per channel). The data from isometric cytograms depicting fluorescence ratio vs. time were analyzed using the 2150 computer to derive the mean fluorescence ratio for each time increment, and this information was used to plot the change in $[Ca^{2+}]_i$ over time following stimulation of cells with ligand.

Analysis of $[Ca^{2+}]_i$ in resting B cells loaded with fluo-3 AM was carried out using an Epics 751 flow cytometer (Coulter) with a Cicero computer system (Cytomation) and a 5 W argon laser (Coherent). The laser excitation wavelength used was 488 nm at 250 mW. Fluorescence emission data were collected through a 525 nm band-pass filter. Fluorescence intensity data were collected during consecutive 5 s increments ($\sim$2,500 cells per increment per channel) and were depicted as a 3-parametric cytogram of fluorescence intensity vs time vs cell number. The data from four consecutive increments (20 s) were analyzed using the Cicero computer to determine the mean fluorescence intensity, and this value was in turn used to plot the change in $[Ca^{2+}]_i$ over time following stimulation with ligand.

In order to convert fluorescence ratio values into actual $[Ca^{2+}]_i$, the Ortho 50H was calibrated using a modification of the procedure described by Grynkiewicz et al. [1985]. We first examined the relationship between the indo-1 fluorescence ratio and the free ionized Ca^{2+} concentration in a panel of graded Ca^{2+} buffers to establish a standard curve. To this end, Ca^{2+}-containing solutions, buffered with EGTA, were constructed on the basis of a computer program developed by Dr. Jan Schmid (National Jewish Center for Immunology and Respiratory Medicine, Denver, CO). Care was taken to match excitation and emission band-pass wavelengths in a Hitachi F-4010 fluorimeter with those used on the Ortho 50H flow cytometer. Following construction of the standard curve relating the change in the 390/490 fluorescence ratio to the concentration of free ionized Ca^{2+} (0.25 nM to 1,000 nM,), the $[Ca^{2+}]_i$ of non-activated and ionomycin-treated (2µM, Sigma) resting B cells was measured on the fluorimeter. The $[Ca^{2+}]_i$ of B cells resuspended in buffer C, pH 7.2) containing 1 µM free Ca^{2+} was found to be 90 and 350 nM in the absence or presence of ionomycin (2 µM), respectively. The same samples were then analyzed on the Ortho System 50H flow cytometer using the resting cells to estabish a 90 nM set point. We could then assign specific $[Ca^{2+}]_i$ to specific channel numbers representing fluorescence ratio. Subsequent analysis of the ionomycin-treated cells verified this calibration method.

Calibration of the Epics 751 for analysis of $[Ca^{2+}]_i$ with fluo-3 was accom-

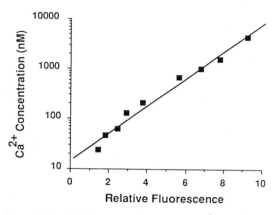

Fig. 6. Relationship of fluo-3 fluorescence intensity to Ca^{2+} concentration. Fluo-3 (cell impermeant, 1 μM) was titrated against varied EGTA-buffered Ca^{2+} solutions, and the intensity of fluorescence emissions monitored using a Hitachi F-4010 fluorimeter (excitation wavelength = 488 nm, emission wavelength = 525 nm). "Relative Fluorescence" is calculated by dividing the fluorescence intensity for fluo-3 in any given Ca^{2+} containing buffer by that obtained for fluo-3 in zero Ca^{2+}.

plished using information derived from the indo-1 calibration procedure described above. First, graded EGTA-buffered Ca^{2+} solutions were used to construct a calibration curve on the Hitachi F-4010 fluorimeter which depicted the relationship between the intensity of the fluo-3 fluorescence signal and the concentration of free ionized Ca^{2+} (Fig. 6). As can be seen, there is a linear relationship between the relative fluorescence of fluo-3 and the concentration of free ionized Ca^{2+}. This allowed us to predict that a given fold increase in fluo-3 fluorescence would reflect a specific molar increase in the concentration of Ca^{2+} and that this relationship should be independent of the absolute intensity of fluo-3 fluorescence (i.e., the relationship is independent of the absolute amount of fluo-3 that is taken up by the cells being studied). Thus, using the above logic, if it is assumed that the basal $[Ca^{2+}]_i$ of resting B cells is 90 nM and stimulation with anti-Ig causes a threefold increase in fluorescence over that of the basal level, then the $[Ca^{2+}]_i$ would have risen to 450 nM. We prepared B cells loaded with fluo-3 in the presence or absence of ionomycin (2 μM) such that their $[Ca^{2+}]_i$ concentrations were 90 and 350 nM, respectively, as previously determined based on indo-I analysis. These samples were then analyzed on the Epics 751 flow cytometer which allowed us to confirm the relationship depicted in Figure 6.

Fluo-3 has been reported to undergo a 40-fold increase in fluorescence intensity upon binding of Ca^{2+} at saturating concentrations [Minta et al., 1989]. However, due to the relatively low affinity of fluo-3 for Ca^{2+}, saturation of the chelator is not achieved until very high Ca^{2+} concentrations are used (>10 μM). Therefore, within the range of physiological $[Ca^{2+}]_i$ (70–1,000 nM) an approximate 2–3-fold increase in fluo-3 fluorescence intensity can be expected.

B. Results

During the past few years, several advances have been made in the techniques for measuring $[Ca^{2+}]_i$. The most widely used indicators for measurement of $[Ca^{2+}]_i$ are the tetracarboxylate fluorescent dyes quin-2, fura-2, and indo-1 [Minta et al., 1989]. All of these dyes are popular by virtue of the fact that they are easily loaded into cells and are well retained following cleavage by cytoplasmic esterases. Of these various Ca^{2+} sensitive fluorescent chromophores, indo-1 is often the indicator of choice due to its superior characteristics [Grynkiewicz, et al., 1985; Rabinovitch et al., 1986]. Indo-1 exhibits a large shift in its fluorescence emission spectrum upon binding of Ca^{2+}, such that measurement of the ratio of fluorescence intensities at two wavelengths allows one to calculate the $[Ca^{2+}]_i$-independent of any variability in intracellular dye concentration or instrument efficiency.

While indo-1 has been a valuable tool for measuring $[Ca^{2+}]_i$, it does exhibit certain characteristics which limit its usefulness. First of all, the existing dyes, including indo-1, require excitation at ultraviolet wavelengths near the cutoff for transmission through glass and plastic. This presents two problems for their use in flow cytometric analysis of $[Ca^{2+}]_i$. First, excitation using UV light compared to light in the visible range is much more expensive, generally requiring tunable water-cooled lasers. Secondly, optical elements used for excitation at UV wavelengths should be made from quartz, glass with enhanced UV transmission or very thin ordinary glass, as opposed to plastic or thick ordinary glass elements, which cause too much absorbance or autofluorescence.

Two other factors relating to the use of indo-1 must be taken into consideration. The first is that UV light, used for excitation, is potentially injurious to cells and may generate a significant amount of autofluorescence. Finally, the UV range overlaps the wavelengths needed to photolyze Ca^{2+} chelators such as nitr-5, nitr-7, and DM-nitrophen, or other caged compounds, causing them to release their natural product. Thus, the existing dyes can not be used with "caged" compounds.

Recently, a new Ca^{2+}-sensitive indicator, fluo-3, has been introduced [Kao et al., 1989; Minta et al., 1989]. Fluo-3 is a derivative of fluorescein and is optimally excited at a wavelength of 488 nm, making it a suitable indicator for use with most argon laser based flow cytometers. Because fluo-3 is analyzed in the visible range, the problems associated with the use of the tetracarboxylate fluorescent indicators are avoided. Interestingly, because fluo-3 has a lower affinity for Ca^{2+} than do the dyes quin-2, fura-2, or indo-1, it exhibits a greater ability to resolve changes in the concentration of Ca^{2+} at micromolar levels or higher. Furthermore, the greater dynamic range in fluo-3's sensitivity to Ca^{2+} is reportedly not accompanied by a decreased ability to measure changes in Ca^{2+} at low levels (10–100 nm) [Minta et al., 1989]. Thus, fluo-3 may be particularly useful for measuring changes in $[Ca^{2+}]_i$ above 1 μM, whereas its predecessors (i.e., indo-1) usually do not resolve changes in $[Ca^{2+}]_i$ above 1 μM.

The major disadvantage of fluo-3 is that, upon binding of Ca^{2+}, there is little or no shift in its excitation or emission spectrum. Since Ca^{2+} binding does not cause a shift in the fluo-3 spectrum, it is not possible to ratio either excitation or emission wavelength pairs, as is the case for indo-1. As previously mentioned, ratioing is an important advantage of indo-1 because it obviates the effect of variable dye uptake on fluorescence emission and it tends to minimize other variables associated with analysis. Essentially, measurement of $[Ca^{2+}]_i$ with indo-1 is independent of absolute changes in fluorescence intensity at a given wavelength. Thus, the use of fluo-3 may present a problem when loading heterogeneous populations of cells which exhibit differences in their size or granularity. Variation in cell size (i.e., cytoplasmic volume) may lead to heterogeneous uptake of dye which is proportional to the size of a given cell. Alternatively, if a population of cells exhibits a large degree of heterogeneity with respect to their granularity, this might result in varied degrees of intracellular compartmentalization of dye in granules. Without the ability to ratio the emission or excitation wavelengths of fluo-3, any heterogeneity in dye uptake or compartmentalization, leading to a quantitative variation in the basal fluorescence intensity within a given population of cells, will compromise the accuracy of determinations made for $[Ca^{2+}]_i$.

We chose to analyze the effectiveness of fluo-3 for measurement of changes in $[Ca^{2+}]_i$ of resting B cells following stimulation with anti-immunoglobulin (anti-Ig), and to compare the data obtained with that derived from cells loaded with indo-1. Initially, optimal loading conditions were established for fluo-3. As can be seen in Figure 7A, the uptake of fluo-3 is dependent upon time when assessed as a function of basal fluorescence intensity at a constant voltage and gain setting. However, when cells, loaded at 37°C for periods of time ranging from 5 to 40 min, were analyzed at varied gain and voltage settings such that their mean basal fluorescence intensity was identical (channel 50 out of a 250 channel linear scale), there was no difference in the change in fluorescence intensity following addition of 2 μM ionomycin to facilitate influx of extracellular Ca^{2+} (Fig. 7B). This suggests that cells take up fluo-3 AM rapidly and that by 10 min they contain a sufficient concentration of fluo-3 to give a maximal increase in fluorescence intensity in the presence of saturating Ca^{2+} (1.2 mM).

Varying the concentration of fluo-3 AM used to load cells, from 100 nM to

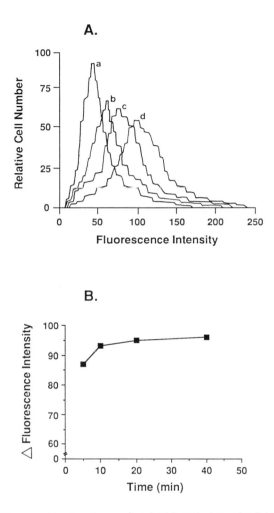

Fig. 7. Effect of increased loading time on fluo-3 AM uptake by resting B lymphocytes. Resting B cells, 5 × 10⁶/ml, were incubated for increasing periods of time in the presence of 1 μM fluo-3 AM in buffer A. After the amount of time indicated, cells were washed 2 times in buffer C and resuspended at 1 × 10⁶/ml in buffer C. **A:** Basal fluorescence intensity of cells loaded for a: 5 min, b) 10 min, c) 20 min, and d) 40 min was measured at a constant voltage and gain setting. **B:** Resting B cells which had been loaded with fluo-3 AM were analyzed at varied voltage and gain settings such that their basal fluorescence intensity exhibited a mean channel number of 50 out of a 250 channel linear scale. Ionomycin, 2 μM, was added to cells, and the peak change in fluorescence intensity measured. "△ Fluorescence Intensity" is equal to the value obtained by subtracting the mean channel fluorescence intensity for cells in the absence of ionomycin from the mean channel fluorescence intensity of cells stimulated with ionomycin.

10 μM, resulted in an increase in the basal fluorescence intensity of resting B cells when analyzed at a constant voltage and gain setting (Fig. 8A). However, when cells loaded with increasing concentrations of fluo-3 were analyzed at varied gain and voltage settings to determine the change in fluorescence intensity which occurred following the addition of ionomycin (2 μM), we observed that the fluo-3 AM concentration used for loading had a distinct effect on the magnitude of the response (Fig. 8B). The maximal change in fluorescence intensity was observed when cells were loaded with 1 μM fluo-3 AM. Either increasing or decreasing the amount of fluo-3 AM resulted in a decrease in the change in fluorescence intensity following addition of ionomycin. At higher concentrations of fluo-3 AM, above 1 μM, dye uptake by cells was increasingly heterogeneous, thus decreasing the efficiency of the readout.

In summary, it was found that cells take up optimal concentrations of fluo-3 AM within 10 min, indicating that after this point, time is no longer a factor. In contrast, the cell density during loading and the concentration of fluo-3 AM used were found to be important factors when trying to optimize loading conditions. For example, we found that loading of various B cell lymphomas (i.e., K46-17 and A20.1) required a much lower cell density, 1×10^6/ml versus 5×10^6/ml for resting B cells, and that the concentration of fluo-3 AM required for optimal loading was often higher, 2–5 μM for lymphomas versus 1 μM for resting B cells. Thus, it is important to optimize loading conditions for each cell type being studied in order to maximize the increase in fluorescence intensity in the presence of saturating Ca^{2+}. Furthermore, optimization of loading conditions will enhance homogeneous uptake of fluo-3 AM by the cell population being studied. It should be noted that the cells being studied here (resting B cells) are very homogeneous with regard to their cytoplasmic volume and granularity, and it is this high degree of homogeneity within the cell population which results in an equivalent uptake of fluo-3 AM by all cells under optimal loading conditions. Unfortunately, as mentioned previously, fluo-3 AM may not be suitable when studying cell populations that exhibit a high degree of variability with respect to their size or granularity which could result in a large variation in the basal fluorescence intensity, thus causing a decrease in sensitivity and accuracy.

Anti-Ig-mediated Ca^{2+} flux was next measured in resting B cells which were stimulated with 0.1 to 25 μg/ml polyclonal sheep anti-mouse Ig antibody following loading with either indo-1 or fluo-3. It was observed that over the range of anti-Ig concentrations tested, both fluo-3 and indo-1 detected qualitatively similar Ca^{2+} flux responses with respect to the kinetics and phenotype of the change in $[Ca^{2+}]_i$ [Fig. 9]. However, fluo-3 and indo-1 did not reveal quantitatively similar increases in $[Ca^{2+}]_i$ over the entire range of anti-Ig concentrations tested. Particularly, at lower anti-Ig concentrations, 0.1 to 1 μg/ml, fluo-3 was not as sensitive as indo-1 under the conditions used for analysis. This

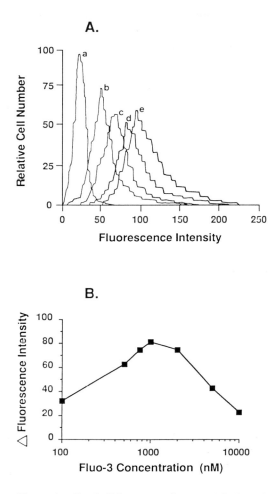

Fig. 8. Effect of increasing fluo-3 AM concentration on uptake by resting B cells. Resting B cells, 5×10^6/ml, were incubated in the presence of varied concentrations of fluo-3 AM in buffer A for 30 min. Following loading, cells were washed two times and resuspended in buffer C at 1×10^6/ml. **A:** Cells incubated in the presence of a: 100 nM, b) 500 nM, c) 1 μM, d) 5 μM, and e) 10 μM fluo-3 AM were analyzed to determine basal fluorescence intensity at a constant voltage and gain setting. **B:** The same cells were then analyzed at varied voltage and gain settings such that their basal fluorescence intensity exhibited a mean channel number of 50 out of 250 channels on a linear scale. Ionomycin, 2 μM, was added and the maximal change in fluorescence intensity measured. "△ Fluorescence Intensity" is equal to the value obtained by subtracting the mean channel fluorescence intensity for cells in the abscence of ionomycin from the mean channel fluorescence intensity of cells stimulated with ionomycin.

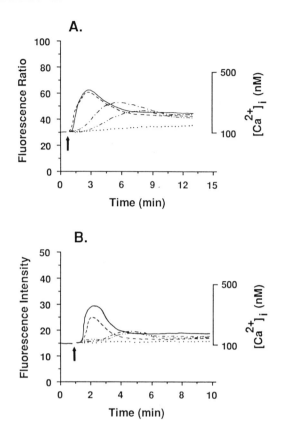

Fig. 9. Measurement of Ca^{2+} flux in resting B cells by indo-1 versus fluo-3 following anti-Ig treatment. Resting B cells, 5×10^6/ml, were incubated in buffer A in the presence of indo-1 (**A**) or fluo-3 (**B**), 5 μM and 1 μM, respectively, for 30 min. Cells were washed and resuspended in buffer C at 1×10^6/ml. Ca^{2+} flux was monitored following addition of 25 μg/ml (————), 10 μg/ml (-----), 1 μg/ml (-·-·-), 0.5 μg/ml (-··-··-), or 0.1 μg/ml (·······) SAMIg (indicated by arrow).

presumably, is due to the inability of fluo-3 to detect small changes in $[Ca^{2+}]_i$, with the degree of sensitivity that is characteristic for indo-1. The relative lack of "sensitivity" observed for fluo-3 is most likely due to its low affinity for Ca^{2+}, and the inability to ratio its fluorescence emission. In contrast, when cells were stimulated with a saturating concentration of anti-Ig, fluo-3 provided a quantitatively similar readout when compared to indo-1. Taken together, these observations would suggest that fluo-3 is satisfactory for analysis of large changes in $[Ca^{2+}]_i$ following maximal stimulation of the cell, but that it is not as effective at measuring small changes in $[Ca^{2+}]_i$.

Thus, it appears that fluo-3 can be used successfully in conjunction with flow cytometric techniques to measure changes in $[Ca^{2+}]_i$. The overriding advantage associated with the use of fluo-3 is that analysis of $[Ca^{2+}]_i$ can be carried out on virtually any type of flow cytometer equipped with an argon-based laser. However, it should be noted that situations which lend themselves to the use of fluo-3 are potentially limited. The characteristics of the cell population being analyzed, the magnitude of the Ca^{2+} response being monitored, and the need for accurate conversion of fluorescence intensity data into values which reflect the actual Ca^{2+} concentration should be considered before fluo-3 is chosen as the Ca^{2+} indicator.

ACKNOWLEDGMENTS

This work was supported by U.S. Public Health Service grants AI 20579 and AI 21768. Dr. Cambier is the endowed Cecil and Ida Green Professor of Cell Biology. Dr. Justement was supported by USPHS training grant AI 00048.

IV. REFERENCES

Aickin CC (1986) Intracellular pH regulation by vertebrate muscle. Annu Rev Physiol 48:349–361.

Berridge MJ (1987) Inositol lipids and cell proliferation. Biochim Biophys Acta 907:33–45.

Boron WF (1986) Intracellular pH regulation in epithelial cells. Annu Rev Physiol 48:377–388.

Boron WF, Roos A (1976) Comparison of microelectrode, DMO, and methylamine methods for measuring intracellular pH. Am J Physiol 231:799–809.

Chandy KG, DeCoursey TE, Cahalan MD, Gupta S (1985a) Ion channels in lymphocytes. J Clin. Immunol 5:1–6.

Chandy KG, DeCoursey TE, Cahalan MD, Gupta S (1985b) Electroimmunology: The physiologic role of ion channels in the immune system. J Immunol 135:787S–792S.

Dennert G, Hyman K, Lesley J, Trowbridge IS (1980) Effects of cytotoxic monoclonal antibody specific for T200 glycoprotein on functional lymphoid cell populations. Cell Immunol 53:350–364.

Gallin EK (1986) Ionic chanels in leukocytes. J Leukocyte Biol 39:241–254.

Gerson DF, Kiefer H, Eufe W (1982) Intracellular pH of mitogen-stimulated lymphocytes. Science 216:1009–1010.

Grinstein S, Furuya S (1986) Cytoplasmic pH regulation in activated human neutrophils: effect of adenosine and pertussis toxin on Na^+/H^+ exchange and metabolic acidification. Biochim Biophys Acta 890:301–309.

Grinstein S, Rotin D, Mason MJ (1989) Na^+/H^+ exchange and growth factor-induced cytosolic pH changes. Role in cellular proliferation. Biochim Biophys Acta 988:73–97.

Grynkiewicz G, Poenie M, Tsien RY (1985) A new generation of Ca^{2+} indicators with greatly improved fluorescence properties. J Biol Chem 260:3440–3450.

Haslett C, Guthrie LA, Kopaniak MM, Johnston RB Jr, Henson PM (1985) Modulation of multiple neutrophil functions by preparative methods or trace concentrations of bacterial lipopolysaccharide. Am J Pathol 119:101–110.

Haugland RP (1989) Handbook of Fluorescent Probes and Research Chemicals. Eugene, Oregon: Molecular Probes, Inc.

Imboden J, Pattison G (1987) Regulation of inositol 1,4,5-trisphosphate kinase activity after stimulation of human T cell antigen receptor. J Clin Invest 79:1538–1541.

Irvine RF, Letcher AJ, Meslor JP (1986) The inositol tris/tetrakisphosphate pathway-demonstration of Ins (1,4,5) P3 3-kinase activity in animal tissues. Nature 320:631–634.

Irvine RF, Moor RM (1986) Micro-injection of inositol 1,3,4,5-tetrakisphosphate activates sea urchin eggs by a mechanism dependent on external Ca^{++}. Biochem J 240:917–920.

Justement LB, Krieger J, Cambier JC (1989) Production of multiple lymphokines by the A20.1 B cell lymphoma after cross-linking of membrane Ig by immobilized anti-Ig. J Immunol 143:881–889.

Kao JPY, Harootunian AT, Tsien RY (1989) Photochemically generated cytosolic calcium pulses and their detection by fluo-3. J Biol Chem 264:8179–8184.

Kolber MA, Quinones RR, Gress RE, Henkart PA (1988) Measurement of cytotoxicity by target cell release and retention of the fluorescent dye bis-carboxyethyl-carboxyfluorescein (BCECF). J Immunol Methods 108:255–264.

Marshak-Rothstein A, Fink P, Gridley T, Raulet OA, Reum MJ, Gefter ML (1979) Properties and applications of monoclonal antibodies against determinants of the Thy-1 locus. J Immunol 122:2491–2497.

Minta A, Kao JP, Tsien RY (1989) Fluorescent indicators for cytosolic calcium based on rhodamine and fluorescein chromophores. J Biol Chem 264:8171–8178.

Molenaar WH (1986) Effect of growth factor on intracellular pH regulation. Annu Rev Physiol 48:363–376.

Molski TFP, Naccache PH, Volpi M, Wolpert LM, Sha'afi RI (1980) Specific modulation of the intracellular pH of rabbit neutrophils by chemotactic peptides. Biochim Biophys Res Commun 94:508–514.

Nasmith PE, Grinstein S (1988) Cytosolic calcium, oxygen consumption and the intracellular pH of stimulated neutrophils. Biosci Rep 8:65–76.

Nishizuka Y (1984) Turnover of inositol phospholipids and signal transduction. Science 225:1365–1370.

Ratcliffe MJH, Julius MH (1982) H-2 restricted T-B cell interactions involved in polyspecific B cell responses mediated by soluble antigens. Eur J Immunol 12:634–641.

Rabinovitch PS, June CH, Ledbetter JA (1986) Heterogeneity among T cells in intracellular free calcium responses after mitogen stimulation with PHA or anti-CD3. Simultaneous use of indo-1 and immunofluorescence with flow cytometry. J Immunol 137:952–961.

Rink TJ, Tsien RY, Pozzan T (1982) Cytoplasmic pH and free Mg^{2+} in lymphocytes. J Cell Biol 95:189–196.

Roos A, Boron WF (1981) Intracellular pH. Physiol Rev 61:296–434.

Simchowitz L (1985) Intracellular pH modulates the generation of superoxide radicals by human neutrophils. J Clin Invest 76:1079–1089.

Simchowitz L, Roos A (1985) Regulation of intracellular pH in human neutrophils. J Gen Physiol 85:443–470.

Thomas JA, Buchsbaum RN, Zimniak A, Racker E (1979) Intracellular pH measurements in Ehrlich ascites tumor cells utilizing spectroscopic probes generated in situ. Biochemistry 18:2210–2218.

Weissman SJ, Punzo A, Ford C, Sha'afi RI (1987) Intracellular pH changes during neutrophil activation. J Leuk Biol 41:25–32.

Noninvasive Techniques in Cell Biology: 375–402
© 1990 Wiley-Liss, Inc.

15. Optical Tweezers: A New Tool for Biophysics

Steven M. Block

Rowland Institute for Science, Cambridge, Massachusetts 02142;
Department of Cellular and Developmental Biology, Harvard University,
Cambridge, Massachusetts 02138

I. INTRODUCTION

Anyone who has seen a science fiction film or TV episode during the last few years is probably familiar with the idea of a *tractor beam,* a beacon of invisible radiation that is supposed to grapple, move, and capture intact spacecraft without inflicting damage. Such a notion remains the stuff of science fiction, but a modern-day counterpart to the tractor beam has recently been invented, called *optical tweezers.* Optical tweezers are more formally described as a ''single beam gradient force optical particle trap,'' or optical trap. Like a tractor beam, an optical trap comprises a beam of intense radiation, in this case from a laser. Like a tractor beam, an optical trap can catch and manipulate objects without touching them and without causing overt damage. But optical tweezers cannot manipulate macroscopic spacecraft: they work instead in the microscopic domain—that of cells, organelles, bacteria, and tiny particles. This chapter will focus on recent progress in using optical tweezers with biological materials, the principles upon which they work, how they are constructed, and on future prospects for their use. The field is young and wide open, with only a handful of articles published at the time of this writing (indeed, there are nearly as many in press as extant). It is arguably too soon for a review in the traditional sense of the word, but a good time to offer the following heuristic treatment of this promising technology.

II. WHAT ARE OPTICAL TWEEZERS?

Optical tweezers were conceived and developed by Arthur Ashkin and colleagues at AT&T Bell Laboratories [Ashkin et al., 1986, 1987; Ashkin and Dziedzic, 1987]. This breakthrough was partly an outgrowth of ongoing research in the use of lasers to cool, stop, and trap atoms and molecules [for reviews, see Ashkin, 1980; Phillips et al., 1988]. Over the years, Ashkin has pioneered a variety of theoretical and experimental approaches to the atom-cooling problem [Ashkin, 1978; Gordon and Ashkin, 1980; Chu et al., 1987; Balykin and Letokhov, 1989]. Today, laser-based atom traps are used routinely to cool atomic beams to micro-Kelvin temperatures, permitting ultra-high resolution spectroscopic measurements in the complete absence of Doppler line-broadening. To put this astounding achievement in perspective, a sodium atom falling under the influence of gravity through a distance of 1 cm gains a kinetic energy equivalent to 200 μK [Thompson, 1988]! The schemes upon which atom traps are based include such colorful names as "optical molasses" and "super molasses," because the forces they produce on neutral atoms are proportional to velocity, in analogy with viscous forces.

The possibility of accelerating or trapping micron-sized particles using the force of radiation pressure was considered by Ashkin some two decades ago [Ashkin, 1970]. At that time, he demonstrated stable trapping of tiny dielectric spheres (latex particles, 0.59–2.68 μm diameter) in an "optical bottle" (optical potential well) generated at the intersection of two weakly focused, counter-propagating laser beams. In a related development, small particles and liquid droplets were levitated against the opposing force of gravity by weakly focused light beams [Ashkin and Dziedzic, 1975, 1980].

Trapping with one laser beam was realized with the advent of the *single-beam gradient force optical trap* [Ashkin et al., 1986]. In this configuration, a strongly focused microbeam is used to create a trapping zone that is stable in all three dimensions. The axial stability of the trap is derived from the large axial gradient in light intensity near the focus, that gives rise to a *gradient force*, or "negative radiation pressure" [Ashkin et al., 1986], capable of pulling a particle back into the trap, even *against the direction of light propagation*. A detailed explanation of this force will be given in the next section. Single-beam particle traps can handle particles with sizes ranging from about 200 μm down to about 25 nm. Conceptually, the single-beam trap, or—as it is aptly dubbed, optical tweezers[1]—is the simplest possible trap. It consists of just one beam brought to a diffraction-limited focus. Such traps are constructed, in practice, by focusing the output of a TEM_{oo} mode laser, that is, one with a simple Gaussian beam profile, through a microscope objective of high numerical aperture.

Ashkin and Dziedzic [1987] were first to train optical tweezers on a biological sample in the form of a suspension of tobacco mosaic virus. From a physicist's point of view, this virus is an asymmetric Rayleigh particle (i.e., a particle with characteristic dimensions much less than the wavelength of light) having a volume of ~500 $Å^3$ and an index of refraction of 1.57. The argon laser which they used to form the trap produced a few hundred milliwatts of power at a wavelength of 514.5 nm (green light). They were successful in monitoring the predicted increase in light scattered at 90° from the trapping region as individual viruses were drawn into the focus and became entrapped. However, after several days of experiments with the virus preparation, they were astonished to find the scattered light signal occasionally go completely off-scale. Their curiosity piqued, they "... noticed the appearance of some strange new particles in diluted samples that had been kept around for several days ... They were quite large compared with Rayleigh particles, on the basis of their scattering of light, and were apparently self-propelled. They were clearly observed moving ... at speeds as high as hundreds of microns per second. They could stop, start up again, and frequently reversed their direction of motion at the boundaries of the trapping beam ... Their numbers increased rapidly as time went by." They had serendipitously discovered just how well the trap was working on bacteria that were contaminating their virus preparation: being physicists, not biologists, one imagines that they had neglected sterile technique!

Encouraged by this discovery, they went on to demonstrate that the trap was capable of capturing and manipulating many species of bacteria, and could hold as many as a half-dozen cells at a time. At low power levels (3–6 mW, measured at the sample), the tweezers produced sufficient trapping force to arrest a swimming bacterium. At higher power levels (~ 100 mW), however, even a short exposure (~1 min) to the intense light was clearly killing the cells. Judging from changes in light scattering and microscopic observations, the bacteria burst and vented their contents into the surrounding medium. The

---

[1]In addition to being a catchy description of the technique, *optical tweezers* is a name uniquely associated with this new technology. I would caution prospective authors against the use of such obvious names as "light traps" or "particle traps." Computer-based literature searches on these keywords turn up, quite literally, thousands of references in journals devoted to subjects such as entomology and public works, respectively. "Light traps" are used by naturalists to catch insects, and "particle traps" are found in sewers!

phenomenon of death by laser irradiation has been termed "opticution" by Ashkin (personal communication).

An obvious way to reduce optical damage at high power levels is to use a laser whose wavelength is less well absorbed by living material. Ashkin and coworkers [1987] soon tried a near-infrared laser, with a line far from the absorption peaks of most cells (Nd:YAG, wavelength 1,064 nm). This permitted apparently damage-free manipulation of many samples, including bacteria, erythrocytes, yeast cells, and even individual organelles within the cell walls of algae such as *Spirogyra*. Power could now be increased to ~100 mW without causing overt damage. Bacteria or yeasts survived long enough to reproduce through several generations, the daughter cells being held by the trap as well. At these higher power levels, bacteria could not only be caught, but towed rapidly through the medium at speeds as high as 500 μm/s, substantially faster than they can swim. This result implies that optical tweezers can grasp an object with a trapping force in the microdyne range.

The enormous potential of this technology was not lost on the biophysical community [Pool, 1988; Block, 1989]. Since the publication of Ashkin's work, several labs, including my own, have undertaken biological experiments with optical traps [Berns, 1989; Berns et al., 1989; Block et al., 1989a,b; Buican et al., 1988, 1989], and many more are underway.[2] Optical tweezers are now under construction in a number of laboratories (e.g., by D.E. Clapham at the Mayo Foundation, and M.P. Sheetz at Washington Univeristy), and an automated cell sorter/manipulator employing trapping technology promises soon to become a commercial product.[3]

III. HOW DO OPTICAL TWEEZERS WORK?

Optical traps work by using radiation pressure generated by light from a laser. A force is imparted to a reflecting or absorbing surface in a vacuum when it is hit by a flux of photons, equal to the rate of change of momentum of the photons. The change in momentum of a single photon upon absorption is $\Delta p = h/\lambda$, the deBroglie relation, where h is Planck's constant and λ is the wavelength. For a light of a fixed power level, P, the force produced by the flux is given by P/c, where c is the speed of light. The force is 2P/c in the case of reflected light, since the change in photon momentum is double for a

---

[2]In view of the publishing sequence of authors' names, one might say that we have now gone from "A" to ":B" in the trapping field! (S. Chu and D. Clapham are doubtless now writing up their work.)

[3]As announced in *Photonics Spectra*, May 1988, p. 60, and *Research & Development*, October 1988, p. 73. The product is being developed by Cell Robotics, Inc., a private firm founded by T. Buican , also of Los Alamos National Laboratories.

reflected beam.[4] Computation shows that a few mW of laser power should produce a μdyne of force (i.e., ~10 pN) when absorbed. *However, this force is not the same "radiation pressure" that makes optical tweezers work!* In fact, this force corresponds to the *scattering force* that occurs when light bounces off an object. Scattering forces tend to push objects along the direction of the beam. The *gradient force* trap, as the name implies, derives its trapping power from a spatial gradient in light intensity, which need not point along the direction of the beam. Gradient forces are generally smaller than scattering forces, but need not always be, especially when the object is nearly transparent. In the single-beam trap, the geometry is so arranged that the gradient force is as *large as possible* and points *everywhere inward,* towards a single, stable trapping zone. The gradient force obtained in this kind of trap is more than an order of magnitude less than the scattering force that would be produced by complete absorption: many mW of power produce a few tenths of a μdyne of trapping force.

Where does the gradient force come from? Two complementary explanations will be offered. Consider first the following argument. When light, which is a high-frequency electromagnetic oscillation, passes through a dielectric material, the time-varying electric field vector polarizes the medium, inducing a set of fluctuating dipoles. Recall that a charge, q, experiences a force in an electric field, E, equal to qE. Similarly, an electric dipole, p, will experience a force in a *spatially varying* electric field—since the separate charges of the dipole find themselves in regions of differing field strength—given by p · ∇E [cf. Purcell, 1985]. The induced dipoles of the transparent material are fixed in the object itself, which thereby experiences a force in the direction of the light gradient. Note, however, that the strength of the induced dipole moment, p, depends both on the polarizability of the medium (which is related to its refractive index, n) and on the strength of the field, introducing another power of E into the field dependence. It can be shown that for a Rayleigh particle of radius r and refractive index n, immersed in a medium of refractive index n_b, the gradient force is given by:

$$F_{grad} = -(\alpha/2) \cdot \nabla E^2$$

where α, the polarizability, equals $r^3 n_b^2 [(n^2 - 1)/(n^2 + 2)]$ [cf. Smith et al., 1981]. So particles that find themselves in a light gradient will experience a force pulling them towards a point where the light is brightest, for example,

---

[4]Strictly speaking, each of these expressions for the force should be multiplied by a factor n_b, equal to the refractive index of the external medium, when the surface is not surrounded by a vacuum. This is because the speed of light in the medium is c/n_b. The correction is of order unity (~1.33 in water).

up into a focus. However, in addition to this gradient force, there is a scattering force, pointing along the beam direction, arising from light that is absorbed or reflected. For a nonabsorbing Rayleigh particle, the scattering force is given by P_{scat}/c, where P_{scat} is the scattered power. Tiny particles scatter light in inverse proportion to the fourth power of the wavelength, a fact that Lord Rayleigh used to account for the blue color of the sky. The exact expression is

$$F_{scat} = \frac{I_0}{c} \; \frac{128\pi^5 r^6}{3\lambda^4} \left\{ \frac{n_b(n^2-1)^2}{(n^2+2)^2} \right\}$$

where I_0, the light intensity, is proportional to E^2 [cf. Ashkin et al., 1986]. Note that F_{scat} and F_{grad} have *different* dependencies on n, n_b, and r, and that F_{scat} depends on the square of the field while F_{grad} depends on the square of the gradient of the field. For tiny particles, and for a sufficiently large ratio $(\nabla E^2)/(E^2)$, a region exists where $-F_{grad}$ is everywhere greater than F_{scat}. This is the trapping zone. This region occurs near the waist of a highly convergent beam, e.g., near the focus of an objective of high numerical aperture. The balance between scattering and gradient forces dictates the range of particle sizes that can be successfully trapped.

The complementary argument explaining the origin of the gradient force is derived from a ray-optic picture. Micron-sized spheres, bacteria, or eukaryotic cells all have dimensions somewhat larger than the wavelength of light, where the Rayleigh approximation, described above, breaks down. For these materials, the more complicated Mie-scattering formulae can be used, but a simpler approach is to consider what happens when light passes through these nearly-transparent, refracting objects. Figure 1 shows two views of a small sphere ($n > n_b$) with light impinging on it: the sphere acts as a convergent lens to bend the light. The sphere on the left is placed in a beam of light coming from directly above; two representative rays through it are shown (large black arrows). The rays are bent by refraction at the interfaces where they enter and leave the sphere. A certain amount of reflection also takes place at these interfaces (small black arrows). Rays of light carry momentum. Before refraction by the sphere, all the rays carry momentum in the longitudinal (z) direction, but none in the transverse (x) direction. However, upon exiting the sphere, the ray on the left has an additional component of momentum to the right, while the ray on the right has an additional component of momentum to the left. By conservation of momentum in the x-direction, these rays must have imparted to the sphere a momentum that is equal and opposite to the momentum change of the rays. If the beam from above were uniform in intensity over its cross-section, the momentum "kicks" from all the rays on the left would identically cancel those on the right. However, if the beam has a light *gradient,* as illustrated by the

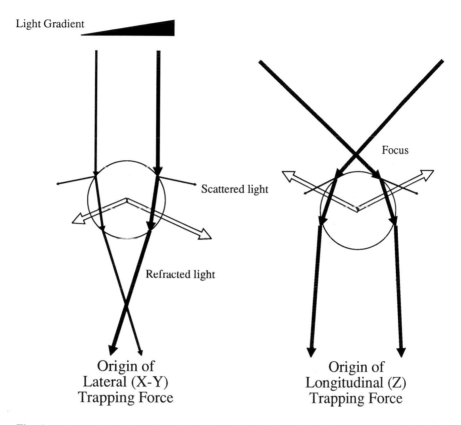

Light Gradient

Scattered light

Refracted light

Focus

Origin of
Lateral (X-Y)
Trapping Force

Origin of
Longitudinal (Z)
Trapping Force

Fig. 1. A ray-optic picture illustrating the origin of the *gradient force* on a small spherical lens. The large and small solid arrows show paths of representative light rays that are refracted or scattered by the lens, respectively. The open arrows show direction of the reaction forces, acting through the center of the lens, that arise from the change in momentum of light rays as they are deflected by the lens. See text for explanation.

wedge-shaped marking, the brighter rays on the right produce the greater reaction momentum. The momentum components due to these rays are shown acting through the center of the sphere (open arrows), and their overall effect is to push the sphere to the right (as well as slightly downward), *towards where the light is brightest*. This is the origin of the *gradient force*. Since the scattering force acts only in the z-direction, the gradient force is entirely responsible for all lateral trapping.

The sphere shown on the right in Figure 1 is located just below the focus of a highly convergent beam. Again, the large black arrows show refracted rays and the small black arrows show (some of) the reflected rays. The light becomes focused by the sphere. Notice that light rays entering from above have a substantial component of lateral (x) momentum and comparatively little longitudinal (z) momentum, while those rays that leave (bottom) are nearly vertical, having gained longitudinal momentum. By conservation of momentum in the z-direction, the sphere will experience a "kick" from these rays upwards (open arrows). Momentum delivered by scattered rays (small arrows) acts downwards, so scattering and gradient forces point in opposite directions. Provided, among other things, that the object is transparent and that the entering light is sufficiently convergent, conditions can be achieved where the gradient force exceeds the scattering force and the trap will be stable. A moment's reflection shows that the position of stable trapping is located somewhere just below the focus: above the focus, the scattering and gradient forces have the same sign. In a similar vein, note that any object with a refractive index *less* than the surrounding medium ($n < n_b$) will be unstable in this trap, since gradient forces will reverse sign.

When object dimensions are comparable with, or exceed, the wavelength of light (the Mie or ray-optic regimes), radiation pressure forces will depend on the exact shape of the object, as well as the shape and direction of the beam. For example, the high local curvature at boundaries of larger objects can give rise to significant gradient forces at these points, accounting for the fact that optical tweezers tend to grasp the edges of bigger cells. This phenomenon makes an ab initio calculation of trapping force a difficult undertaking for all but the simplest geometries. Unfortunately, most cells fall into this category. At present, no analysis has been published of the trapping potential (say) for various simple shapes in the vicinity of the beam waist of a Gaussian laser beam. Such an analysis would be of practical interest in estimating the depth and shape of actual trapping wells. For now, a rule of thumb based on experiment might be

$$F_{trap} \approx 0.03(n_b P/c)$$

as a rough limit on the force obtained with a micron-sized sphere trapped by an infrared laser, i.e., about 3% of the equivalent photon pressure on an absorbing surface.

I would like to emphasize that optical tweezers work by virtue of the fact that trapped particles (or living materials) *refract* light, and not because they absorb or reflect it. Too much absorption can lead to excessive heating and/or opticution. A typical trapping laser can generate many milliwatts of light. When brought to a diffraction-limited focus just one micron across, this power reaches a density of *megawatts per square centimeter*, a rather impressive figure!

Fortunately, the small size of the trap works to advantage, here. The following analysis is based on a back-of-the-envelope calculation by Edward M. Purcell (personal communication). Imagine a sphere 1 μm in diameter, thermally insulated from its environment, with a thermal conductivity equal to that of water (e.g., a bacterial cell), and placed at the focus of a milliwatt laser. Even if just 1 part in 10^3 of the incident radiation were absorbed, this microwatt of heat would raise the temperature of the particle past the boiling point in less than a millisecond—the sphere would be vaporized! Imagine, instead, that this same irradiated sphere is immersed in water. The physics of the problem is now changed: the sphere is water-cooled. The equation for thermal conduction has the same form as the diffusion equation, and is readily solved for spherical geometry. The steady-state temperature rise, ΔT, for a sphere in water is given by

$$\Delta T = \frac{3P}{8\pi\kappa r}$$

where r is the radius, κ is the thermal conductivity of water, and P is the power delivered to the sphere. Putting in numbers, one discovers that the same microwatt of absorbed power will only raise the temperature $\sim 1°C$ at steady-state, a quite tolerable number. The crucial point is that the power disposition is confined to an exceedingly small volume which can shed heat rapidly to its surroundings.

In the light of this result, optical damage is seen to be a more complex phenomenon than nonspecific heating: resonant absorption by chromophores in the material and specific photochemistry are likely culprits. The peak trapping force is linear with the light flux, and therefore the highest possible power is desired. It makes sense to use a wavelength that is not only minimally absorbed, but one that is poorly coupled to deleterious photochemistry. Such practical considerations will be taken up in the next section.

IV. CONSTRUCTION OF OPTICAL TWEEZERS

Building a trapping microscope with a single-beam trap is a straightforward task. Briefly, 1) a laser beam must be sent through the back of the objective in a manner that does not interfere with normal function; 2) the beam must be so arranged as to completely fill (or slightly overfill) the back aperture of this lens, so as to be maximally convergent; and 3) the beam must be brought to a diffraction-limited focus at the specimen plane, in coincidence with the sample being viewed. As a further convenience, some form of external x-y-z control over the position of the laser focus is desirable, so that the specimen and laser beam can be moved independently. Finally, there should be a means of visualizing the laser spot against the microscopic image, so that it can be positioned and aligned. These considerations do not unduly constrain the basic

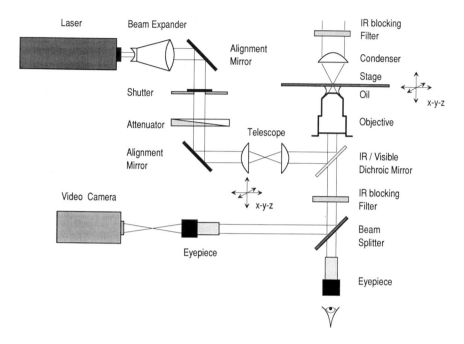

Fig. 2.   Schematic drawing of optical tweezers (not to scale). The diagram highlights design elements of an optical trapping microscope. This figure displays an inverted microscope configuration, but a similar arrangement of elements serves for an upright microscope. Light from a CW laser, operating in TEM_{oo} mode, passes through a beam expander and reflects from an alignment mirror mounted at 45°. A shutter and variable attenuator are used to modulate the laser output. After reflection from a second 45° mirror, the beam passes through a telescope formed by two plano-convex relay lenses (alternatively, two laser aplanats). The first lens of this telescope is mounted on an x-y-z micropositioner, as indicated by the arrows. This arrangement allows the laser focus inside the microscope to be moved about independently of the specimen. The relative powers of the two lenses are chosen so as to have the beam fill the entrance pupil at the rear of the microscope objective. (To avoid an external focus, the first lens could be replaced by a plano-concave lens of shorter focal length, essentially producing a Galilean beam expander at this position.) Light enters the microscope body and reflects from a dichroic mirror that reflects infrared and transmits visible light. It is focused by an objective of high numerical aperture (here shown as an oil-immersion objective) onto a specimen mounted on the microscope stage. Separate x-y-z micropositioners on the stage (arrows) manipulate the specimen. Laser light transmitted through the specimen is captured by an infrared blocking filter placed after the microscope condenser. A small fraction of the laser light is also retro-reflected from the coverglass over the specimen and makes its way back down the microscope through the dichroic mirror. This light is attenuated by another infrared blocking filter before passing on to the video camera. The small spot formed by this light is superposed on the microscope image formed by the conventional light path through the microscope. Visualized in this way, the spot can be used to align the optics, check beam quality, and to position the trapping zone. Note on alignment: In practice, the longitudinal (z) position of the external relay lens is adjusted to bring the z-position of the trapping

design, and several practical implementations now exist. Figure 2 shows one such design, which is simple and efficient, delivering as much of the laser light to the sample as possible. To conserve laser power, the beam is introduced into the microscope via a dichroic mirror mounted in the epi-illuminator head. Wherever possible, laser lenses are anti-reflection coated. The heart of the design is a telescope (Fig. 2, legend) that produces a nearly parallel beam of light. Changing the separation of the telescope lenses makes the beam slightly divergent or convergent and thereby moves the focal point up and down relative to the specimen plane. Small lateral displacements of the first telescope lens, to a first-order approximation, generate small x-y translations of the beam. This moves the laser spot about within the specimen plane. In some microscopes (e.g., Zeiss Universal), a lens that is part of the normal epi-condenser system can serve as the second lens of the telescope.

A note on microscope objectives. The design in the figure assumes that the objective is *infinity corrected,* that is, that the back focal plane of the lens is located at infinity. This is true of all "ICS" objectives in the new Zeiss Axio-series microscopes. For the older Zeiss inverted microscope (IM-35 series), a divergent lens located in the headstage below the objective seat effectively moves this focal point to infinity. In nearly all other microscopes (Zeiss Standard or Universal, Nikon, Leitz, etc.), the back focal plane is a standard distance, 160 mm, from the rear entrance pupil of the objective, as specified by the Royal Society for Microscopy. For these microscopes, the light entering the objective must appear to diverge from a point 160 mm back in order to be properly focused. This may be accomplished by adjusting the focal length and placement of the telescope lenses.

The size of the beam as it enters the objective pupil is set by the combined powers of the beam expander and telescope. Since this pupil is typically 4–9 mm in diameter, filling it may require a beam with a $1/e^2$ diameter ranging up to 1 cm. This is the only aperture in the system: in all other cases, the optical elements are substantially larger than the beam in order to remain diffraction-limited. Experience has shown that spatial filtering of the laser is not necessary.

The microscope should be outfitted with a videocamera sensitive to near infrared, as well as visible, light. For most applications, CCD cameras work extremely well in this capacity. For low-light-level work, I have found that a silicon diode head (Extended-red Ultricon) works better than either CCD or Newvicon cameras. A videocamera is essential for imaging the (invisible) beam, assessing the quality of the laser spot, and performing optical alignment.

---

zone into coincidence with the specimen plane (i.e., the focus of the microscope). This is done with a small particle (e.g., a 0.2 μm glass bead) in the trap. Thereafter, the stage z-position is used exclusively to move the trap up and down. Either the x- and y-position of this lens, or of the stage, can be used to move the trap relative to any object in the specimen plane.

Other designs exist. Ashkin and Dziedzic [1987] introduced their laser beam into a Zeiss Universal microscope through one of the eyepiece tubes of the trinocular head. A portion of the laser light is lost in this configuration, due to beam splitters in the microscope that carry light to the other eyepiece and to the videocamera. However, one virtue of this design is that it does not require an epi-illuminator head, nor does it require a custom dichroic mirror. Beam translation is again achieved through movement of an external relay lens (in this case, a divergent lens), and the beam diameter is adjusted by placing the laser at a considerable distance (>1 m) from the microscope, taking advantage of the intrinsic laser beam divergence.

Figure 3 shows two actual trapping microscopes based on the same general design. The photos show the relative position of various optical elements that are mounted on a bench beside each microscope. In addition to other components, it is convenient to have a fast shutter in the system in order to close off the laser beam. Our shutter is normally operated by a momentary footswitch; this safety feature helps to prevent accidental exposure. An additional safety feature is to provide infrared blocking filters or attenuators (either dielectric mirrors or KG-series Schott glass) in all parts of the visible light path. A variable attenuator in the laser beam sets the light level. Mirrors serve to perform the necessary beam translation and rotation to align the complete system.

In one of the systems, the first lens of the telescope is moved by an electromechanical servo system using deflection coils. This can be used to steer the x-y position of the beam automatically via command voltages [Block et al.,

---

Fig. 3.   Photographs of two light-trapping microscopes. **a**: Optical tweezers based on a water-cooled conventional CW laser and a Zeiss Axiovert inverted microscope. A portion of the laser with its beam expander can be seen at the far left. The laser is carried on rails mounted to the same air table as the rest of the apparatus. An optical bench placed at right angles to the laser, seen in the middle, holds additional components, including mirrors, electronic shutter and dual-wedge compensating attenuator. The x-y-z micropositioner holding the external relay lens that forms the first half of the telescope is indicated by the arrow. The second lens of the telescope is mounted on a custom slider located just below the nosepiece of the microscope. This slider, which also holds the dichroic mirror, is visible just behind the camera port (the black tube projecting from the left side of the microscope body). A videocamera, normally mounted on a rail behind the camera port, has been removed for the picture. **b**: Optical tweezers based on a solid-state diode-pumped microlaser and a Zeiss Universal microscope. The laser is seen on the far right, mounted to an optical bench that holds additional components (as above). An electromechanical servo system, indicated by the arrow, produces small deflections of a beam holding the first lens of the telescope. This beam, seen as the narrow aluminum tube leading toward the microscope, carries a laser aplanatic lens at its distal end. Movement of the lens is damped by a dashpot mounted on a second (wider) aluminum tube, seen running along the side of the microscope orthogonal to the first tube. A 45° mirror, mounted behind the epi-illuminator head, sends the light into the microscope. Some of the electronics that control the servo system can be seen at the bottom of the instrument rack.

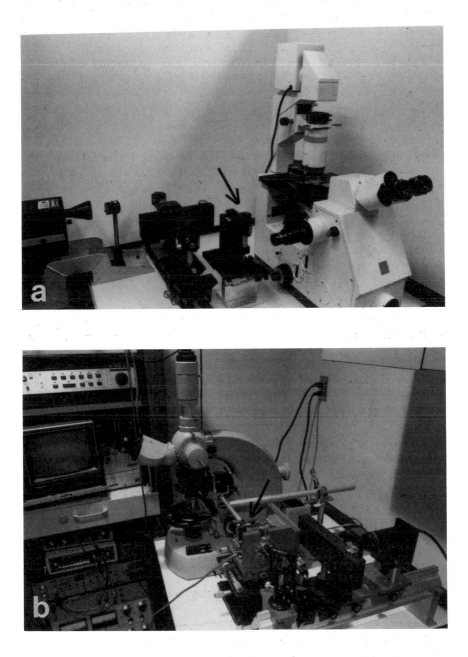

1989b]. Alternative steering options include deflecting the beam with either galvanometer mirrors or with acousto-optic modulation (AOM) devices based on Bragg cells. Both these methods are likely to be faster than approaches based on lens motion—low-inertia galvanometer mirrors can modulate in msec, AOM's in μsec—but are more costly. Beam-steering permits manipulation despite the presence of fixed constraints, such as microelectrodes or probes, near the sample area. The alternative to steering the beam, of course, is to move the sample. All current systems use stage movement for the z-control, since the working distance of the objective is fixed. An advantage to moving the sample in the other two dimensions, as well, is that nearly all microscopes come equipped with x-y control on their stages; no further equipment is required. Moreover, stage movement is the only reasonable means of moving trapped specimens long distances (mm, or even cm). For finer control, motorized microscope stages are available commercially, and—over smaller distances—piezoelectric transducers mounted to the sample cell are an effective means of achieving electronic control with rapid response (Ashkin, personal communication). In practice, it is convenient to have both stage and beam controlled independently, as different experimental situations will demand one or the other.

The first trapping microscope used brightfield optics [Ashkin et al., 1987]. However, one can obtain higher contrast and definition with either phase contrast or Nomarski differential interference contrast (DIC) optics. Trapping microscopes using both these methods have been successfully constructed in my lab. Phase objectives contain a ring inside that phase-retards and *reduces the intensity* of the annulus of light passing through it. Extreme care must be taken to insure that excessive heating of this ring does not occur at the power levels used, or the objective may be destroyed. Phase objectives work well only with relatively low-power beams that *do not* overfill the back aperture. DIC optics require that the laser beam be sent through the Nomarski slider; this has not proved to be a problem even at the highest powers used (~500 mW). Here, the full back aperture of the lens is available. I have used water and glycerol immersion objectives (with and without cover glass), as well as oil immersion objectives (with cover glass), in powers from $40\times$ to $100\times$, and numerical apertures from 1.25 to 1.40, all with success. Because it is restricted to objectives of lower numerical aperture, darkfield microscopy is probably not very fruitful for single-beam trapping. Darkfield optics uses the condenser at the high aperture, but limits light into the objective to a narrower cone. Image enhancement by video microscopy, either analog or digital, can further improve contrast in conjunction with any optical arrangement.

A straightforward elaboration of the single-beam trap is to have two independently adjustable trapping beams [see Ashkin et al., 1987]. A number of intriguing experimental possibilities open up with this approach, taking advantage of the ability of two traps to orient biological particles in space while levitating them. These possibilities have hardly been explored. The second trapping beam can be sent into the microscope in one of several ways. The beam can be brought in "from below," through a condenser of high numerical aperture (e.g., as used with double oil immersion optics), or through a second objective, used in place of the normal condenser. Alternatively, the second beam can be brought in "from above," through the same trapping/viewing objective as the first, using beam splitters [Ashkin and Dziedzic, 1980].

Optical manipulation need not be limited to single-beam gradient traps. While such traps are conceptually simple, easy to build, and offer versatile three-dimensional control, they require an objective of high numerical aperture. These lenses typically have very high magnification and short working distance, a disadvantage for some preparations, e.g., cell or tissue culture work. High numerical aperture, however, is only required for axial stability; even weakly focused microbeams can produce strong lateral trapping. Ashkin's original particle trap design [1970] provided axial stability by means of two coaxial, counter-propagating beams; beam waists were positioned to produce a trapping region near their intersection. Buican and coworkers [1988, 1989] have adapted this design to an optical-trapping cell sorter with versatile capabilities (Fig. 4). It is certainly the most ambitious trap to date, incorporating a micromachined chamber with various compartments, fluidics, a solid-state laser, 3-D servomechanisms, video processing, etc., all controlled by computer via a graphics interface. The design permits use of relatively long-working-distance, low-magnification objectives. There are certain advantages to cell sorting or manipulation by optical tweezers. The chamber can be entirely enclosed, reducing the possibility of contamination; cell manipulation is relatively gentle and noninvasive; the process can be automated; positional accuracy is extremely high; the volumes handled remain small, etc. While not as fast as conventional cell sorters that divert fluid droplets containing cells, optical approaches may afford a greater opportunity for detailed cell examination and/or measurement.

In constructing optical tweezers, the choice of a laser is clearly important. Recall that the argon green line at $\lambda = 514.5$ nm caused optical damage to biological specimens, so Ashkin and colleagues [1987] switched to an infrared Nd:YAG laser at $\lambda = 1064$ nm. This was a natural choice, since visible light is heavily abosrbed by natural chromophores such as hemoglobins, cytochromes, flavins, chlorophylls, and the like. Longer infrared wavelengths, e.g., the 10.6 μm CO_2 line, are heavily absorbed by water, but biological materials are nearly transparent to the near infrared. Moreover, conventional microscope lens design is optimized for visible light, and will not handle well any wavelength too far outside the visible spectrum. The characteristic size of the "grasp" of optical tweezers is governed by the width of the diffraction-limited beam waist, namely, $\sim\lambda$ for a Gaussian beam at high numerical aperture. It

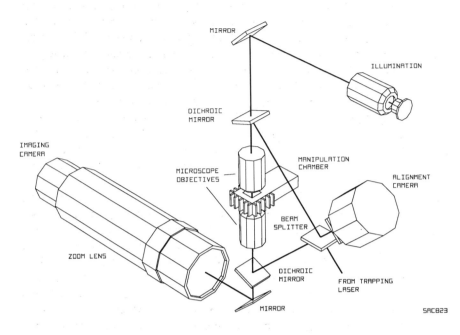

Fig. 4. A simplified diagram of OCAM (Optical-trapping Cell Analyzer & Manipulator), a prototype instrument developed by Tudor Buican and colleagues [Buican et al., 1988, 1989; derived from an earlier 2-D trapping device, Buican et al., 1987]. Biological materials are introduced into a manipulation chamber through any of a number of ports, which communicate with a complex series of interconnecting compartments and channels micromachined into the chamber (not shown); the compartments can be filled with different things. The x-y-z position of the chamber is adjusted by servoed micropositioners (not shown), and the entire system is controlled by computer. The light beams remain stationary. The system uses two counter-propagating beams of light to effect axial (z) trapping. The two-beam design allows each beam to be only weakly focused, permitting the use of long working-distance objectives of relatively low magnification. Two microscope objectives, mounted above and below the chamber, are used both as an objective/condenser pair for the trans-illuminating light (i.e., as a microscope) and as focusing elements for the trapping beams. Laser light, entering from the lower right-hand corner (laser not shown), is divided by a beam splitter and sent through each objective via a dichroic mirror. An alignment camera, aimed at the beam splitter, is employed to ensure that the two beams are strictly coaxial. A commercial zoom lens mounted on a video camera (lower left corner) serves as an eyepiece with variable magnification to pick up the secondary image of the microscope. This camera is connected to a video frame processor and the rest of the computer system. (Diagram courtesy of Dr. Tudor Buican.)

makes sense to keep the wavelength as short as possible. These considerations still allow considerable latitude in the choice of wavelength. Continuous-wave lasers in the red and near infrared, especially those capable of producing half a watt of power or thereabouts, are limited to a number of fixed wavelengths. The Nd:YAG line at 1,320 nm has been explored briefly by Ashkin, while a diode laser line near 800 nm has been tried by Buican (personal communication). Unfortunately, no data exist for optical damage at these wavelengths compared with 1,064 nm. Newer high-intensity sources in the spectral range 740–1,300 nm are rapidly becoming available, including a variety of laser diodes and tunable titanium sapphire (Ti:Al_2O_3) lasers. Perhaps one of these devices will prove best-suited for biological work?

The current choice of 1,064 nm is not without its problems. Ashkin and coworkers [1987] have reported that yeast and bacteria reproduce in traps up to ~80 mW power, and that cells of an unknown strain of bacteria maintain their motility. However, they noted some loss of flexibility in the membranes of erythrocytes trapped for tens of minutes. We have found that cells of *E. coli* (a gram-negative, facultative aerobic bacterium) lose motility after several minutes' exposure to a trap of ~50 mW [Block et al., 1989b]. Interestingly, cells of a strain of *Streptococcus* (V4051, a gram-positive bacterium that does glycolysis only) remain motile in the same trap for hours (S.M. Block, D.F. Blair, H.C. Berg, unpublished data). This difference is certainly not due to heating in the usual sense of the word, as shown earlier, but may possibly be due to direct photogeneration of singlet molecular oxygen by the 1,064 nm light. Singlet oxygen is known to responsible for many so-called *photodynamic effects* in biology [cf. Bellus, 1979]. It is conceivable that adding chemical quenchers of singlet oxygen to the medium may afford longer-lasting protection from damage. Alternatively, another laser wavelength might be required.

At present, two types of Nd:YAG laser are used for optical tweezers. The first is the conventional water-cooled continuous-wave laser (manufactured, for example, by CVI or Quantronix). These lasers operate from a high-intensity lamp as a pump source. They produce power in the range 0.1–100 W (TEM_{oo} mode), or even higher, depending upon size and cost, and have noise levels on the order of 2% with stabilized power supplies. They are available with both polarized and unpolarized output. The second type is the diode-pumped solid-state microlaser, which uses laser diodes to excite the Nd:YAG crystal, typically in an end-pumped configuration. Microlasers are generally cooled with a Peltier device, and have extremely high wall-plug efficiency (as much as 30%). Currently, microlasers are limited in output power by the availability of high-power laser diode arrays. Diode lasers are becoming more powerful all the time, and commercial microlaser models can now be purchased with power outputs in the range of 10–350 mW (manufactured, for example, by Amoco or Adlas), with 0.5-1 W promised soon. Experimental models have reached well beyond a watt by now. Microlasers are small enough to be held in the hand, extremely convenient, rugged, and stable, and have noise levels that are an order of magnitude better than conventional lasers (0.1%–0.2%). Moreover, they do not require large power supplies, cooling water, nor do they need alignment or adjustment. For now, the cost per watt of a microlaser is substantially more than a conventional laser, but there is every reason to expect that microlasers will soon become more affordable, dropping below the cost of a conventional model in the power range under 5 W.

In principle, solid-state (GaA1As) laser diodes themselvves might provide direct power for use as trapping lasers, rather than serving as pump sources for crystal lasers. Their characteristic wavelengths, 680–880 nm, lie in a potentially good range for trapping. They are small, stable, easy to set up, and inexpensive. However, there are still technical drawbacks to these devices. Currently, they are available packaged as low-power single-mode chips—probably still too weak to serve as useful trapping lasers—or as high-power multi-mode arrays. Unlike the "clean" TEM_{oo} beams available from crystal lasers, all diode lasers produce highly astigmatic beams with enormous divergence, typically over 30°, in the widest dimension. This makes it difficult to derive good beams from them. Effort is being spent in the laser industry to design custom aspheric collimators and correcting lenses that will clean up diode laser beams to the point where they can be focused close to the diffraction limit, as required for good trapping. T. Buican has experimented with diode lasers in a 2-D trapping arrangement (personal communication). There is reason to hope that high-power laser diodes may eventually prove quite important in trapping technology.

V. MANIPULATIONS WITH OPTICAL TWEEZERS

What can be done with optical tweezers? As shown earlier, there is a practical limit to the amount of force that can be produced. Roughly speaking, this limit is in the region of a few μdyne (a few tens of pN). To put this in perspective, 1 μdyne is the towing force required to move a sphere 1 μm in diameter through water at just over 500 μm/s. Forces of this magnitude are sufficient to catch motile bacteria or sperm cells, move eurkaryotic cells through fluid, and displace certain vesicles and organelles within larger cells [Ashkin et al., 1987; Block et al., 1989b; Buican et al., 1989]. They can bend or twist most cytoskeletal polymers and assembles, e.g., axonemes, bacterial flagella, filapodia, stereocilia, microtubules, actin cables, etc. [Block et al., 1989b and unpublished data]. They can overcome the torque of a bacterial rotary motor. They are probably sufficient to activate some mechanoreceptive cells. Forces of a few μdyne are *not sufficient,* for example, to perform microdissection, or to replace a suction electrode in holding cells for microinjection. Nor are they sufficient to pull apart cytoskeletal assemblies, such as thin filaments or microtubules (unpublished data). They can move isolated chromosomes, but cannot completely arrest moving chromosomes during mitosis (Berns et al., 1989). It remains an open question as to whether they can slow down or arrest molecular motors, such as myosin, dynein, and kinesin, in various in vitro motility assays [Block et al., 1989c]. The sliding force associated with a single myosin or dynein crossbridge has been estimated to be in the range 0.2–0.9 μdyne [cf. Kamimura and Takahashi, 1981], so there might be sufficient force provided that the number of crossbridges is small.

A potpourri of examples will help to convey the flavor of what can be accomplished with optical tweezers. Figure 5 shows a single, motile bacterium that has been captured and isolated by a trap. The cell remains actively motile for a time after it is trapped, spinning rapidly about its long axis as a result of the continued movement of its flagella. Figure 6 shows yeast cells, which are not motile, that have been trapped at a low power level over a period of several hours. They continue to bud and reproduce, the daughter cells remaining in the trap alongside one another. Figure 7 shows a rounded rat macrophage that has been lifted off the bottom of an enclosed chamber by a two-beam trap in a cell sorter (OCAM, Fig. 4). Figure 8 illustrates the unique capacity of optical tweezers to grasp vesicles *inside* intact cell membranes, in this case through the organelle wall of a chloroplast from *Nitella*. In a similar fashion, optical tweezers can also manipulate entire organelles inside cells [see Ashkin et al., 1987]. Figure 9 shows the Nomarski image of a vesicle being moved across the diameter of the micropile, a part of the *Drosophila* embryo. Figure 10 demonstrates beam-steering with optical tweezers, used in this case to rotate a cell body tethered to the coverglass by a single flagellum. Counterclockwise circular motion of the trap spins the cell, building up torsional strain in the flagellum. When the trap is turned off, the cell body rotates clockwise as the flagellum unwinds. Measurements of this kind can be used to determine the elastic properties of the flagellum. The trapping force generated by optical tweezers can be calibrated, for example, against Stokes' drag [Block et al., 1989a,b] so that the device can be used as a *microtensiometer* to make direct force measurements. Figure 11 demonstrates the use of optical tweezers to affect the motion of single mitotic chromosomes on the metaphase plate in a cultured cell, as viewed by computer-enhanced video microscopy.

VI. FUTURE PROSPECTS

What further developments lie in store for optical tweezers? Clearly, additional experiments need to be done exploring aspects of optical damage in a variety of preparations and over a range of powers and wavelengths. Optimum conditions have surely not yet been found: peak force could be improved by an order of magnitude, perhaps more, in certain experimental preparations. Lasers (especially microlasers and diode lasers) continue to evolve in the direction of becoming more compact, more convenient, as well as less expensive. The simple design of optical tweezers will make it feasible to consider outfit-

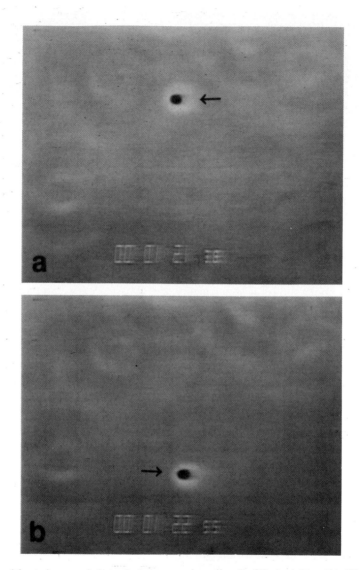

Fig. 5. A bacterium caught by optical tweezers. A motile cell of *Escherichia coli* (wild-type for motility and chemotaxis, strain AW405) swam at ~30 μm/s into the trap and was caught and viewed in phase contrast optics. The cell's shape is that of a cylinder with hemispherical endcaps, roughly 2 μm in length by 0.5 μm in diameter and it is seen here end-on. **a:** The cell (arrow) becomes aligned by the trap with its long axis perpendicular to the specimen plane, i.e., parallel to the trapping beam. The cell's several flagella (not visible) continue to turn, causing the body to spin rapidly, like a top, about its long axis. **b:** The trap has been moved in the specimen plane at about 50 μm/sec to a new location. The cell continues to be held by the tweezers (arrow).

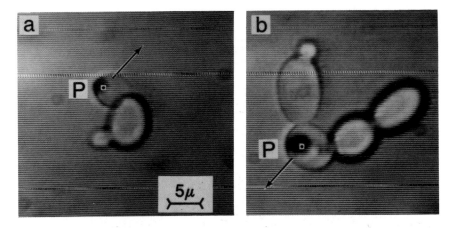

Fig. 6. Yeast cells maintained in an optical trap. The center of the trap is indicated by the dot beside the letter "P." Originally, two cells were caught by the tweezers. **a**: The same cells after ~0.5 h. One of the cells has divided, and all three are now held in the trap, which is being moved in the direction shown by the arrow. An additional bud has appeared on one cell near the trapping zone. **b**: The same preparation, ~3 h later. The original two cells have divided and six cells are now held in the trap. The cells are being dragged at several µm/sec in the direction shown by the arrow, so that viscous forces bring the individual cells up into focus in the specimen plane. (Photo courtesy of Dr. Arthur Ashkin.)

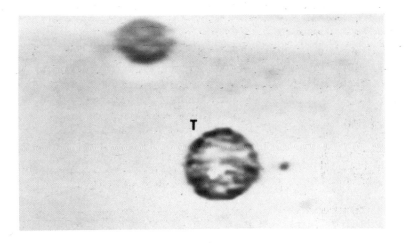

Fig. 7. Optical trapping of eukaryotic cells from cell culture. The photo shows a rat alveolar macrophage (marked T) suspended in a trap based on two counter-propagating beams (cf. Fig. 4). A second cell, top, is seen lying on the bottom of the chamber. The ability to manipulate relatively large cells in culture opens avenues for many potential applications, including cell sorting, cell fusion, cell-cell/cell-substrate interactions, investigations of cell motility, etc. (Photo courtesy of Dr. Tudor Buican.)

Fig. 8.   Successive images of a chloroplast manipulated by optical tweezers, showing how the trap can be used to "reach inside" intact cell membranes. Individual chloroplasts from the fresh-water alga *Nitella axillaris* were exposed by microdissection of a giant internodal cell. The chloroplasts, normally 6-8 μm in diameter, were swollen to ~10 μm with a buffer of low osmolarity. **a:** A transparent chloroplast (arrow), containing a number of heavily pigmented grana, viewed in phase contrast optics. **b:** The light trap is trained through the outer chloroplast wall onto the pigmented grana. They coalesce into a single clump under the trapping forces. **c:** The trap has been moved to the right, moving the clump of grana to the point where it bumps into the chloroplast wall. **d,e:** The trap continues to move to the right, carrying the entire chloroplast with it.

Fig. 9.   Successive images of a vesicle inside a *Drosophila* embryo manipulated by optical tweezers. The photos show the *micropile,* a short (~30 μm-long) appendage at the rostral end of the blastoderm that corresponds to the site of sperm entry, as viewed by Nomarski optics. A portion of the main body of the embryo can be seen at the bottom of each photograph. This embryo is ~2 h old, at a stage just prior to cellularization. **a:** A micron-sized vesicle is visible near the blastoderm wall inside the micropile. **b–e:** The vesicle is trapped and moved from right to left across the diameter of the micropile. The initial and final positions are indicated by arrows.

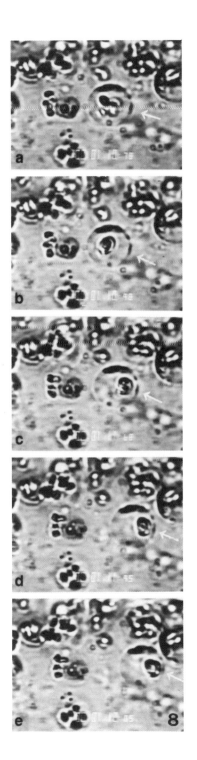

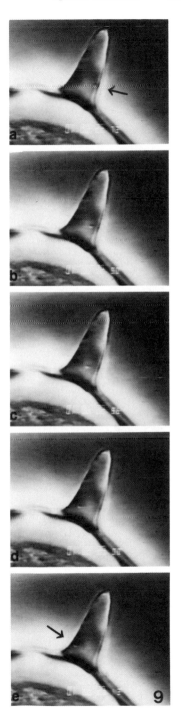

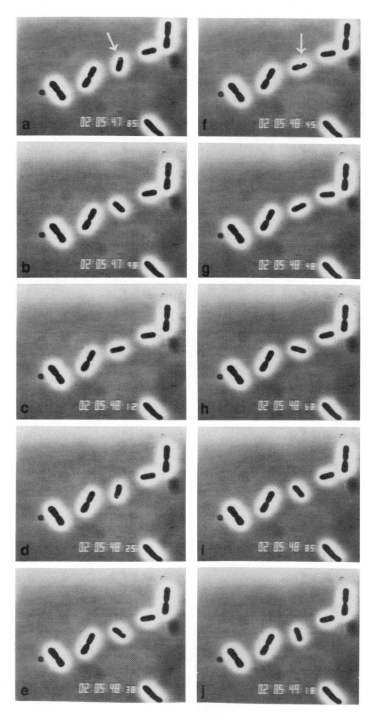

Fig. 10.   Optical tweezers used to rotate a tethered bacterial cell. Cells of *Streptococcus* (strain V4051) were tethered by a their flagella to the glass coverslip and viewed in phase contrast. Their rotary motors were arrested by treatment with a protonophore that depletes the transmembrane proton gradient. This treatment causes the motors to "lock up" partially, so that turning the cell body builds up elastic strain (torsion) in the flagellar filament by which the cell is tethered. The trap is used to grab one end of the cell, opposite to its point of tethering. Circular motion of the trap, generated by beam-steering optics, causes rotation of the cell body. The figure shows successive images of the cell at various time intervals. **a**: A tethered cell has been trapped at one end. The trap itself is visible as a small white spot (arrow) at the top of the image. This spot is due to reflection of the laser light from the glass/water interface, located just over the cell. **b–f**: The trap is rotated counterclockwise, turning the cell at about 2 Hz at constant angular velocity about its point of attachment. The trap position is indicated by the arrow in panel (f). The cell rotates nearly one full revolution. **g**: The laser light is turned off, releasing the cell from the trap. **h–j**: The cell body rebounds clockwise throug ~90°, as the flagellum unwinds. The angular position of the cell relaxes exponentially from the point of release to a new equilibrium value. From the decay time of this relaxation, the torsional stiffness of the flagellum can be calculated [Block et al., 1989a,b].

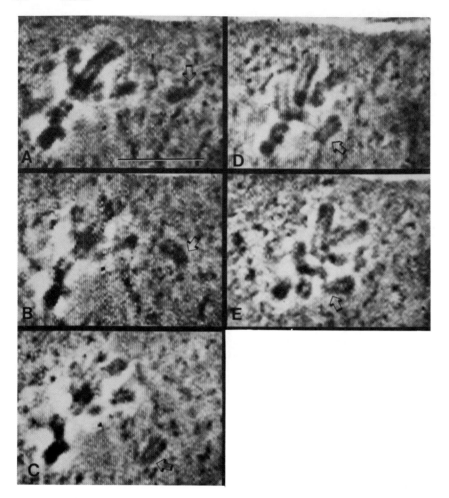

Fig. 11.   Use of optical tweezers on individual chromosomes in mitosis. A light trap was trained on a chromosome from a PTK2 cell in metaphase. The application of trapping forces to chromosomes in prometaphase or early metaphase appears to cause otherwise stationary chromosomes to start moving, and subsequently to induce relatively high-speed motion the chromosome in a direction opposite to that of the applied force [Berns et al., 1989]. The calibration bar is 10 μm. **A:** The trap is applied to a single centrophilic chromosome at the spot marked by the arrow, on the side of the chromosome farthest from the spindle pole (not visible). **B:** The chromosome reorients within a few seconds of exposure to the trapping force and begins to move rapidly poleward, pulling out of the trap. **C:** The chromosome has moved to the spindle pole within 1 min. The trap is then moved from its former location and applied to the side of the chromosome closest to the spindle pole. **D–E:** Within 2–3 sec after the second application of the trap, the chromosome begins to move away from the pole, stopping only when it reaches the metaphase plate. This cell subsequently completed normal mitosis. (Photo courtesy of Dr. Michael Berns.)

ting any microscope used for cell examination with trapping capability. For now, the price of the laser and its associated optics alone can equal, or even exceed, the price of a good research microscope.

In addition to using optical tweezers as a microtensiometer for in vitro motility assays of molecular motors (above), a host of other uses are now being explored or contemplated. A short sampling among them: A. Ashkin [1989a,b] has been measuring the microviscosity and viscoelastic properties of cytoplasm in plant cells, mapping out physical aspects of the intracellular milieu. D.E. Clapham's group (personal communication) has launched an ambitious project to build a multi-laser system that will incorporate confocal scanning microscopy (requiring a visible light laser), ultraviolet excitation (for fluorescent indicator dyes such as fura-2 and photoreleasable compounds such as caged calcium, etc., requiring a UV laser), and optical tweezers (requiring an IR laser), all in a single microscope. T.N. Buican's group (personal communication) continues to develop automated cell sorters based on trapping principles, and has proposed a miniaturized version of their instrument, the Cytometry Workstation, for placement in earth orbit aboard the Space Station.

It would seem that the sky is indeed the limit for this new technology: perhaps "tractor beam" was not so far off, after all?

ACKNOWLEDGMENTS

Thanks are due to Mike Berns, Tudor Buican, Dave Clapham, and especially Art Ashkin for providing me with material in advance of publication. I thank Ed Purcell and Howard Berg for helpful discussions; Howard Berg, Art Ashkin, and Linda Turner for comments on the manuscript; and Jay Scarpetti and Mary Ann Nilsson for photographic work. This research was supported by the Rowland Institute for Science, Cambridge, MA 02142.

VII. REFERENCES

Ashkin A (1970) Acceleration and trapping of particles by radiation pressure. Phys Rev Lett 24:156–159.
Ashkin A (1978) Trapping of atoms by resonance radiation pressure. Phys Rev Lett 40:729–732.
Ashkin A (1980) Applications of laser radiation pressure. Science 210:1081–1088.
Ashkin A, Dziedzic JM (1974) Stability of optical trapping by radiation pressure. Appl Phys Lett 24:586–588.
Ashkin A, Dziedzic JM (1975) Optical levitation of liquid drops by radiation pressure. Science 187:1073–1075.
Ashkin A, Dziedzic JM (1987) Optical trapping and manipulation of viruses and bacteria. Science 235:1517–1520.
Ashkin A, Dziedzic JM (1980) Observation of light scattering from nonspherical particles using optical levitation. Appl Optics 19:660–668.
Ashkin A, Dziedzic JM (1989a) Optical trapping and manipulation of single living cells using

infra-red laser beams. In Berichte der Bunsen-Gesellschaft für Physikalische Chemie Int J Phys Chem, Darmstadt, West Germany.

Askin A, Dziedzic JM (1989b) Internal cell manipulation using infrared laser traps. Proc Natl Acad Sci USA 86:7914–7918.

Ashkin A, Dziedzic JM, Bjorkholm JE, Chu S (1986) Observation of a single-beam gradient force optical trap for dielectric particles. Optics Lett 11:288–290.

Ashkin A, Dziedzic JM, Yamane T (1987) Optical trapping and manipulation of single cells using infrared laser beams. Nature 330:769–771.

Balykin VI, Letokhov VS (1989) Laser optics of neutral atomic beams. Phys Today 42(4):23–28.

Bellus D (1979) Physical quenchers of singlet molecular oxygen. Adv Photochem. 11:105–205.

Berns MW (1989) Optical trapping with a laser microbeam: A new method to study cell motility (abstract). J Cell Biol 107,6; part 2:453a.

Berns MW, Wright WH, Tromberg BJ, Profeta GA, Andrews JJ, Walter RJ (1989) Use of a laser-induced optical force trap to study chromosome movement on the mitotic spindle. Proc Natl Acad Sci USA 86:4539–4543.

Block SM (1989) Laser light traps for biology. Phys Today 42,1:S17-S19.

Block SM, Blair DF, Berg HC (1989a) Investigations of the bacterial rotary motor using optical tweezers (abstract). Biophys J 55,2; part 2:258a.

Block SM, Blair DF, Berg HC (1989b) Compliance of bacterial flagella measured with optical tweezers. Nature 338:514–518.

Block SM, Goldstein LSB, Schnapp BJ (1989c) Using optical tweezers on kinesin-coated beads moving along microtubules (abstract). J Cell Biol 109;4 Part 2:81a.

Buican TN, Neagley DL, Morrison WC (1988) OCAM—A robotic optical trapping cell analyzer and manipulator (abstract). Cytometry (suppl 2):88.

Buican TN, Neagley DL, Morrison WC, Upham BD (1989) Optical trapping, cell manipulation, robotics. In New Technologies in cytometry, Society of Photo-optical Instrumentation Engineers (SPIE) Proceedings, v. 1063. SPIE, Bellingham, WN. (in press).

Buican TN, Smyth MJ, Crissman HA, Salzman GC, Stewart CC, Martin JC (1987) Automated single-cell manipulation and sorting by light trapping. Appl Optics 26:5311–5316.

Chu S, Prentiss MG, Cable AE, Bjorkholm JE (1987) Laser cooling and trapping of atoms. In Svanberg S, Persson W (ed): Laser Spectroscopy VIII. Springer Series in Optical Sciences, volume 55. Berlin: Springer-Verlag, pp 58–63.

Gordon J, Ashkin A (1980) Motion of atoms in a radiation trap. Phys Rev A21:1606–1617.

Kamimura S, Takahashi K (1981) Direct measurement of the force of microtubule sliding in flagella. Nature 293:566–568.

Phillips WD, Gould PL, Lett PD (1988) Cooling, stopping and trapping atoms. Science 239:877–882.

Pool R (1988) Trapping with optical tweezers. Science 241:1042.

Purcell EM (1985) Electricity and Magnetism. Berkeley Physics course, Volume 2. New York: McGraw-Hill, Inc.

Smith PW, Ashkin A, Tomlinson WJ (1981) Four-wave mixing in an artifical Kerr medium. Optics Lett 6:284–286.

Thompson R (1988) Hot favourites for atom cooling. Nature 335:588–589.

Index

403

total internal reflection fluorescence combined with, 120, 124
Fluorescence signals, from microinjected cells
analysis of, 186-188
detection of, 185
Fluorescence tracking, fluorescence aggregates for, 131
Fluorescent analogs, microinjected cytoskeletal, 179-184
Fluorescent fluid phase labels, endocytic pathway assessed via, 162-164
Fluorescent probes. *See specific probes*
Fluorescent ligand conjugates, endocytic pathway assessed via, 156-162
Fluorescent lipid membrane probes, endocytic pathway assessed via, 164
Fluorine-nuclear magnetic resonance (^{19}F-NMR), BAPTA used in for Ca^{2+} studies inside whole cells, 4
Fluorophores. *See specific dyes*
FMLP. *See f*-Methionyl-leucyl-phenylalanine
^{19}F-NMR. *See* Fluorine-nuclear magnetic resonance
N-Formyl-methionyl-leucyl-phenylalanine, membrane potential studies and, 331
FPR. *See* Fluorescence photobleaching recovery
Free radicals
optical measurement of, 233
photochemistry of photobleaching and, 146
rate of formation of, 217
Frog skin epithelium, cell volume imaging and, 248-249
Front Face Accessory, fluorescence spectra recorded via, 340
Fructose-1,6 diphosphate, rhodamine-labeled aldolase binding to actin and, 227
Funaria hygrometrica, ionic currents around, 279
Fura-2
as Ca^{2+} indicator, 3-4, 22, 34-38, 40-44, 49, 255, 260-261, 266, 283, 300, 333-337, 367, 402
emission spectra of, 11
excitation spectra of, 335
quantum yield of, 16
spatial resolution and, 40-43
structure of, 5-6
temporal resolution and, 43-44
Fura-2 analog

structure of, 9
synthesis of, 16

Gating, ion channels and, 355
GDH. *See* Glutamate dehydrogenase
Gelsolin, as microinjected fluorescent analog, 181, 192-193
Gibbs–Donnan equilibrium, inward diffusion of ions and, 240
Glass, metal-coated, fluorescence emission and in total internal reflection fluorescence, 116-117
Glass substrate, fluorescence emission near in total internal reflection fluorescence, 115
Glucose, single cell metabolism and, 229
Glucose deoxygenase/glucose, in deoxygenation, 111
Glucose oxidase, oxidative metabolism studies and, 344
Glutamate dehydrogenase (GDH), mitochondrial matrix spacing and, 218
Glycerol, in optical configurations for microscope in total internal reflection fluorescence, 99, 101, 108-109
Glycocalyx matrix, label interaction with, in fluorescence photobleaching recovery, 131
Glycolysis, cytosolic redox reactions and, 217
Gold, colloidal, in single particle labeling, 131
Goos–Hanchen shift, characterization of, 112
Gradient force, atom trapping studies and, 376, 379-382
Gramicidin
membrane potential studies and, 331
Na^+ activity and, 267
Gramicidin D, calibration curves for K^+/Na^+ indicators constructed using, 16
Granulocytes, ion transport in, 354
Griffithsia pacifica, ionic currents around, 279

H^+
cell volume imaging and, 244
fluorescent dye indicators available for, 3
intracellular levels and distributions of, 239
ionic currents and, 280, 284
single cell fluorescence measurements and, 265-267